Series in BioEngineering

More information about this series at http://www.springer.com/series/10358

J. Paul Robinson · Andrea Cossarizza
Editors

Single Cell Analysis

Contemporary Research and Clinical Applications

 Springer

Editors
J. Paul Robinson
Purdue University Cytometry Laboratories,
  Department of Biomedical Engineering
Purdue University
West Lafayette, IN
USA

and

Purdue University Cytometry Laboratories,
  Department of Basic Medical Sciences,
  Weldon School of Biomedical
  Engineering
Purdue University
West Lafayette, IN
USA

Andrea Cossarizza
Chair of Pathology and Immunology,
  Department of Medical and Surgical
  Sciences for Children and Adults
University of Modena and Reggio Emilia
  School of Medicine
Modena
Italy

ISSN 2196-8861   ISSN 2196-887X   (electronic)
Series in BioEngineering
ISBN 978-981-13-5151-8   ISBN 978-981-10-4499-1   (eBook)
DOI 10.1007/978-981-10-4499-1

Printed on acid-free paper

This Springer imprint is published by Springer Nature
The registered company is Springer Nature Singapore Pte Ltd.
The registered company address is: 152 Beach Road, #21-01/04 Gateway East, Singapore 189721, Singapore

Preface

Negotiating the variety of biological applications available for a single-cell technology can be a daunting task in the twenty-first century. There was a time when flow cytometry offered an opportunity to analyze single cells by studying their phenotype or their cell cycle and provide an easy-to-achieve solution to well-known problems. But times have changed. The applications and opportunities for flow cytometry have never been more significant than they are today. New instrumentation, faster, more sensitive, more available, has moved the field. This is no more evident than in the opening chapter of this book, where a detailed review of cellular microenvironments promises to open a new era in understanding how therapeutic agents must be designed to accommodate the knowledge that can be obtained only using the techniques outlined in this book. This chapter is undoubtedly the most comprehensive review of microenvironment cytometry available today.

For three decades there have been proposals to move flow cytometry into the clinical microbiology laboratory, but complexity and cost have prevented any widespread adoption from a clinical perspective. This has changed with a chapter on rapid, low cost implementation using microtiter-plate analysis (now available on virtually every commercial instrument), whereby critical data such as antibiotic sensitivity can be achieved in a few hours on a routine basis. Similarly, technology advances in high throughput have made some approaches to drug screening almost routine, producing valuable data in a short period of time, both reducing the cost and decreasing the time to result. These processes demand more opportunities for analytical data reduction; a section of this book addresses this rapidly changing environment. Two other vital areas of single-cell analysis demand attention: standards/quality control and instrument sensitivity, and assay validation. These are addressed in two ways; a discussion of approaches to ensure that data are verifiable and quantifiable, as well as assay validation, and a discussion of the very sensors from which the signals we use are derived. In this unique chapter on photon detection, we bring the most up-to-date discussion of sensor technology available, with suggestions of how next-generation sensors may transform single-cell analysis, and in particular the analysis of very small particles where every photon counts. While the most modern technology available is presented, so too is a bird's eye review of how the field of cytometry became a field to itself and how the very tools we have today evolved from the original discoveries.

In the clinical environment, for more than 30 years flow cytometry has been crucial to quantify circulating CD4+ T lymphocytes in patients with HIV infection, a parameter required first to start antiretroviral therapy, then to follow its efficacy. Over these decades, spectacular advancements in all the fields related to this technology have dramatically improved the capability to finely diagnose a large number of human diseases, and to identify the phenotype and function of new cell populations. So, several chapters in this volume focus on real-world applications in clinical environments. From the critical aspects of running multicenter studies to analysis of rare cells and the impact on patient diagnostics, important factors are identified. Further, the tremendous opportunities for evaluating the very detailed aspects of metabolic function of mixed populations of cells is one of the most challenging but highly rewarding features of single-cell analysis using flow cytometry. Indeed, this technology has allowed the opening of a completely new field that studies how metabolic changes play a role in determining the differentiation, maturation and functionality of immune cells.

This book covers a range of research and clinical applications and brings together a *state-of-the-art* focus on the uniqueness of cytometric tools.

West Lafayette, USA   J. Paul Robinson
Modena, Italy   Andrea Cossarizza

Acknowledgements

We gratefully acknowledge the editorial support of Gretchen Lawler, who assisted us in reviewing and editing each chapter. We are also very grateful to Carmen Gondhalekar, who painstakingly edited and modified every figure in the book to comply with the publisher's requirements. This was no easy accomplishment, as getting authors to provide high-quality figures is not an easy task!

Contents

Contributors

Sara De Biasi Department of Surgery, Medicine, Dentistry and Morphological Sciences, School of Medicine, University of Modena and Reggio Emilia, Via Campi, Modena, Italy

Gary Bonhert Tempero Pharmaceuticals, A GSK Company, Cambridge, MA, USA

Andrea Cossarizza Chair of Pathology and Immunology, Department of Medical and Surgical Sciences for Children and Adults, School of Medicine, University of Modena and Reggio Emilia, Via Campi, Modena, Italy

Sofia Costa-de-Oliveira Department of Microbiology, University of Porto and Faculty of Medicine, Porto, Portugal

Scott Davis Tempero Pharmaceuticals, A GSK Company, Cambridge, MA, USA

Rachel J. Errington School of Medicine, Institute of Cancer and Genetics, College of Biomedical and Life Sciences, Cardiff University, Cardiff, UK

Lara Gibellini Department of Surgery, Medicine, Dentistry and Morphological Sciences, School of Medicine, University of Modena and Reggio Emilia, Via Campi, Modena, Italy

Carmen Gondhalekar Weldon School of Biomedical Engineering and College of Veterinary Medicine, West Lafayette, IN, USA

Cherie Green Flow Cytometry Biomarkers Development Sciences, Genentech, Inc., a Member of the Roche Group, South San Francisco, CA, USA

Victoria Griesdoorn School of Medicine, Institute of Cancer and Genetics, College of Biomedical and Life Sciences, Cardiff University, Cardiff, UK

Guadalupe Herrera Cytometry Service, Central Research Unit (UCIM), Incliva Foundation, The University of Valencia and University Clinical Hospital, Valencia, Spain

Jonathan Hill Surface Oncology, Cambridge, MA, USA

Robert A. Hoffman Livermore, CA, USA

Rob Jepras GlaxoSmithKline Medicines Research Centre, Stevenage, UK

Beatriz Jávega Laboratory of Cytomics, Joint Research Unit CIPF-UVEG, The University of Valencia and Principe Felipe Research Center, Valencia, Spain

Andrew Lake Surface Oncology, Cambridge, MA, USA

Anis Larbi Biology of Aging Laboratory, Singapore Immunology Network, Agency for Science Technology and Research (A*STAR), Singapore, Singapore; A*STAR Flow Cytometry, SIgN Immunomonitoring Platform Facility, A*STAR, Singapore, Singapore; Faculty of Medicine, University of Sherbrooke, Sherbrooke, QC, Canada; Department of Biology, Faculty of Sciences, ElManar University, Tunis, Tunisia

Virginia Litwin Hematology/Flow Cytometry, Covance Central Laboratory Services, Indianapolis, IN, USA

Steve Ludbrook GlaxoSmithKline Medicines Research, Stevenage, UK

Alicia Martínez-Romero Laboratory of Cytomics, Joint Research Unit CIPF-UVEG, The University of Valencia and Principe Felipe Research Center, Valencia, Spain; Cytomics Technological Service, Principe Felipe Research Center, Valencia, Spain

Milena Nasi Department of Surgery, Medicine, Dentistry and Morphological Sciences, School of Medicine, University of Modena and Reggio Emilia, Via Campi, Modena, Italy

José-Enrique O'Connor Laboratory of Cytomics, Joint Research Unit CIPF-UVEG, The University of Valencia and Principe Felipe Research Center, Valencia, Spain; Department of Biochemistry and Molecular Biology, Faculty of Medicine, University of Valencia, Valencia, Spain

Metul Patel GlaxoSmithKline Medicines Research Centre, Stevenage, UK

Jordi Petriz Josep Carreras Leukaemia Research Institute, Badalona, Barcelona, Spain

Cidália Pina-Vaz Department of Microbiology, University of Porto and Faculty of Medicine, Porto, Portugal

Marcello Pinti Chair of Pathology and Immunology, Department of Life Sciences, School of Medicine, University of Modena and Reggio Emilia, Via Campi, Modena, Italy

J. Paul Robinson The SVM Professor of Cytomics, Professor of Biomedical Engineering, Purdue University, West Lafayette, USA

Acácio Gonçalves Rodrigues Department of Microbiology, University of Porto and Faculty of Medicine, Porto, Portugal

Francisco Sala-de-Oyanguren Ludwig Institute for Cancer Research, Département D'Oncologie Fondamentale, Faculté de Biologie et Médecine, Université de Lausanne, Épalinges, Switzerland

Poonam Shah GlaxoSmithKline Medicines Research Centre, Stevenage, UK

Ana Silva-Dias Department of Microbiology, University of Porto and Faculty of Medicine, Porto, Portugal

Ana Pinto Silva Department of Microbiology, University of Porto and Faculty of Medicine, Porto, Portugal

Oscar F. Silvestre School of Medicine, Institute of Cancer and Genetics, College of Biomedical and Life Sciences, Cardiff University, Cardiff, UK; Department of Nanophotonics, INL—International Iberian Nanotechnology Laboratory, Braga, Portugal

Paul J. Smith School of Medicine, Institute of Cancer and Genetics, College of Biomedical and Life Sciences, Cardiff University, Cardiff, UK; OncoTherics Limited, Shepshed, Leicestershire, UK

Rita Teixeira-Santos Department of Microbiology, University of Porto and Faculty of Medicine, Porto, Portugal

Alessandra Vitaliti BioMarker Development, Novartis Pharma AG, Basel, Switzerland

Gregory Wands Tempero Pharmaceuticals, A GSK Company, Cambridge, MA, USA

Lili Wang Biosystems and Biomaterials Division, National Institute of Standards and Technology, Galthersburg, MD, USA

Masanobu Yamamoto Basic Medical Sciences, Purdue University, West Lafayette, IN, USA; CTO, Miftek Corporation, West Lafayette, IN, USA

Microenvironment Cytometry

Paul J. Smith, Victoria Griesdoorn, Oscar F. Silvestre
and Rachel J. Errington

Abstract It is becoming self-evident that biological systems—from microbes to human tissues—create local microenvironments that can foster survival or impose Darwinian selection that creates unwanted phenotypes. Understanding the dynamic nature of cancer-cell microcommunities in such situations presents a challenge from the conceptual level of cell attractors to the practical level of tracking discrete molecular events and relevant changes by cytometry. There has been a significant increase in the attractiveness of therapeutic strategies that seek to disrupt the ability of neoplastic cells to exploit their microenvironments and present drug-resistant and pro-metastatic phenotypes. Aligned with this is the demand for 3D cellular systems that can mimic aspects of the microenvironment. The chapter discusses methods of cell encapsulation, in particular the application of hollow-fiber technology. Strategies also need to take into account patterns of cellular plasticity and the dynamic heterogeneity in target molecule presentation, especially in co-culture systems. The focus here is on the cytometry of cellular communities within their microenvironments with examples of targeted therapeutic agents. A critical feature of cellular microenvironments, arising from inadequate vascularity, is reduced

P.J. Smith (✉) · V. Griesdoorn · O.F. Silvestre · R.J. Errington
School of Medicine, Institute of Cancer and Genetics, College of Biomedical
and Life Sciences Cardiff University, Heath Park, Cardiff CF14 4XN, UK
e-mail: smithpj2@cf.ac.uk; paul@oncotherics.com

V. Griesdoorn
e-mail: vicky@vandalon.nl

R.J. Errington
e-mail: erringtonrj@cf.ac.uk

P.J. Smith
OncoTherics Limited, 56 Charnwood Road, Shepshed, Leicestershire LE129NP, UK

O.F. Silvestre
Department of Nanophotonics, INL—International Iberian Nanotechnology Laboratory,
Avenida Mestre José Veiga s/n, 4715-330 Braga, Portugal
e-mail: oscar.silvestre@inl.int

© Springer Nature Singapore Pte Ltd. 2017
J.P. Robinson and A. Cossarizza (eds.), *Single Cell Analysis*,
Series in BioEngineering, DOI 10.1007/978-981-10-4499-1_1

1

oxygenation levels. The chapter highlights some of the cellular responses to hypoxia, while research experience suggests that the targeting of hypoxic cells in tumor regions has therapeutic potential.

Keywords Neoplasms · Metabolism · Tumor microenvironment · Hypoxia · Flow cytometry · Microscopy · Fluorescence · Mitosis · Polyploidy · Single-cell analysis · Hollow fiber · Bioreactor · Anthraquinones · Antineoplastic agents · DNA topoisomerase type II · Drug resistance · Drug screening assays

1   Microcommunities and Microenvironments

1.1   *Introduction*

In 1828 the Scottish philosopher Thomas Carlyle (1795–1881), when translating the German word 'Umgebung' in a work by Goethe, coined the term 'environment' to convey the sense of "nature, conditions in which a person or thing lives" [1]. The term "cellular microenvironment" is therefore self-descriptive and recognizes the local milieu within which a functionally defined population of cells—a micro-community—resides, or indeed through which individual cells may transit. However, this can also be extended to the composition of cell types and biological constituents within that localized physical environment.

Cell-to-cell heterogeneity, arising by stochastic processes in apparently isogenic populations, is a common feature throughout the natural world and not least in microbial systems [2]. Indeed, the application of flow cytometry for the detection and separation of diverse microbial populations allows for a system-level charac-terization of unknown bacterial phyla in environmental studies [3] and assessment of environmental threats [4]. In cytometric studies, cellular components are often utilized as reporters of microenvironmental states and such profiling is used to interpret the state of the microenvironment itself—not least the presence of per-turbing influences. In some cases, there is a technological bridge that for example enables a soluble factor such as a cytokine to be reportable though cell-based capture. Clearly, there is a frequent requirement to validate by parallel analyses. Further, reliance is often placed on a genotypically defined cellular subset to functionally qualify as reporters, without anticipating how the dynamics of adap-tation or variation can influence reporter performance.

An unavoidable feature of cancer is its intrinsic potential to generate or even thrive in conditions of low oxygen tension (hypoxia)—an overriding microenvi-ronmental factor that can influence cancer-cell metabolism and gene expression, through to complex changes in behavior [5]. Even in the early stages of tumor proliferation, hypoxia can develop through an inadequate or temporal-lag in vas-cular supply (see Fig. 1). This pervading physical influence exists in many tumor

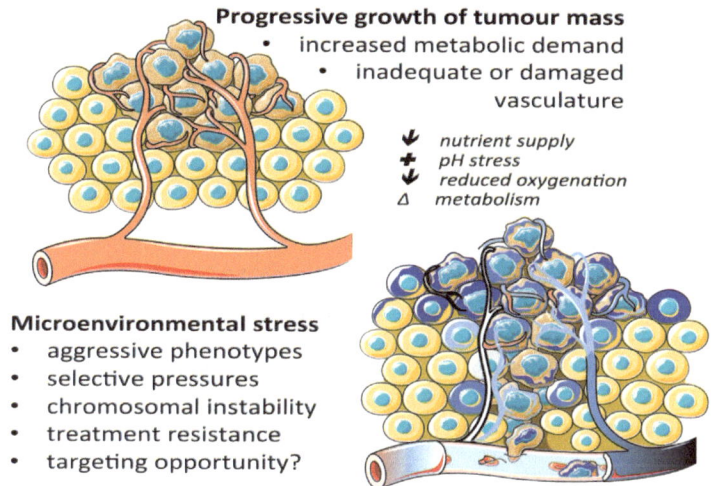

Fig. 1 The generation of microenvironments in the tumor milieu

settings, where the definition of "normoxia" in surrounding tissues is an important issue [6]. The origins of hypoxia are readily attributed to the changed metabolism of neoplastic cells and the confused vascular organization accompanying progressive growth—resulting in both diffusion-limited hypoxia and acute changes in oxygen tension.

This chapter will briefly address, within the context of cytometry, the nature of cellular heterogeneity and plasticity within cellular microcommunities; provide examples of in vitro model systems with respect to cell-cell interactions and target-molecule expression; address encapsulation systems for creating boundary-defined in vitro and in vivo models. The primary theme is how neoplastic cells respond to the pervasive microenvironmental influence of tumor hypoxia, noting progress towards hypoxia-targeted anti-cancer therapeutics.

1.1.1 Boundaries and Niches

The limiting boundaries of a microenvironment are determined by the nature or extent of chemical and physical states. Boundaries may be relatively well defined in physical terms or exist as homeostatically maintained gradients within which cells may experience varying degrees of stress or advantage. In the case of tumors, this dynamic and complex environment encompasses normal cells, neoplastic cells, elements of the immune system, blood vessels, soluble factors, a plethora of molecules, and an extracellular matrix with biophysical, mechanical, and biochemical properties specific for each tissue. Cells respond to or contribute their own influences on the local environment. Even within such a milieu there can exist particular tumor zones or niches [7] with specialized environmental characteristics where, for example, stem cells may be retained in undifferentiated but self-renewable states.

The pervading physical microenvironment is critically important in tumor progression and in attempts to recapitulate cellular complexity for in vitro studies. Tumor hypoxia co-opts the stroma and potentiates progression by multiple mechanisms, including the promotion of a functionally immunosuppressive microenvironment. One route is by stimulating Treg cell function and increasing expression of immune checkpoint molecules such as PD-L1 and CTLA4 on tumor cells [5]. Here the cytometric challenge is to capture the dynamic changes that lead to treatment resistance or indeed may reveal targeting opportunities.

1.1.2   Tumor Microenvironment [TME]

The cellular microenvironment surrounding neoplastic and pro-metastatic cells is of specific interest since it can both drive progression and frustrate therapeutic intent. Critical interactions are apparent at the levels of structural proteins, cellular components, and a range of non-tumor cells, including fibroblasts and infiltrating inflammatory cells [8]. Tumor cell–stromal cell interactions affect tumor-cell signaling and the balance between survival and barrier-breaching proliferation [9]. It is not surprising that such dynamic changes can affect drug sensitivity or create immunosuppressive states. Critically, it is also apparent that the range of mechanisms involved in driving interactions will depend on the host tissue, whether that be the primary site or a metastatic niche [9]. A striking example is pancreatic ductal adenocarcinoma (PDA), a common and lethal malignancy with little progress in survival rates over the last four decades. A contributing factor to the failure of systemic therapies is the tumor stromal content—a desmoplasia with growth of fibrous or connective tissue [10]. The PDA stroma supports tumor growth and promotes metastasis while presenting a physical barrier to drug delivery as well as creating a reduced oxygen environment.

The TME is complex, with high levels of site-specific variations in vivo. As a result, there has been difficulty in generating physiologically mimetic models in vitro. The TME literature is extensive, with a 7-fold increase in peer-reviewed publications in the last decade compared with the previous (source: US National Library of Medicine, National Institutes of Health.). The literature has explored multiple aspects of therapeutic intent, including stemness characteristics, stromal interactions, and insights for immunotherapy [11–14]. Immunotherapy concepts have even progressed to "reprogramming" of the TME to enable adoptive cellular therapy [15]. The TME is not fixed, changing upon treatment and continuing to provide mechanisms for resistance [16] and recurrence [17]. The essential problem with cancer progression and anti-cancer therapy is the capacity of cancer cells to migrate and establish distant metastases [18], especially to preferred sites and receptive microenvironments. Cytometry has revealed the heterogeneity evident in tumor cell populations, demanding an understanding of the behavior of defined microcommunities that may share a local milieu undergoing change but have different capacities for response and adaptation [19, 20].

There is a continuing rise in the capacity of cytometry to generate high-dimensionality data from gene-expression profiling [21] to biomarker screening [22], all supported by options for automated analysis [23]. This will enable a re-visualization of the dynamic states in cellular microcommunities by considering them as networks that reflect the probability of change and the nature of the states attained. This is captured in the concept of "attractor networks" [24]. Viewed in this manner, a case can be made for the inevitability of the neoplastic state, contesting the more convenient view of linear oncogenic pathways [25]. Here genetic and chromosomal instability is accompanied by epigenetic changes and the phenotypic plasticity of tumor cells.

2 Plasticity of Cellular Communities

2.1 Landscapes and Attractor States

Genetics alone cannot explain human variation and disease. This is also true for the dynamic plasticity shown by tumor cells under different microenvironmental influences. The popular term *epigenetics* embodies a partial explanation of both situations. It was first introduced by Haldane and Waddington [26, 27] to describe "the causal interactions between genes and their products, which brings the phenotype into being." More recently, epigenetics has been used to define heritable changes in gene expression that are not due to any alteration in DNA sequence [28]. As an embryologist and theoretical biologist, Waddington (1905–1975) is credited with developing the wider concept of the epigenetic landscape, envisioned as a series of ridges and valleys through which a cell may pass to reach a given state of stability. Perturbations to the highly dynamic signaling networks in cells [24] act to differentially "attract" cells within that epigenetic landscape towards new malignant signaling [24] and aggressive phenotypic states [29]. Essentially, cells can be considered as attractor networks whose basic property is to adopt a set of distinct, stable, self-maintaining states [30]. Accordingly, cancer development and progression within defined boundaries can be viewed as the outcomes of "cancer network attractors."

The operation of cancer network attractors presents a view of how heterogeneous cellular communities can change within the prevailing epigenetic landscape defined at one level by a pervading microenvironmental influence. In this view, attempts to "target" that resident population based on the network configuration of an "average" cancer cell are clearly suboptimal, since the population comprises moving targets creating an intrinsic capacity for treatment failure or relapse [31]. Exploration of these concepts lends itself to computational modeling and the analysis of rich gene-expression datasets from different cancer types to seek shared attractors [32]. Experimentally these concepts can be tested through the capacity of various cytometric techniques to track changes in population heterogeneity, to create a census for gene expression [33], and to monitor in high dimensionality molecular features at the single-cell level. There are well-recognized attractors, including epithelial–mesenchymal transitions (EMTs) [34], malignant versus non-malignant states [35], and mitotic chromosomal instability (CIN) [32].

2.2   Cell Transitions and Interactions: The Glycome

The epithelial-to-mesenchymal transition and subsequent capacity of cells to migrate is a striking example of cellular plasticity. It is relevant to the more adverse behaviors of tumor cells and has an extensive literature, but will not be reviewed here. However, it is important to note that although transitions may demand that the investigator qualify the cells as having distinct features that mark the steps in transition, it has been suggested that there is more likely to be a general continuum of morphological varieties collectively realized as "transitions," with a wide array of downstream migratory behaviors [36]. Accordingly, cytometry-based analyses must take this heterogeneity and continuum into consideration. Fundamentally, it is a change in the capacity of cells to interact with each other, the microcommunity and the surrounding environment that functionally defines transition. Cellular interactions at the level of surface-surface contacts are at the first level determined by the physical properties. For cell-cell interactions, the variable expression of a complex array of charged molecules with glycans plays a central role in determining those properties. This cell-surface glycome comprises various forms of carbohydrate chains with signatures that can be used to identify both cell types and their developmental or indeed pathological states [37].

The case has been made for glycome as the third group of bioinformative macromolecules to be elucidated after the genome and proteome [38]. Functioning microcommunities require mediators to modulate interactions. Glycosylation of a protein, as distinct from the non-enzymatic attachment of sugars via glycation, involves the addition of a glycosyl group to various potential targets moieties (arginine, asparagine, cysteine, hydroxylysine, serine, threonine, tyrosine, or tryptophan), generating a glycoprotein. Such post-translational modifications of proteins extend the chemical range of the amino-acid components, with glycosylation altering protein folding and stability in addition to modifying the regulatory functions.

The repertoire of glycans bearing glycosidic linkages to proteins and lipids undergo a stepwise incorporation of saccharides into glycan chains and branches during their transit through the secretory process. The n-dimensional nature of the array of molecules generated presents challenges for analysis [39]. Cytometric approaches are typically blind to this complexity or to the context of a specific modification. Rather, there is a focus on the expression of key molecules relevant to cell states and their dynamic of expression as cells modify behavior and function. Typically, these are detected using antibody probes.

In tumor cell populations, the potential for wide variation in the physical manifestation of the surface glycome—sometimes referred to as the glycocalyx according to the organism under study—provides options for improving Darwinian fitness. Critically, there is widespread evidence that beyond exploiting a niche environment, the surface glycome pattern is modulated during tumor progression and spread, providing signatures for staging and prognosis. Accordingly, this has attracted efforts to therapeutically interfere with the glycocalyx. Again the plethora of cell-surface glycosylation can mask the impact of therapeutic targeting [40].

2.3   Cellular Interactions in a Microcommunity: Transitions in NCAM Polysialylation

Changes in the polysialylation of neural cell adhesion molecule (NCAM) is an example of developmentally controlled glycosylation that enables behavioral transitions through changes in membrane-membrane apposition [41]. PolySia-NCAM is a key prognostic indicator in tumors of neural-crest origin. Such cells re-express polySia-NCAM, and its occurrence correlates with aggressive and invasive disease and poor clinical prognosis in different cancer types, including SCLC, pancreatic cancer, and neuroblastoma. Conceptually the tumor cells can be regarded as reiterating aspects of developmental roles for the polysialylation of NCAM in permitting neural-cell plasticity [42]. Down-regulation of polysialylation occurs in normal adult tissues [42]. Polysialylated NCAM expression also presents a legitimate target for modulation in cancer [40] and in engineering repair of the central nervous system [43].

Recent evidence suggests that the key enzyme polysialyltransferase ST8SiaII is a druggable target, and pharmacological inhibition can interfere with the migratory activity of neuroblastoma cells [44]. In SCLC it is thought to regulate NCAM-mediated cell-surface interactions, imparting anti-adhesive and pro-migratory properties to cells [40, 44]. In line with this view, established SCLC cultures expressing polySia-NCAM [CD56] typically show adhesion-independent growth, forming multicellular aggregates that can be used as model systems of micro-tumors.

Live-cell cytometry using SCLC 3D culture models can exploit a novel method of polySia-NCAM detection based on polySia's being a substrate for processive endosialidase cleavage by bacteriophage-borne glycosyl hydrolases. Certain endosialidase mutants have lost the cleaving capacity while retaining high-affinity substrate binding. Fusion of these inactive enzymes with eGFP yields convenient tools for the detection of polySia on live cells [45, 46]. The EndoNGFP probe rapidly binds to polySia molecules and has been used in combination with the non-toxic far red–fluorescent viability dye DRAQ7™ to allow a real-time approach for assessment of polySia expression in co-culture systems (Fig. 2). FCM profiling using EndoNGFP has indicated an influence of SCLC microcommunity composition independent of substrate adherence potential [47]. Co-culture of micro-aggregate culture forms (polySia^{hi}) with adherence transition forms (polySia^{lo}) was found to increase polySia expression in the low-expressing fraction with a concomitant reduction of expression in the high-expression fraction (Fig. 2). Using a viability dye it could be shown that this was not due to any cell death, given that that loss of viability results in the loss of polySia expression per se (Fig. 2). These microcommunity changes occurred within a relatively short time period (6 days), showing the capacity for even simple co-cultivation conditions to significantly shift target presentation profiles.

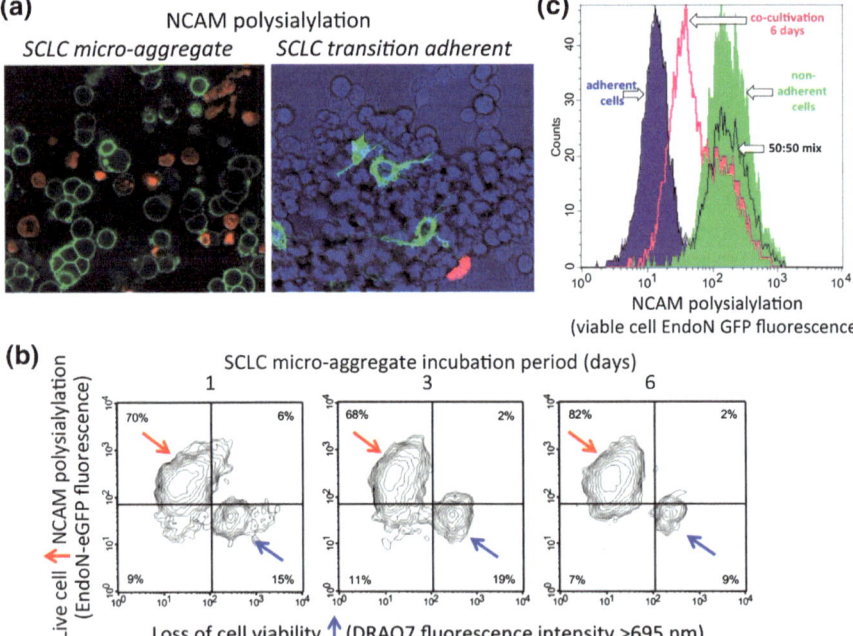

Fig. 2 Dynamic changes in NCAM polysialylation tracked in small-cell lung carcinoma (SCLC) micro-aggregates and cells undergoing plasticity changes to a transition adherent form. Panel **a**, *left* confocal section image of EndoNGFP binding (*green*) in a background of non-viable cells (DRAQ7 positive; *red*). Panel **a**, *right* merged transmission image (*blue*) with EndoNGFP-positive cells (*green*) showing dendritic-like morphology and a low background of non-viable cells (DRAQ7 positive; *red*). Panel **b** flow cytometric detection of viable EndoNGFP-positive cells showing heterogeneity in NCAM polysialylation maintained during 3D micro-aggregate culture with a persistent background of non-viable cells. Panel **c** co-culture generates a novel NCAM polysialylation profile within the population not attributable to cell death or predicted by admixture alone

3 Chromosomal Instability (CIN) and the Microenvironment

A well-recognized attractor is the route to mitotic chromosomal instability (CIN) [32]. CIN and aneuploidy are known to furnish the gross steps in genetic variation that allow tumor cells to generate surviving lineages adapted to changes in environments such as nutrient fluctuations and hypoxia [48]. This inherent capacity for variation is also readily observed in established cell-line models such as HeLa, generating sub-clonal responses to hypoxic stimuli [49]. In contrast, normal stromal cells are genetically stable under hyperglycemic and ischemic conditions [50]. Given that the propensity of tumor cells to undergo mitotic chromosomal instability is strongly associated with tumor grade, there is a need to understand the nature of the routes to CIN in cellular microcommunities. Of particular interest here is the application of FCM, microscopy, and time-lapse microscopy to map occult routes to CIN that underpin shifts in heterogeneity.

Programmed physiological routes can lead to polyploid cells with multiples of the diploid chromosome number [51]. Conceptually this can promote greater resilience or adaptation for chronic stress and injury [52]. However, under aberrant control it may contribute to CIN in cancer [53]. Classically CIN arises from mitotic anomalies of chromosome non-disjunction promoting cleavage furrow regression [54] and adaptation to a protracted activation of the spindle assembly checkpoint with mitotic slippage [55]. Cancer cells under stress are able to breach cell-cycle control restraints and progress into tetraploidy by readily observable anomalies in mitotic progression. Critically, tetraploid cells are relatively more resistant to DNA-damaging agents and are genetically unstable [56]. It is likely that physiological stresses within the microenvironment can encourage CIN [48], providing a base-line activity for CIN generation that becomes a more frequent event upon drug challenge. The occult route to tetraploidy via endocycle entry, by-passing mitosis [53], requires a collection of cytometric methods to both track and validate.

3.1 Cytometric Analysis of Chromosomal Instability

By combining fluorescent DNA stains with proliferation-tracking dyes and immunostaining for mitotic cells, it is possible to elucidate the division history and cell-cycle position within even an asynchronous population. Furthermore, by using imaging FCM, resolution of at least four main mitotic phases can be achieved [57]. Analysis of cell-division abnormalities, arising either stochastically or within failed attempts to resolve a cell-cycle stress response, has been actively studied to identify the origins of CIN. However, CIN-type chromosome mis-segregation does not adequately explain the presence of cells with triploid or near-tetraploid chromosome numbers in many cancers, including osteosarcoma [58]. Rather, it is suggested that tetraploidy presents an important intermediate stage to progressive polyploidization and wider CIN [58] through downstream cell-division abnormalities [59]. Tetraploid cells can mis-segregate chromosomes owing to supernumerary centrosomes and the increased chromosomal mass, generating aneuploid subclones [58, 60]. It is unclear how a route to tetraploidization via initial entry into an endocycle might allow cancer cells to also evade the consequences of anti-cancer drug–induced cell-cycle stress and contribute to resistance of recovering lineages with ongoing CIN [56]. A reasonable explanation is that the bypass of mitosis effectively avoids mitotic catastrophe [60]. The promotion of mitotic failure may offer therapeutic advantages in its own right [61].

The initial tetraploidization step can occur either by *endomitosis*, in which the cell does not engage cytokinesis, or by *endocycling*, in which the S and Gap phases alternate but with no intervening cell division [53]. A non-physiological route to tetraploidization is virus-induced cell-cell fusion [62]. In flow cytometric analyses it is important to distinguish in vitro cell-cell fusion events from aggregated cells [63]. The process of cell-cell fusion generating heterokaryons is clearly a potential route to CIN and tumor progression [64]. Interestingly, fusion can be considered as a "collision" of networks; this has been modeled computationally [30].

3.2 Mitotic Bypass

Endocycle entry in response to chemotherapeutic stress bypasses mitotic checkpoints and enhances the probability of survival by evading first-cycle mitotic catastrophe [60]. In the absence of stress resolution, cells undergoing endocycle entry may delay in a tetraploid G_1 state [59], potentially misconstrued as cell-cycle exit in G_2 [60]. G_2 arrest of U-2 OS cells by a DNA-damaging DNA topoisomerase II poison typically generates cells with high levels of cyclin B1 expression [65]. In contrast, using a cyclin B1–eGFP reporter [66] and stress induced by the DNA topoisomerase II catalytic inhibitor ICRF-193, mitotic bypass can be tracked by fluorescence time-lapse imaging (Fig. 3) and the quantification of cytoplasmic versus nuclear fluorescence (Fig. 4). ICRF-193 [meso-4,4-(2,3-butanediyl)-bis (2,6-piperazinedione)] is a complex-stabilizing inhibitor of DNA topoisomerase II, inducing arrest at a late cell-cycle decatenation checkpoint. Checkpoint failure results in the frequent inability of sister chromatids to separate properly in anaphase [67]. Effective tracking of mitotic bypass requires time-lapse imaging, confirming that under ICRF-193–induced stress, U-2 OS cells frequently do not attempt mitosis and that destruction of the cyclin-B1 signal is cytoplasmic.

FCM allows the tracking of drug-induced population shifts, through the destruction of cyclin B1 expression in G_2, identified by simultaneous DNA content analysis using the live-cell DNA dye DRAQ5^{TM} [68]. FCM further provides a cytometric signature distinguishing initial endocycle entry from downstream re-entry into an endocycle S phase in live cells (Fig. 5).

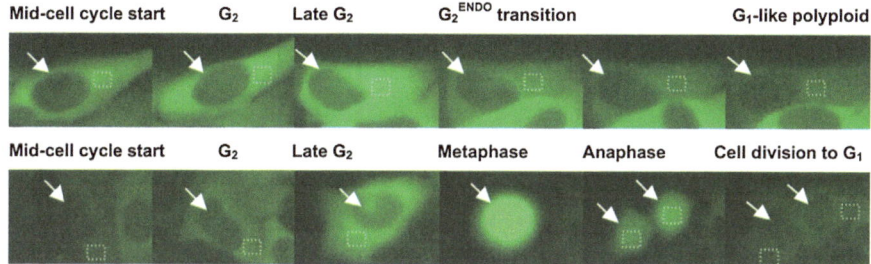

Fig. 3 Typical sequence of cyclin B1 destruction dynamics in U-2 OS cyclin B1-eGFP cells captured by fluorescence time-lapse microscopy. *Top*, typical sequence of cyclin B1 destruction in 2 μg/mL ICRF-193-treated cell, indicating mitotic bypass. *Bottom*, typical cyclin B1 dynamics in U-2 OS cyclin B1-eGFP solvent control cells showing commitment to mitosis. Both sequences shown were obtained over a period of 10 h. White arrows indicate the cell nucleus and white region-of-interest (ROI) areas represent exemplar location of the ROIs to extract fluorescence intensities over time, placed in the cytoplasm bordering the nucleus (see Fig. 4)

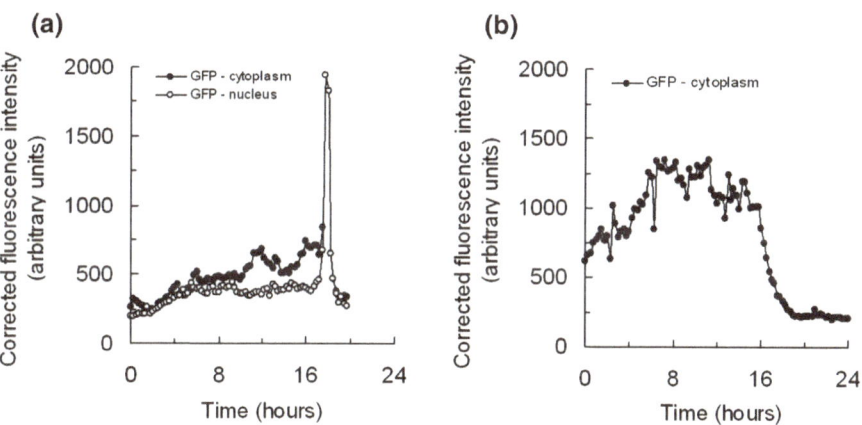

Fig. 4 Typical nuclear and/or cytoplasmic cyclin B1–eGFP fluorescence profiles for U-2 OS cyclin B1–eGFP cells undergoing normal mitosis and mitotic bypass. Plots show frame-by-frame mean fluorescence intensity over time with profiles of solvent control cells undergoing normal mitosis (panel **a**) and mitotic bypass (panel **b**) after exposure to 2 μg/mL ICRF-193. Intensity profiles were corrected against the background fluorescence. The maximal nuclear intensity during mitosis is an artefact due to cell rounding and the altered temporal pace through which a cell undergoes mitosis

3.3 Mitotic Bypass and the p53 Network

Attractor landscape analysis of p53 network dynamics and its regulation can identify potential therapeutic strategies for treating cancer [69]. Complete DNA repair or stress resolution does not appear to be a precondition for cell-cycle resumption following checkpoint activation [70]. In its tumor-suppressor role, p53

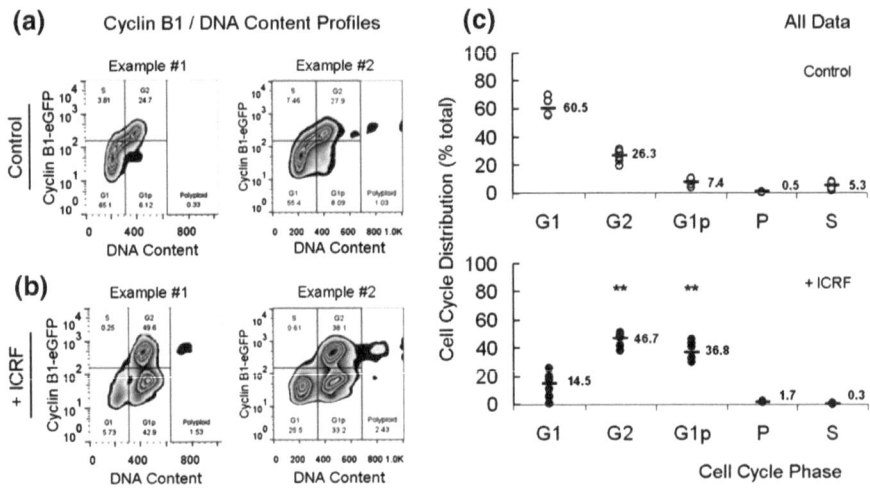

Fig. 5 Cell-cycle progression of U-2 OS cyclin B1–eGFP cells after 24-hour exposure to ICRF-193 in FCM. Panel **a**, typical cyclin B1–eGFP versus DNA content (DRAQ5) profile plots for U-2 OS cyclin B1–eGFP solvent control populations (0.1% v/v DMSO treatment) at 24 h after exposure. Panel **b**, same for 2 μg/mL ICRF-193–treated cells. Panel c, cell-cycle distribution plots for collection of typical cyclin B1–eGFP versus DNA content profiles showing solvent control populations (*top*) and 2 μg/mL ICRF-193–treated cells (*bottom*). $G_1p = G_1$ polyploid; P = polyploid. Cross-bars and number inserts indicate mean values (* = $p < 0.05$; ** = $p < 0.01$) (n = 6)

protein functions can invoke either cell-cycle arrest or progression to apoptosis in response to DNA damage. Induction of p53 is achieved largely through uncoupling the p53-Mdm2 interaction, leading to elevated p53 levels (review: [71]). Attractor landscape analysis of the cellular response to DNA damage, validated by single-cell imaging of a fluorescent p53 reporter in the MCF7 breast-cancer cell line, was found to simulate the effect of the Mdm2 inhibitor nutlin-3 in having limited efficacy in triggering cell death. Nutlin-3 within the p53 functional background produced a state consistent with oscillatory p53 dynamics and cell-cycle arrest [69]. Typically, checkpoint breaching, for example under DNA decatenation stress induced by the DNA topoisomerase inhibitor ICRF-193 [72], drives p53 dys-functional cells towards mitotic failure. This event generates CIN via mitotic spindle breakage or the rupture of linked chromatids during condensation or seg-regation [73].

Previous studies have shown that human osteosarcoma U-2 OS cells (p53wt) exposed to ICRF-193 enter into a G_1-like polyploid cell-cycle phase via mitotic bypass with evidence of concomitant p53 stabilization and increased p21 (waf1) expression [60, 74]. U-2 OS cells are reported to show attenuated p21 (waf1) responses through the effects of the enhanced Mdm2 negative regulation of wtp53 and a downstream reduced rate of p21(waf1) mRNA translation [75, 76], raising the hypothesis of an underlying role for moderated p53 stabilization in favoring endocycle entry [56].

In a recent study using U-2 OS cells (Griesdoorn et al., unpublished data), nutlin-3 (1–5 μM) was found to impose moderate p53 stabilization across the cell cycle, deplete S-phase, and generate a G_1 polyploid subpopulation with no evidence of further cell-cycle progression until released from drug treatment. Cells that breached arrest or avoided mitotic bypass escaped through mitosis with minimal failure rates (Griesdoorn et al., unpublished data). The hypothesis is that non-genotoxic p53 stabilization per se presents a finite probability of promoting endocycle entry in cells not arrested proximal to an endocycle commitment stage in G_2. An implication is that oscillatory changes in the p53-Mdm2 loop bring with them a finite possibility of entry into CIN and survival. This is in line with the normal ability of p53 to alter the consequences of activation in response to the level of environmental stress [77]. For example, during starvation or oxygen stress, pathways are directed towards conservation and survival, while as environmental stress becomes persistent or more intense then the system switches to cell-death activation [77].

4 In Vitro Microenvironment Models and Hollow Fiber Technology

4.1 3D Models: Multicellular Tumor Spheroids

One limitation of multicellular tumor spheroids (MCTSs) is that during long-term culture there is a reliance upon the passive diffusion of both nutrients and O_2 into the spheroid mass. This causes the center of spheroids of diameters >600 μm to become necrotic [78]. Monitoring cell "health" with real-time reporting in MCTS systems is essential to assess the physiological state of the microcommunity within drug screens. For example, the cell-viability probe DRAQ7 can provide a sensitive, real-time readout of cell death induced by a variety of stressors such as hypoxia, starvation, and drug-induced cytotoxicity [79]. Figure 6a and b shows the detection of the necrotic center of a MCTS by confocal imaging using DRAQ5, which stains both live and dead cells. Figure 6c shows the tracking of spheroid cell health with the viability dye DRAQ7 and the appearance of non-viable cells during culture, using the Celigo® bench-top, microwell plate–based, bright-field and fluorescent imaging system.

Recent attempts to solve this drawback of static cultures for anti-cancer drug screens have included in vitro perfused, three-dimensional (3D) spheroid models based on the TissueFlex system that establishes controlled O_2, nutrient, and metabolite gradients [80]. In the cancer-cell attractor conceptual model, heterogeneity within a clonal cell population represents dynamic states clustered around a gene-expression attractor point. This highlights the importance of incorporating tumor heterogeneity into in vitro 3D tumor models that comprise competing or interacting phenotypes within an appropriate microenvironment [81]. Further, given

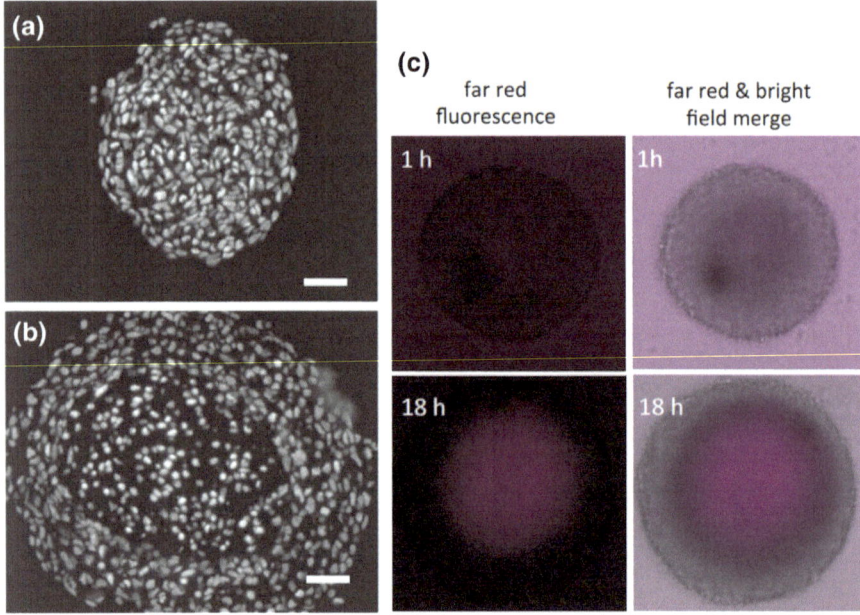

Fig. 6 Monitoring cell health in 3D MCTSs. **a** and **b** Far-red fluorescence images of confocal sections of spheroid human small-cell lung-cancer cell line adherent subline (AP78) of NCI-H69. Image shows nuclear labeling (live-cell penetrant dye DRAQ5; bar = 100 μm) and the development of a compromised center to the spheroid. **c** Real-time monitoring of the appearance of necrotic cores in U87MG spheroids (500 cells, 4-day formation of MCTS; stained with the non-toxic viability nuclear dye DRAQ7; 1.5 μM). Images obtained on Celigo Image Cytometer; courtesy of Nexcelom Bioscience

that tumor cell spread is a three-dimensional migration process in which cells attempt to survive selective pressures and colonize microenvironmental niches, it is reasonable to attempt to reiterate physical aspects of the barriers within in vitro models. For example, Liu et al. constructed a 3D microfabricated chip incorporating low planar areas and isolated square plinths (>100 microns) to assess cancer-cell movement potential. Highly metastatic PC-3 prostate-cancer cell movement was assessed by confocal microscopy; cells were found to readily occupy the plinths, whereas non-metastatic cells appeared to be inefficient at such an activity [82], possibly owing to the imposition of contact inhibition signals.

Changes in multicellular aggregate and spheroid cell populations can be assessed by enzymatic disassembly and FCM [35]. Continuous changes in cell states within their own attractor, having the effect of driving repopulation, can also be tracked using fluorescent-dye tracing [35]. Following changes in populations in situ can further exploit the increasing number of imaging platforms for 3D culture systems and linked techniques [83]. The physical location of a cell with an influential phenotype within a microcommunity can also have an impact upon the whole cellular system. For example, a recent study has shown that in a co-culture 3D in vitro spheroid model, the location of ABCG2-expressing cells (nominally drug-resistant owing to enhanced efflux; [84]) can act to control drug exposure within the whole tumor microenvironment [85]. Such micropharmacokinetic influences can be analyzed by FCM and confocal imaging, particularly if the drugs involved have convenient spectral properties [86] that also allow for compartmental modeling at the single-cell level [87–89].

4.2 Encapsulation Systems

Recreating controllable microenvironments in ex vivo model systems is a means of defining the critical influences on cell proliferation and behavior. Cell encapsulation aims to entrap viable cells within the confines of matrices or semi-permeable membrane barriers. Encapsulation physically isolates a cell mass from an outside environment and aims to maintain normal cellular physiology within the permeable barrier. These methods have been developed based on the promise of therapeutic usefulness for tissue transplantation [90] and also for testing the in vivo activity of chemotherapeutic drugs. Transplanted encapsulation systems result in cells being protected by the barrier from immune rejection, potentially allowing transplantation (allo- or xenotransplantation) without the need for immune suppression. Four aspects are critical in the development or success of the encapsulation approach, namely the platform permeability, mechanical properties, immune protection, and biocompatibility [91].

4.2.1 Microencapsulation and Macroencapsulation

Numerous cellular encapsulation techniques have been developed over the years, generally divided into two classes. *Microencapsulation* involves small spherical capsules (0.3–1.5 mm), most of them produced from hydrogel matrices, such as alginates. Examples involve microfluidic technologies for the production of perfectly spherical alginate microspheres with a fine size range [92]. *Macroencapsulation* involves a larger planar or cylindrical geometry, in which live cells are physically isolated from directly interacting with host tissue by enclosure between two or more selectively permeable flat sheet membranes or within the lumen of a semi-permeable tubular membrane of hollow fibers (HFs) [93–95]. These encapsulation platforms rely on the host animal's own homeostatic mechanisms for the control of pH, metabolic waste removal, electrolytes, and nutrients inside the encapsulation microenvironment [91]. HF technologies have found a range of applications, including cell filtration and bioreactor culture systems. The HF implant technology [94] is of particular note given that it is readily amenable to analysis by cytometric approaches, including FCM.

4.2.2   HF Bioreactor (HFB) Cell-Culture Systems

HF-assisted culture aims to continuously mimic optimal conditions when attempting to scale up cell-culture capacity, for example in expanding human embryonic stem populations [96] or to meet the demands of tissue-engineering applications for regenerative medicine [97]. HFBs also offer high culture surface area-to-volume ratio with the potential to enable high levels of cellular production with less population variation. The concept that HFB designs seek to embrace is that in attempting to balance cell production and system performance there is a fundamental need to provide a microenvironment that mimics that of a complex tissue, given the expected limitations in oxygen and nutrient delivery [98].

Several culture bioreactor designs for mammalian tissue growth meet the requirements of a wide variety of cell/tissue types and applications [99], including HF bioreactors. Generically a HF bioreactor consists of a closed vessel, normally a cylindrical module, filled with medium and mammalian cells into which a bundle of semi-permeable HFs is inserted. The HFs provide nutrients (i.e., glucose, serum, O_2) to the cells that grow in the extra-fiber space and eliminate their metabolic waste by-products (i.e., lactic acid and CO_2) through a constant flow of pumped medium. This efficient exchange results in increased cell densities and higher yields of secreted product, allowing long culture times and maintaining higher viability and cell morphology. This type of bioreactor is widely applied for the production of high-value biological molecules/cells for pharmaceutical or research use, such as cytokines [100], monoclonal antibodies [101], recombinant proteins and viruses [102], hepatocyte cultures [103], and extracorporeal hepatic assist devices [104–106].

4.2.3   HF Microenvironment Control

HF bioreactor design could also have the cells growing attached to the HF inner wall, with the nutrients in the same manner supplied through the fibers and the waste disposal performed by the outer-fiber medium. This type of culture config-uration has been applied to specific endothelial-cell models, where the cells attached to the inner wall surface are subject to a uniform shear stress that is directly proportional to the inner fiber flow rate, this being vital to maintain cell phenotype [107, 108]. It is also possible to have two different types of cells growing in the bioreactor in co-culture models, one type inside the HF lumen and another in the extra-fiber space, for example as a model of the in vitro blood-brain barrier [109].

The challenge for HF-assisted culture is particularly complex for adherent cells that may require a supportive matrix where design needs to mimic blood capillary geometry and where the transfer rate across the fiber wall is independent of the shear stresses experienced by the cellular community [97]. In attempts to under-stand the dynamic processes within HFBs there have been extensive attempts to model fluid flow, mass transport, and cell distribution, especially when cell behavior is driven by nutrient and chemotactic gradients [110]. Simulating the growth of cell aggregates (rat cardiomyocytes) along the outer surface of a single-fiber HFB, in which local oxygen concentration is flow-rate and pressure dependent, has been used to predict optimal conditions for growth and shear-stress tolerance (0.05 Pa) [111].

A critical factor for HFB performance is the retention or indeed acquisition of desired characteristics by the cellular community during the expansion phase. In particular, multilayer constructs for tissue engineering, sustained within a HFB system, may even allow for critical cell migration between layers [98]. HFB approaches have been used to explore capacity for novel immunotherapeutic approaches, such as the transplantation of human bone-marrow stromal cells, where preservation of immunophenotype and the absence of differentiation during device operation are critical [112]. A three-dimensional perfusion culture approach, achieved within a miniaturized HFB design comprising oxygen and medium capillaries, has been used to reduce demand for cells while permitting the demonstration of serum-free culture of primary human hepatocytes and efficient expression in the majority of P450 (CYP) isoenzymes, transport proteins, and enzymes of phase II metabolism [113]. Further, integrating video microscopy with a miniaturized HFB has enabled real-time observation of cell-population responses to imposed variations in medium, pH, and oxygenation for validation studies with primary cells (HUVEC, human hepatocytes) and cell lines (HuH7, THP-1) in mono- and co-culture formats [114].

4.2.4 HF Implant Assay

In essence, the HF platform represents an attempt to establish half way between truly in vivo systems and the in vitro assays. It should be reiterated that the HF platform is not intended to replace more detailed biological models, such as transgenic or knockout animal models that allow insights into biology and pathogenesis; neither can it mimic the interactions of tumor cells with the host tissue, as they relate to stromal or immunological responses [115]. Although various models using human tumors xenografted into immune-compromised mice have proved invaluable in chemotherapeutic-agent development and assessment, they are accompanied by various limitations, including high costs associated with large-scale screening, time, and the numbers of mice required. The empirical dosing and development of pharmacokinetic assays for each compound evaluated in xenograft models acts to greatly reduce the rate at which compounds progress to the clinic [94, 115, 116].

To address this problem, Hollingshead et al. [94] developed the HF assay. The HF assay is an in vivo test that involves the short-term growth of tumor cells within biocompatible modified-polyvinylidene fluoride (mPVDF) hollow fibers, permeable to substances with a molecular weight <500 kD, surgically implanted in mice prior to treatment. To address the growing demand for candidate drug testing in traditional xenograft models, the US National Cancer Institute (NCI) established a format for the hollow-fiber assay (HFA). Earlier studies indicated its predictive value in that drugs that show efficacy in HFAs generally also show anti-tumor activity in xenograft studies [115, 117, 118]. Overall, the HFA constitutes a simpler and cost-effective screening method to run [115, 118–120].

In brief, the methodology of the NCI HFA (as described by the Biological Testing Branch of the Developmental Therapeutics Program; Division of Cancer Treatment and Diagnosis) comprises a standard panel of 12 tumor cell lines heat-sealed within mPVDF hollow fibers and incubated in tissue-culture medium for 24–48 h prior to implantation and subsequent drug administration. The standard panel comprises NCI-H23, NCI-H522, MDA-MB-231, MDA-MB-435, SW-620, COLO 205, LOX, UACC-62, OVCAR-3, OVCAR-5, U251, and SF-295. HFAs can be used to track both cell behavior and perturbation by pharmacological agents when maintained either in vitro or as an encapsulated xenograft in vivo. The HFA has been used for the discovery of novel anticancer agents from fungi [121] and for natural product screening [122], while assay modifications have demonstrated its potential for analyzing anti-tumor vasculature agents [123, 124]. In the latter case, neovasculature arises as a consequence of the presence of tumor cells within HFs subcutaneously embedded on the dorsal flanks of mice. Furthermore, the HF assay can be carried out in vitro as an alternative with the potential to vary the external conditions or the sequence of compound addition and removal (see Fig. 7).

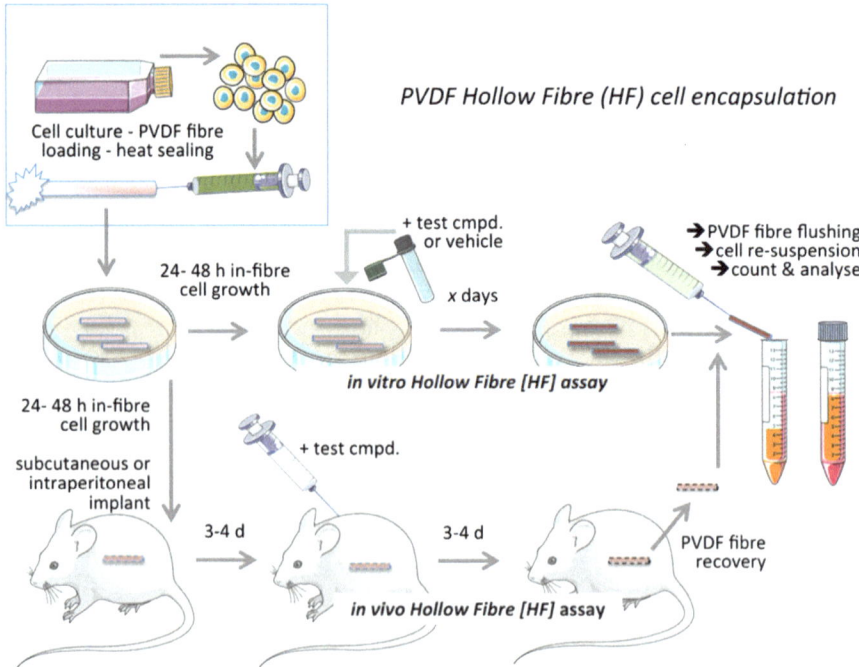

Fig. 7 Hollow-fiber (HF) cell encapsulation: outline procedure for in vitro cell culture and implant assay (for overview see Sharma et al. [118])

4.2.5   HFA Cytometry

The HF assay format is flexible, allowing analysis of cell populations in situ (by bioluminescence or fluorescence imaging of reporter cells), or direct examination of recovered HFs. Figure 8 shows SEM images of cells within sectioned HFs and the arrest of cells following Taxol exposure.

A clear cytometric advantage of the HFA is that the cellular encapsulated microcommunities can be retrieved from in vitro or in vivo contexts for FCM analysis of short-term pharmacodynamic endpoints, including viability, cell-cycle perturbations, and apoptosis [125]. To assess performance of the U-2 OS human osteosarcoma cell line in HFs we have used a subline, noted above, stably transfected with cyclin B1-fused eGFP (U2OS-GFP) [66, 126, 127] that has been used to track pharmacodynamic responses by FCM [60, 128]. A typical analysis of the calculated experimental cell density in the HF was found to be $\sim 1.2 \times 10^5$ cell/cm^2 HF surface area, while a 2D control culture had a density of $\sim 7.2 \times 10^4$ cell/cm^2 100-mm TC dish area. FCM showed that the population cell-cycle distribution was G1 = 64%; S = 13%; G2/M = 23% (sd_max. ±2.2, average of 3 HF) and G1 = 53%; S = 19%; G2/M = 28% (CON TC dish), suggesting the absence of an overt cell-cycle delay but a reduction in the number of cells in cycle upon encapsulation related to cell density. Figure 9 shows a typical flow cytometric analysis of U-2 OS cells recovered from encapsulation in HFs.

In earlier studies, many groups have explored different pharmacodynamic and pharmacokinetic end-point measurements in HFA systems, including DNA damage induction, apoptosis, cell-cycle analysis, and gene expression [125, 129–136]. The advantage of being able to retrieve a "pure" cell population from the HF, without host-cell contamination, was particularly evident in studies on prostate-cancer cells [130, 135], where the LNCaP HF model provided a mean of obtaining a pure population of LNCaP cells during the different stages of progression to androgen for gene-expression analysis.

Hollingshead et al. initially demonstrated detection of encapsulated biology without cell retrieval. This involved using tumor cells transfected with luciferase to provide a bioluminescent reporter-based dynamic assay [137] with further additional developments [119, 120, 138, 139]. Bioluminescence could be detected through the wall of the HF on subcutaneous implants, and the capacity of cells within an HF to support bioluminescence demonstrated that there was sufficient oxygenation of the contents of the fiber in spite of the lack of vascularization. Furthermore, the bioluminescence in subcutaneous fibers following administration of the luciferin substrate demonstrated the speed with which small-molecule substrates can be distributed systemically. This cytometric approach has the capacity to perform time-series imaging of many animals simultaneously, using short exposure times (~ 1 min) and sensitive imaging systems, and can exploit luciferase construct reporters for specific promoters.

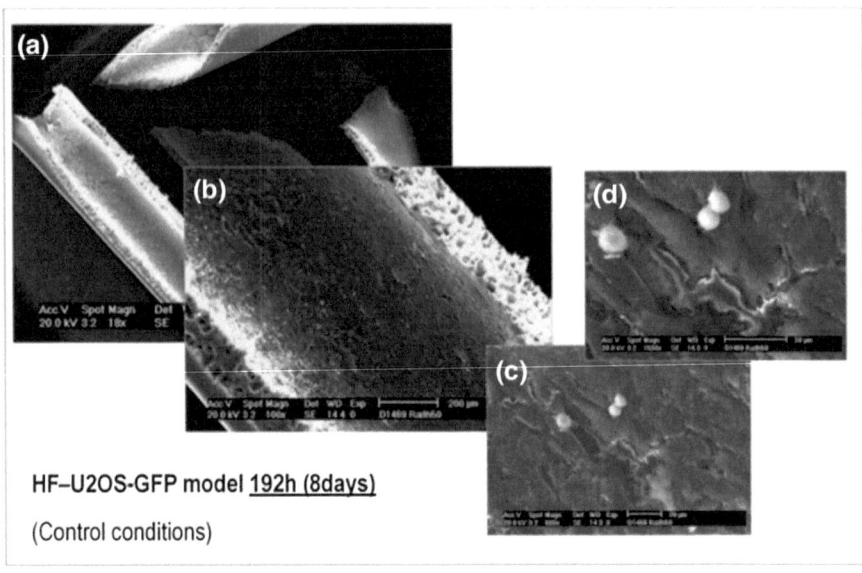

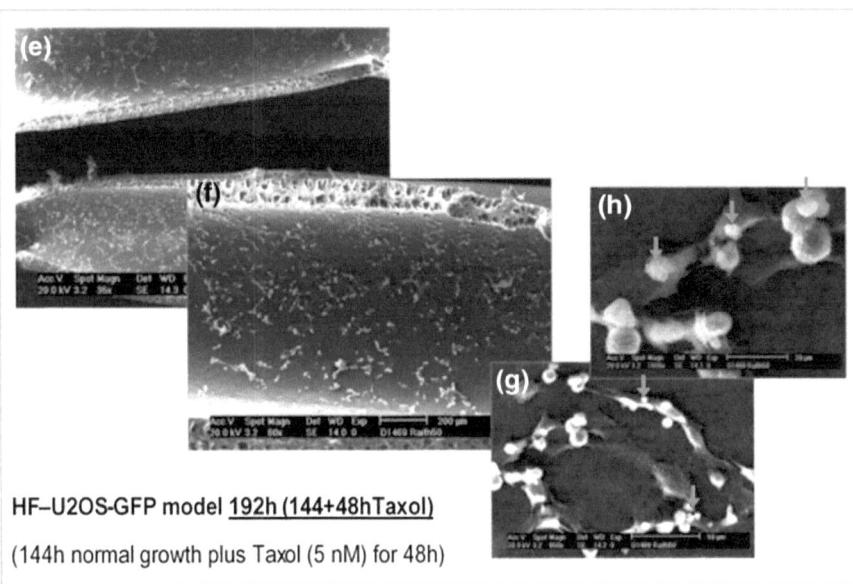

Fig. 8 Scanning electron microscopy images of longitudinally cut open HF–U2OS-GFP in-vitro cell culture at 192 h (8 days). Progressive zoom-in images of a typical HF (outer diameter 1.2 mm) loaded at $\sim 1.0 \times 10^6$ cells/ml (*top panel*) Normal-growth un-treated control cells at 192 h (8 days) and (*bottom panel*) 5 nM Taxol added at 144 h (day 6) and incubated for 48 h to demonstrate the blocking of cell division/mitotic arrest (*arrowed*) in-fiber

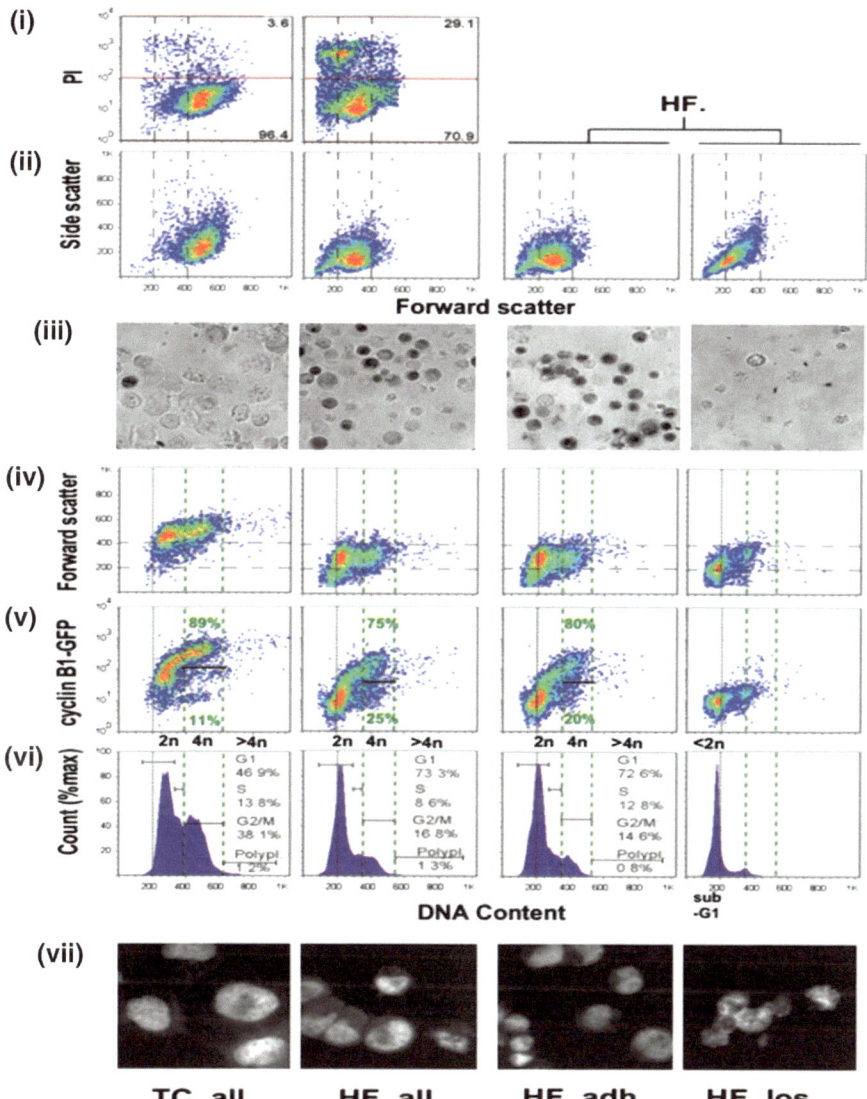

Fig. 9 Flow and image cytometric analysis of cells retrieved from the HF–U2OS-GFP model versus standard tissue culture after 192 h (8 days). The panels show the cell fraction analyzed versus the assay performed. Cell fraction (columns): (TC all) low confluency (~75% CF) standard tissue culture all cell content (first); (HF all) All cells isolated from the hollow fiber (second); (HF adh) Adherent-layer component requiring trypsinization to remove cells, ~90% all HF content (third); (HF_los) Flush-out component, comprising the "loose" cellular material, ~10% all HF content (fourth). Assay type (rows): *i* Identification of damaged/dead cells [propidium iodide (PI)]; *ii* forward- and side-scatter analysis; *iii* bright-field transmission image of the same samples in the plots (field: 128 × 128 μm); *iv* forward scatter versus DNA content (DRAQ5); *v* cyclin B1–GFP plots versus DNA content (DRAQ5), with detail of the G2/M fraction with high/low signal; *vi* cell-cycle histograms with compartment manual analysis (G1, S, G2/M, >4n Polyploidy); *vii* fluorescent image of fixed cells labeled with DRAQ5 to depict nucleus shape and chromatin status (field: 64 × 64 μm). *Vertical* and *horizontal dotted lines* on each panel provide a means for visualizing the assay across the row, thus comparing each cell fraction with every other. Data shown represent typical output, n = 3 × 3 (independent experiments x individual HF)

5 Microenvironment and Hypoxia

5.1 Modeling the Hypoxic Microenvironment

A problem arises when attempting to re-capitulate the relevant oxygen gradients within in vitro or in vivo tumor models. This issue is particularly relevant to oxygenation levels in micro-clusters of cells or MCTS models that attempt to mimic early metastatic entities. The current consensus is that the prognostic importance of isolated tumor cells (ITCs) and micrometastases (MICs) and metastases sized more than 2 mm can be regarded as a continuum. A micrometastasis is classified as a tumor deposit "greater than 0.2 mm but not greater than 2.0 mm in largest dimension." For example, the presence of metastatic disease is measured by the size of the largest contiguous metastasis for multiple foci in sentinel lymph node biopsy with <200 cells (ITC) versus >200 cells (micrometastasis) [140]. This sets pre-clinical models of micrometastases into perspective with the potential to represent hypoxic microenvironments [141]. Using previous descriptions of oxygen consumption and diffusion in three-dimensional tumor spheroids, Grimes et al. have developed a method of estimating rates of oxygen consumption and a diffusion limit of 232 ± 22 μm, suggesting that even micrometastases are subject to a degree of hypoxic stress [142].

5.2 Stromal-Tumor Hypoxia

With recent novel treatment concepts focusing on immunotherapy it is critical to view hypoxia as a potential barrier to achieving therapeutic intent because of the restriction of immune plasticity, and hypoxic cells initiating a sequence of events that promote the differentiation and expansion of immune-suppressive stromal cells. Essentially there is a remodeling of the microenvironment into an immunologically privileged site [143]. One approach to dealing with this challenge is to identify therapeutic targets within hypoxia-associated pathways. This brings with it the complexity of validation, FCM being well placed to assess the changing balance and competence of immune-system cells captured from microenvironment-defined locations [143].

Translating the above cellular complexities into in vivo model systems is a problem, since they are frequently generated in immune-compromised hosts. Further, they typically present as treatment-naïve systems, while in reality the clinical experience is that novel treatments are usually attempted in previously treated patients. For example, it is now appreciated that significant vascular, stromal, and immunological changes are imposed by irradiation [17], changing the perceived setting for the molecular targeting of hypoxia in radiotherapy [144].

Likewise, a study of prostate-cancer xenografts indicates that bicalutamide-induced hypoxia can select for cells that show malignant progression [145]. This places a premium on the detection and validation of the degree of hypoxia in model systems and on non-invasive tools to predict treatment outcome [146].

5.3   Hypoxia Detection

Hypoxia, a hallmark of most solid tumors, is a negative prognostic factor owing to its association with an aggressive tumor phenotype and therapeutic resistance. Given its prominent role in oncology, accurate detection of hypoxia is important, as it impacts prognosis and could influence treatment planning. A variety of approaches have been explored over the years for detecting and monitoring changes in hypoxia in tumors, including biological markers and noninvasive imaging techniques. A number of bioreductive hypoxia markers (based on 2-nitroimidazole chemistry) are in use, including pimonidazole, CCI-103F, EF5, and F-misonidazole [147]. Expression of cell-surface carbonic anhydrases as markers of hypoxia in vivo have been investigated using specific antibody detection, with evidence that fluorescently labeled CAIX-specific antibody is a robust indicator of hypoxia and correlates with pimonidazole staining patterns [148]. Currently, positron emission tomography (PET) is the preferred exploration method for non-invasive imaging of tumor hypoxia, exploiting the ability of tracers to report intracellular oxygenation levels. In this area, not all tumors are accessible to PET analysis at the resolution required, while the performance of different hypoxia-sensing radiotracers and their pharmacokinetic properties can be limiting [149]. ^{18}F-FAZA, a nitroimidazole PET tracer, is a promising imaging technique to guide hypoxia-targeted cancer therapy. It has been compared with other methods (e.g., measuring pO$_2$ using OxyLite, EPR oximetry, and ^{19}F-MRI) to qualify the ability of the ^{18}F-FAZA PET image intensities to report tumor oxygenation [150].

5.4   Cellular Responses to the Hypoxic Tumor Environment

Microenvironmental cues, including hypoxia, conspire to modulate the behavior of tumor cells within a stressed stromal setting, with shared roles for the hypoxia-inducible factor (HIF) family of transcription factors [5]. Under hypoxia, tumor cells enhance their co-option of stromal cells [9] into a complex response that leads to functional immunosuppression, enhanced tumor aggressiveness, and tumor spread (summarized in brief in Fig. 10).

This is a wide subject beyond the scope of this chapter, but it is clear that cytometry has a leading role in identifying the temporal heterogeneity and dynamic population content at such tumor sites. For example, tissue-resident macrophages demonstrate heterogeneity and uncertain provenance, and reside within a varying

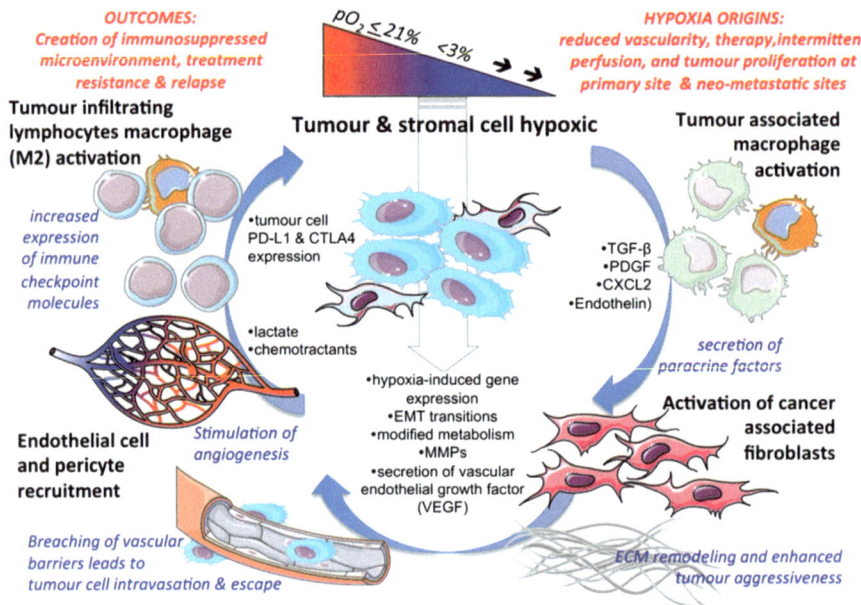

Fig. 10 The cellular and extracellular matrix recruitment cascade initiated by tumor and stromal cells within hypoxic microenvironments, leading to functional immunosuppression, enhanced tumor aggressiveness, and spread (reviews: [5, 9, 151])

signaling environment under normal and inflammatory conditions [152]. They can even demonstrate proliferative potential in vivo, as shown for peritoneal macrophages using FCM [153]. There are an increasing number of applications of FCM in tumor model systems to track the expression profiles of blood and tissue-infiltrating monocytes/macrophages [154].

Functionally, tumor-associated macrophages (TAMs) can be considered as phagocytic cells operating within a pro-inflammatory tissue. Under hypoxia there is a shift in the cellular metabolic profile, with an increased production of lactic acid, further propagating the TAM phenotype [152]. HIF-1α activation of glycolytic genes fosters this attempt to metabolically adapt to hypoxia via an increased conversion of glucose to pyruvate and subsequently to lactate, involving the shunting of glucose metabolites from mitochondria to glycolysis [155]. The production of lactic acid by tumor cells contributes to the signaling environment, leading to increased vascular endothelial growth factor (VEGF) expression and the M2-like polarization of TAMs, again mediated by HIF-1α [156]. This appears to be a critical step in the hypoxia cascade, given that the potent cytokine VEGF causes downstream recruitment of endothelial cells and pericytes from the surrounding vasculature.

It is recognized that the extracellular matrix (ECM) constrains not only tissue structure but also the integrity of mature blood vessels. The hypoxia cascade further overlays the ability of tumor cells to remodel the ECM. In neoplasia the ECM contributes to the modified behavior and migration potential of tumor cells, not least during epithelial to mesenchymal transitions. Hypoxia activates the expression of TET1, which also serves as a co-activator of HIF-1α transcriptional regulation to modulate HIF1α downstream target genes and promote EMT [157]. In a hypoxia study using FCM to analyze multiple endothelial-cell (EC) markers together with in situ analysis of collagen deposition, it has been suggested that ECM deposition and assembly by ECs is regulated by hypoxia-inducible factors 1α and 2α [158]. Critically, hypoxia eventually generates a persistent immunosuppressive microenvironment [143] by stimulating Treg cell function and increasing tumor-cell expression of immune checkpoint molecules (e.g., PD-L1 and CTLA4) [5].

The central features of the cellular responses to hypoxia appear to be preserved in multiple tumor types. Microenvironmental influences operate in vivo in small-cell lung carcinoma (SCLC), as supported by the finding that biopsy samples show a correlation between the expression of HIF-1α and VEGF [159]. The authors suggest that a HIF-1α/VEGF angiogenic pathway operates in vivo in SCLC, similar to that in non-SCLC [159]. Interestingly, despite its central role in angiogenesis, there has been only modest success for VEGF-targeted therapies, perhaps reflecting the complexity of the hypoxic milieu [160]. Next-generation targeted therapies are increasingly focused on immune modulation. It is likely that a constant challenge to their success will be the cellular response cascade persistently generated by the chemo-, radio-, and potentially immuno-resistant hypoxic fraction of tumors. One approach is to effectively target the environment itself to efficiently ablate cells (tumor, reactive TAMs, and CAFs) at these critical locations. This is the rationale behind the use of hypoxia-activated prodrugs.

6 Targeting Hypoxia

6.1 The Therapeutic Paradigm

The radiobiology and radiotherapy communities have collectively recognized the role of tumor hypoxia in cellular responses to radiation since the influential publications of Gray et al. [161, 162]. In developing approaches to deal with tumor hypoxia, a problem arises from our historical view of the categorization of tumor types based the organ of origin. This is a categorization to which we can readily relate. However, whilst there are clearly multiple histological cancer types and sub-types, a modern categorization with respect to treatment options is based on molecular pathology—a concept embodied in precision medicine and targeted therapies. Thus, cancers can be categorized and therefore treated, regardless of their site of origin, based, for example, on the inhibition of shared molecular drivers involved in cancer initiation and progression irrespective of their site of origin (e.g., EGFR, RAF, Myc, ALK fusion, or through an interference with functional p53 up-regulation) [163]. These cell-based targets are readily accessible to cytometric methodologies for profiling shared molecular pathways. Hypoxia is a common driver of tumor progression and metastasis, shared by multiple cancer types. Thus, exploiting hypoxia to location-target can be extended to multiple cancer types where the principle can also be used avoid normoxic tissue effects. The paradigm therefore is to target cells—tumor, stromal, or immune—that occupy hypoxic microenvironments to prevent the appearance of resistance and immunosuppression-generating cell-response loops [164–166]. Leading candidates in this area of microenvironment targeting include hypoxia-activated prodrugs and hypoxia response inhibitors that target carbonic anhydrase IX [146].

Targeting hypoxia in multiple tumor types presents a "generic" approach to the control of tumor progression and metastasis irrespective of the tissue of origin, since the microenvironment is the tissue-differentiating principle and not the tumor type. This also allows for drug screening in different in vitro and in vivo models to establish proof of principle. Clinical acceptance of this view is evidenced by the progression of the anthraquinone-based AQ4N (Banoxantrone) to a Phase II clinical trial in glioblastoma, and the more recent progression of TH-302 (Evofosfamide) to Phase III clinical trials in soft-tissue sarcoma and pancreatic ductal adenocarcinoma (PDAC). It is increasingly apparent that the fundamental mechanism of "activation" of a drug under hypoxia is an important factor for targeting, together with the properties of the pharmacophore within a complex cellular environment.

6.2   Hypoxia-Activated Prodrugs: HAPs

Hypoxia-activated prodrugs (HAPS) are a diverse group of chemicals that, crucially, are reduced in single-electron steps by several enzymes, including cytochrome P450 reductase [165]. Five different chemical entities have been shown to be capable of selectively targeting hypoxic cells, and these include nitro (hetero)-cyclic compounds, aromatic N-oxides, aliphatic N-oxides, quinone, and transition-metal complexes [167]. An ideal HAP should combine certain fundamental properties, including the ability to penetrate through to the target cells during the window of pharmacokinetic availability, and undergo preferential activation under pO_2 levels that differentiate normal from target tissues [6]; the activated product should have significant cytotoxic potential in both proliferating and non-proliferating cell subpopulations found within hypoxic regions. Further properties start to address the wider potential clinical value, and these include low toxicity and adverse-effect profile, persistence of the activated drug at target to contribute to downstream maintenance of cytotoxic action, and bystander effects [168].

Those HAPs designed for clinical applications have previously been allocated to two broad classes [169]. Class I HAPs (like the benzotriazine N-oxides tirapazamine and SN30000), are activated under relatively mild hypoxia. In contrast, Class II HAPs (such as the nitro compounds PR-104A or TH-302) are maximally activated only under extreme hypoxia. On reduction some produce nitrogen mustard alkylating agents, which bind covalently to DNA. Recent HAPs of this type are TH-302 [170] and PR104 [167]. Alternatively, reduction of some HAPs (e.g., tirapazamine; 3-amino-1,2,4-benzotriazine 1,4-dioxide; TPZ) generates toxic hydroxyl radicals that cause DNA double-strand breaks. Typically, none of these HAPs forms a persistent cytotoxin since the reactive intermediates are (i) lost by further reduction, (ii) consumed through binding of the activated functional group, or (iii) have the one-electron reduction rapidly reversed in the presence of moderate levels of oxygen. This 'redox cycling' is associated with production of cytotoxic superoxide radicals. Thus, for these HAPs, hypoxia-selective cell killing is dependent on the reduced radical that accumulates in hypoxic cells being more toxic than the superoxide formed in normoxic cells.

6.3   Unidirectional HAPs (uHAPs) and Multilevel Targeting

Unidirectional HAPs (AQ4N and OCT1002) are a distinct class of HAP. The first-in-class comprise di-N-oxides of tertiary amine anthraquinones and are reduced to a cytotoxin through an irreversible two-step process, each involving two-electron reduction. Consequently, they are defined as "unidirectional HAPs." This difference is crucial, as it results in unique characteristics of these prodrugs. Different proteins catalyze this reduction, including several cytochrome P450s, nitric oxide synthase, and in principle any Fe^{2+} hemoprotein, these reactions being inhibited by oxygen. AQ4N and OCT1002 are electrically neutral but have an unequal distribution of charge across the two functional N-oxide groups; this allows for extensive prodrug distribution, including into the brain and poorly-vascularized tumors. In distinct contrast, the corresponding reduced drugs (AQ4 and OCT1001) are tertiary amine intercalators that bind non-covalently to DNA with very high affinity. Thus, they remain tightly bound for many days in the hypoxic cell in which they are formed. The reduced drugs AQ4 and OCT1001 remain localized and cytotoxic irrespective of any subsequent changes in oxygen, causing cell death by inhibition of DNA processing and topoisomerase II activity. AQ4N has shown considerable efficacy in a range of preclinical tumor models with clinical proof of concept [171]. Both AQ4N [145] and OCT1002 [172] demonstrate significant activity in prostate-cancer (PrCa) xenografts.

Under tumor-relevant hypoxic conditions the two sequential $2e^-$ reductions of uHAPs yield the targeted toxic metabolites. The concept is shown in Fig. 11 together with the hypoxia dependency for activation of OCT1002, the induction of cell-cycle arrest, and DNA damage signaling. These stable reduction products have very high affinity for DNA and the targeting of topoisomerase II, which can effect a long-term inhibition of both DNA replication and cell-cycle traverse. The stability of AQ4N in oxygenated tissues, its independence from flavoprotein reductases for

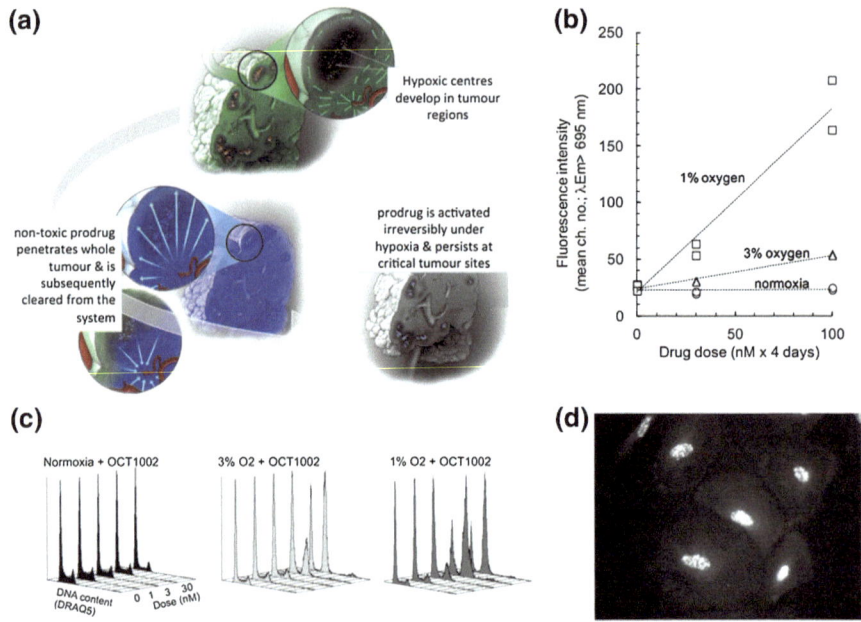

Fig. 11 **a** Schematic representation of the unidirectional hypoxia-activated pro-drug concept. **b** Reciprocity between pO_2 level exposure and accumulation/retention of OCT1001. Flow cytometric detection of hypoxic activation of OCT1002 to a persistent intracellular metabolite detected by cellular retention of the far-red fluorescent OCT1001. Human lung adenocarcinoma A549 cells were incubated for 4 days in the constant presence of OCT1002 prior to collection and analysis by FCM. **c** Hypoxia-dependent late cell-cycle arrest in A549 cells exposed to OCT1002 × 4 days. **d** A549 cells treated with 100 nM OCT1002 in 1% hypoxia for 5 days and probed for long-term DNA damage effects. Immunofluorescence image shows the detection of nuclear γH2AX phosphorylation

conversion, and AQ4N's specific inhibition of the DNA/Topo II–driven tumor proliferation distinguish it from HAPs, including TH-302, a non-specific DNA-alkylating agent based on the oxazaphosphorines, like cyclophosphamide [165]. Since the active form of the uHAP is generated within the cell, it turns tumor cells into "drug factories"—side-stepping drug-resistance mechanisms associated with uptake and efflux.

7 The Continuing Challenge

Given the physical and cellular complexity of the microenvironment, cytometry is a key generic technology for probing heterogeneity while offering routes to explore its exploitation. The example of uHAP development, along with other anti-cancer therapies that aim to exploit the different milieus that tumor cells either create or are attracted to, will constantly require novel probes and new cytometric methods. Currently, the central challenge is to associate critical bioactivities in cellular microenvironments with the differential expression of immune and stress response pathways in different tumor types. The downstream aim is to provide a rational approach to precision medicine–based combination therapies. Here the potential to disrupt the immunosuppressive micro-environment by hypoxia targeting is highly promising.

Acknowledgements and Declarations The authors acknowledge the contributions of Dr. Robert Falconer (Bradford University, UK; support by Yorkshire Cancer Research) and Dr. Emeline Furon (Cardiff University, UK, EPSRC Case award; Protec'Som SAS, Valognes, France) on NCAM-polysialylation studies. The authors also acknowledge the contributions of Mrs. Marie Wiltshire (Cardiff University), Mrs. Sally Chappell (Cardiff University; Malaghan Institute of Medical Research, New Zealand), and Dr. Peter Giles (Cardiff University) for flow cytometry, imaging, and bioinformatics support, respectively. Authors PJS and RJE declare that they are non-executive directors of Biostatus Ltd, the commercial supplier of DRAQ5 & DRAQ7. PJS is a director of Oncotherics Ltd, developer of uHAP technology. VG & ORS have no conflicts to declare. We thank Oncotherics Ltd for permission to reproduce artwork in Fig. 11.

References

1. Jessop R (2012) Coinage of the term environment: a word without authority and carlyle's displacement of the mechanical metaphor. Lit Compass 9:708–720
2. Lidstrom ME, Konopka MC (2010) The role of physiological heterogeneity in microbial population behavior. Nat Chem Biol 6:705–712. doi:10.1038/nchembio.436
3. Ferrari BC, Winsley TJ, Bergquist PL, Van Dorst J (2012) Flow cytometry in environmental microbiology: a rapid approach for the isolation of single cells for advanced molecular biology analysis. Methods Mol Biol 881:3–26. doi:10.1007/978-1-61779-827-6_1
4. Sgier L, Freimann R, Zupanic A, Kroll A (2016) Flow cytometry combined with viSNE for the analysis of microbial biofilms and detection of microplastics. Nat Commun 7:11587. doi:10.1038/ncomms11587
5. LaGory EL, Giaccia AJ (2016) The ever-expanding role of HIF in tumour and stromal biology. Nat Cell Biol 18:356–365. doi:10.1038/ncb3330
6. McKeown SR (2014) Defining normoxia, physoxia and hypoxia in tumours-implications for treatment response. Br J Radiol 87:20130676. doi:10.1259/bjr.20130676
7. Perez-Velazquez J, Gevertz JL, Karolak A, Rejniak KA (2016) Microenvironmental niches and sanctuaries: a route to acquired resistance. Adv Exp Med Biol 936:149–164. doi:10.1007/978-3-319-42023-3_8
8. Broxterman HJ, Georgopapadakou NH (2007) Anticancer therapeutics: a surge of new developments increasingly target tumor and stroma. Drug Resist Updat 10:182–193. doi:10.1016/j.drup.2007.07.001
9. McMillin DW, Negri JM, Mitsiades CS (2013) The role of tumour-stromal interactions in modifying drug response: challenges and opportunities. Nat Rev Drug Discov 12:217–228. doi:10.1038/nrd3870
10. Feig C, Gopinathan A, Neesse A, Chan DS, Cook N, Tuveson DA (2012) The pancreas cancer microenvironment. Clin Cancer Res 18:4266–4276. doi:10.1158/1078-0432.CCR-11-3114, 18/16/4266 [pii]

11. Crespo J, Sun H, Welling TH, Tian Z, Zou W (2013) T cell anergy, exhaustion, senescence, and stemness in the tumor microenvironment. Curr Opin Immunol 25:214–221. doi:10.1016/j.coi.2012.12.003

12. Hamada S, Masamune A, Shimosegawa T (2013) Novel therapeutic strategies targeting tumor-stromal interactions in pancreatic cancer. Front Physiol 4:331. doi:10.3389/fphys.2013.00331

13. Pitt JM, Marabelle A, Eggermont A, Soria JC, Kroemer G, Zitvogel L (2016) Targeting the tumor microenvironment: removing obstruction to anticancer immune responses and immunotherapy. Ann Oncol 27:1482–1492. doi:10.1093/annonc/mdw168

14. Smyth MJ, Ngiow SF, Ribas A, Teng MW (2016) Combination cancer immunotherapies tailored to the tumour microenvironment. Nat Rev Clin Oncol 13:143–158. doi:10.1038/nrclinonc.2015.209

15. Beavis PA, Slaney CY, Kershaw MH, Gyorki D, Neeson PJ, Darcy PK (2016) Reprogramming the tumor microenvironment to enhance adoptive cellular therapy. Semin Immunol 28:64–72. doi:10.1016/j.smim.2015.11.003

16. Csermely P, Korcsmaros T (2013) Cancer-related networks: a help to understand, predict and change malignant transformation. Semin Cancer Biol 23:209–212. doi:10.1016/j.semcancer.2013.06.011

17. Barker HE, Paget JT, Khan AA, Harrington KJ (2015) The tumour microenvironment after radiotherapy: mechanisms of resistance and recurrence. Nat Rev Cancer 15:409–425. doi:10.1038/nrc3958

18. Ratajczak MZ, Suszynska M, Kucia M (2016) Does it make sense to target one tumor cell chemotactic factor or its receptor when several chemotactic axes are involved in metastasis of the same cancer? Clin Transl Med 5:28. doi:10.1186/s40169-016-0113-6

19. Smith PJ, Khan IA, and Errington RJ (2009) Cytomics and cellular informatics—coping with asymmetry and heterogeneity in biological systems. Drug Discov Today 14:271–277. doi: 10.1016/j.drudis.2008.11.012, S1359–6446(08)00404-2 [pii]

20. Smith PJ, Falconer RA, Errington RJ (2013) Micro-community cytometry: sensing changes in cell health and glycoconjugate expression by imaging and flow cytometry. J Microsc 251:113–122. doi:10.1111/jmi.12060

21. Marr C, Zhou JX, Huang S (2016) Single-cell gene expression profiling and cell state dynamics: collecting data, correlating data points and connecting the dots. Curr Opin Biotechnol 39:207–214. doi:10.1016/j.copbio.2016.04.015

22. Atkuri KR, Stevens JC, Neubert H (2015) Mass cytometry: a highly multiplexed single-cell technology for advancing drug development. Drug Metab Dispos 43:227–233. doi:10.1124/dmd.114.060798

23. Mair F, Hartmann FJ, Mrdjen D, Tosevski V, Krieg C, Becher B (2016) The end of gating? An introduction to automated analysis of high dimensional cytometry data. Eur J Immunol 46:34–43. doi:10.1002/eji.201545774

24. Coghlin C, Murray GI (2014) The role of gene regulatory networks in promoting cancer progression and metastasis. Future Oncol 10:735–748. doi:10.2217/fon.13.264

25. Huang S (2011) On the intrinsic inevitability of cancer: from foetal to fatal attraction. Semin Cancer Biol 21:183–199. doi:10.1016/j.semcancer.2011.05.003

26. Haldane JB, Waddington CH (1931) Inbreeding and linkage. Genetics 16:357–374

27. Waddington CH (1939) Introduction to modern genetics. Macmillan, New York

28. Esteller M (2009) Epigenetics in biology and medicine. In: Esteller M (ed) C. Press, UK

29. Creixell P, Schoof EM, Erler JT, Linding R (2012) Navigating cancer network attractors for tumor-specific therapy. Nat Biotechnol 30:842–848. doi:10.1038/nbt.2345

30. Koulakov AA, Lazebnik Y (2012) The problem of colliding networks and its relation to cell fusion and cancer. Biophys J 103:2011–2020. doi:10.1016/j.bpj.2012.08.062

31. Huang S, Kauffman S (2013) How to escape the cancer attractor: rationale and limitations of multi-target drugs. Semin Cancer Biol 23:270–278. doi:10.1016/j.semcancer.2013.06.003

32. Cheng WY, Ou Yang TH, Anastassiou D (2013) Biomolecular events in cancer revealed by attractor metagenes. PLoS Comput Biol 9:e1002920. doi:10.1371/journal.pcbi.1002920

33. Smith PJ (2016) Cytometric routes to single cell transcriptomics. Cytometry A 89:424–426. doi:10.1002/cyto.a.22869

34. Sun L, Fang J (2016) Epigenetic regulation of epithelial-mesenchymal transition. Cell Mol Life Sci 73:4493–4515. doi:10.1007/s00018-016-2303-1

35. Li Q, Wennborg A, Aurell E, Dekel E, Zou JZ, Xu Y, Huang S, Ernberg I (2016) Dynamics inside the cancer cell attractor reveal cell heterogeneity, limits of stability, and escape. Proc Natl Acad Sci USA 113:2672–2677. doi:10.1073/pnas.1519210113

36. Campbell K, Casanova J (2016) A common framework for EMT and collective cell migration. Development 143:4291–4300. doi:10.1242/dev.139071

37. Bennun SV, Hizal DB, Heffner K, Can O, Zhang H, Betenbaugh MJ (2016) Systems glycobiology: integrating glycogenomics, glycoproteomics, glycomics, and other 'omics data sets to characterize cellular glycosylation processes'. J Mol Biol 428:3337–3352. doi:10.1016/j.jmb.2016.07.005

38. Hirabayashi J, Arata Y, Kasai K (2001) Glycome project: concept, strategy and preliminary application to *Caenorhabditis elegans*. Proteomics 1:295–303. doi:10.1002/1615-9861(200102)1:2<295:aid-prot295>3.0.co;2-c

39. Nie H, Li Y, Sun XL (2012) Recent advances in sialic acid-focused glycomics. J Proteomics 75:3098–3112. doi:10.1016/j.jprot.2012.03.050, S1874-3919(12)00201-1 [pii]

40. Falconer RA, Errington RJ, Shnyder SD, Smith PJ, Patterson LH (2012) Polysialyltransferase: a new target in metastatic cancer. Curr Cancer Drug Targets 12:925–939

41. Rutishauser U (1992) NCAM and its polysialic acid moiety: a mechanism for pull/push regulation of cell interactions during development? Dev Suppl 99–104

42. Rutishauser U (2008) Polysialic acid in the plasticity of the developing and adult vertebrate nervous system. Nat Rev Neurosci 9:26–35. doi:10.1038/nrn2285

43. El Maarouf A, Rutishauser U (2010) Use of PSA-NCAM in repair of the central nervous system. Adv Exp Med Biol 663:137–147. doi:10.1007/978-1-4419-1170-4_9

44. Al-Saraireh YM, Sutherland M, Springett BR, Freiberger F, Ribeiro Morais G, Loadman PM, Errington RJ, Smith PJ, Fukuda M, Gerardy-Schahn R, Patterson LH, Shnyder SD, Falconer RA (2013) Pharmacological inhibition of polysialyltransferase ST8SiaII modulates tumour cell migration. PLoS ONE 8:e73366. doi:10.1371/journal.pone.0073366

45. Jokilammi A, Ollikka P, Korja M, Jakobsson E, Loimaranta V, Haataja S, Hirvonen H, Finne J (2004) Construction of antibody mimics from a noncatalytic enzyme-detection of polysialic acid. J Immunol Methods 295:149–160. doi:10.1016/j.jim.2004.10.006, S0022-1759(04)00367-9 [pii]

46. Jakobsson E, Schwarzer D, Jokilammi A, Finne J (2012) Endosialidases: versatile tools for the study of polysialic acid. Top Curr Chem. doi:10.1007/128_2012_349

47. Smith PJ, Furon E, Wiltshire M, Chappell S, Patterson LH, Shnyder SD, Falconer RA, Errington RJ (2013) NCAM polysialylation during adherence transitions: live cell monitoring using an antibody-mimetic EGFP-endosialidase and the viability dye DRAQ7. Cytometry A 83:659–671. doi:10.1002/cyto.a.22306

48. Giam M, Rancati G (2015) Aneuploidy and chromosomal instability in cancer: a jackpot to chaos. Cell Div 10:3. doi:10.1186/s13008-015-0009-7

49. Frattini A, Fabbri M, Valli R, De Paoli E, Montalbano G, Gribaldo L, Pasquali F, Maserati E (2015) High variability of genomic instability and gene expression profiling in different HeLa clones. Sci Rep 5:15377. doi:10.1038/srep15377

50. Sharma S, Bhonde R (2015) Mesenchymal stromal cells are genetically stable under a hostile in vivo-like scenario as revealed by in vitro micronucleus test. Cytotherapy 17:1384–1395. doi:10.1016/j.jcyt.2015.07.004

51. Raslova H, Kauffmann A, Sekkai D, Ripoche H, Larbret F, Robert T, Le Roux DT, Kroemer G, Debili N, Dessen P, Lazar V, Vainchenker W (2007) Interrelation between polyploidization and megakaryocyte differentiation: a gene profiling approach. Blood 109:3225–3234. doi:10.1182/blood-2006-07-037838

52. Pandit SK, Westendorp B, de Bruin A (2013) Physiological significance of polyploidization in mammalian cells. Trends Cell Biol 23:556–566. doi:10.1016/j.tcb.2013.06.002
53. Fox DT, Duronio RJ (2013) Endoreplication and polyploidy: insights into development and disease. Development 140:3–12. doi:10.1242/dev.080531
54. Shi Q, King RW (2005) Chromosome nondisjunction yields tetraploid rather than aneuploid cells in human cell lines. Nature 437:1038–1042. doi:10.1038/nature03958
55. Musacchio A, Salmon ED (2007) The spindle-assembly checkpoint in space and time. Nat Rev Mol Cell Biol 8:379–393. doi:10.1038/nrm2163
56. Rello-Varona S, Vitale I, Kepp O, Senovilla L, Jemaa M, Metivier D, Castedo M, Kroemer G (2009) Preferential killing of tetraploid tumor cells by targeting the mitotic kinesin Eg5. Cell Cycle 8:1030–1035. doi:10.4161/cc.8.7.7950
57. Filby A, Day W, Purewal S, Martinez-Martin N (2016) The analysis of cell cycle, proliferation, and asymmetric cell division by imaging flow cytometry. Methods Mol Biol 1389:71–95. doi:10.1007/978-1-4939-3302-0_5
58. Davoli T, de Lange T (2011) The causes and consequences of polyploidy in normal development and cancer. Annu Rev Cell Dev Biol 27:585–610. doi:10.1146/annurev-cellbio-092910-154234
59. Storchova Z, Kuffer C (2008) The consequences of tetraploidy and aneuploidy. J Cell Sci 121:3859–3866. doi:10.1242/jcs.039537
60. Smith PJ, Marquez N, Wiltshire M, Chappell S, Njoh K, Campbell L, Khan IA, Silvestre O, Errington RJ (2007) Mitotic bypass via an occult cell cycle phase following DNA topoisomerase II inhibition in p 53 functional human tumor cells. Cell Cycle 6:2071–2081. 4585 [pii]
61. Mc Gee MM (2015) Targeting the mitotic catastrophe signaling pathway in cancer. Mediators Inflamm 2015:146282. doi:10.1155/2015/146282
62. Duelli D, Lazebnik Y (2007) Cell-to-cell fusion as a link between viruses and cancer. Nat Rev Cancer 7:968–976. doi:10.1038/nrc2272
63. Gomez-Icazbalceta G, Ruiz-Rivera MB, Lamoyi E, Huerta L (2015) FRET in the analysis of in vitro cell-cell fusion by flow cytometry. Methods Mol Biol 1313:217–227. doi:10.1007/978-1-4939-2703-6_16
64. Mohr M, Zaenker KS, Dittmar T (2015) Fusion in cancer: an explanatory model for aneuploidy, metastasis formation, and drug resistance. Methods Mol Biol 1313:21–40. doi:10.1007/978-1-4939-2703-6_2
65. Gong J, Traganos F, Darzynkiewicz Z (1995) Discrimination of G2 and mitotic cells by flow cytometry based on different expression of cyclins A and B1. Exp Cell Res 220:226–231. doi:10.1006/excr.1995.1310
66. Thomas N, Kenrick M, Giesler T, Kiser G, Tinkler H, Stubbs S (2005) Characterization and gene expression profiling of a stable cell line expressing a cell cycle GFP sensor. Cell Cycle 4:191–195
67. Nakazawa N, Mehrotra R, Arakawa O, Yanagida M (2016) ICRF-193, an anticancer topoisomerase II inhibitor, induces arched telophase spindles that snap, leading to a ploidy increase in fission yeast. Genes Cells 21:978–993. doi:10.1111/gtc.12397
68. Smith PJ, Blunt N, Wiltshire M, Hoy T, Teesdale-Spittle P, Craven MR, Watson JV, Amos WB, Errington RJ, Patterson LH (2000) Characteristics of a novel deep red/infrared fluorescent cell-permeant DNA probe, DRAQ5, in intact human cells analyzed by flow cytometry, confocal and multiphoton microscopy. Cytometry 40:280–291
69. Choi M, Shi J, Jung SH, Chen X, Cho KH (2012) Attractor landscape analysis reveals feedback loops in the p 53 network that control the cellular response to DNA damage. Sci Signal 5:ra83. doi:10.1126/scisignal.2003363
70. Shaltiel IA, Krenning L, Bruinsma W, Medema RH (2015) The same, only different—DNA damage checkpoints and their reversal throughout the cell cycle. J Cell Sci 128:607–620. doi:10.1242/jcs.163766
71. Meek DW (2015) Regulation of the p53 response and its relationship to cancer. Biochem J 469:325–346. doi:10.1042/bj20150517

72. Luo K, Yuan J, Chen J, Lou Z (2009) Topoisomerase IIalpha controls the decatenation checkpoint. Nat Cell Biol 11:204–210. doi:10.1038/ncb1828

73. Iwai M, Hara A, Andoh T, Ishida R (1997) ICRF-193, a catalytic inhibitor of DNA topoisomerase II, delays the cell cycle progression from metaphase, but not from anaphase to the G1 phase in mammalian cells. FEBS Lett 406:267–70

74. Smith PJ, Chin SF, Njoh K, Khan IA, Chappell MJ, Errington RJ (2008) Cell cycle checkpoint-guarded routes to catenation-induced chromosomal instability. SEB Exp Biol Ser 59:219–242

75. Chang LJ, Eastman A (2012) Differential regulation of p21 (waf1) protein half-life by DNA damage and Nutlin-3 in p53 wild-type tumors and its therapeutic implications. Cancer Biol Ther 13:1047–1057. doi:10.4161/cbt.21047

76. Chang LJ, Eastman A (2012) Decreased translation of p21waf1 mRNA causes attenuated p53 signaling in some p53 wild-type tumors. Cell Cycle 11:1818–1826. doi:10.4161/cc. 20208

77. Humpton TJ, Vousden KH (2016) Regulation of cellular metabolism and hypoxia by p 53. Cold Spring Harb Perspect Med 6. doi:10.1101/cshperspect.a026146

78. Friedrich J, Seidel C, Ebner R, Kunz-Schughart LA (2009) Spheroid-based drug screen: considerations and practical approach. Nat Protoc 4:309–324. doi:10.1038/nprot.2008.226

79. Akagi J, Kordon M, Zhao H, Matuszek A, Dobrucki J, Errington R, Smith PJ, Takeda K, Darzynkiewicz Z, Wlodkowic D (2013) Real-time cell viability assays using a new anthracycline derivative DRAQ7(R). Cytometry A 83:227–234. doi:10.1002/cyto.a.22228

80. Wan X, Li Z, Ye H, Cui Z (2016) Three-dimensional perfused tumour spheroid model for anti-cancer drug screening. Biotechnol Lett 38:1389–1395. doi:10.1007/s10529-016-2035-1

81. Fong EL, Harrington DA, Farach-Carson MC, Yu H (2016) Heralding a new paradigm in 3D tumor modeling. Biomaterials 108:197–213. doi:10.1016/j.biomaterials.2016.08.052

82. Liu L, Sun B, Pedersen JN, Aw Yong KM, Getzenberg RH, Stone HA, Austin RH (2011) Probing the invasiveness of prostate cancer cells in a 3D microfabricated landscape. Proc Natl Acad Sci USA 108:6853–6856. doi:10.1073/pnas.1102808108

83. Costa EC, Moreira AF, de Melo-Diogo D, Gaspar VM, Carvalho MP, Correia IJ (2016) 3D tumor spheroids: an overview on the tools and techniques used for their analysis. Biotechnol Adv 34:1427–1441. doi:10.1016/j.biotechadv.2016.11.002

84. Smith PJ, Wiltshire M, Chappell SC, Cosentino L, Burns PA, Pors K, Errington RJ (2013) Kinetic analysis of intracellular Hoechst 33342–DNA interactions by flow cytometry: misinterpretation of side population status? Cytometry A 83:161–169. doi:10.1002/cyto.a. 22224

85. Curran S, Vantangoli MM, Boekelheide K, Morgan JR (2015) Architecture of chimeric spheroids controls drug transport. Cancer Microenviron 8:101–109. doi:10.1007/s12307-015-0171-0

86. Errington RJ, Ameer-Beg SM, Vojnovic B, Patterson LH, Zloh M, Smith PJ (2005) Advanced microscopy solutions for monitoring the kinetics and dynamics of drug-DNA targeting in living cells. Adv Drug Deliv Rev 57:153–167. doi:10.1016/j.addr.2004.05.005

87. Evans ND, Errington RJ, Shelley M, Feeney GP, Chapman MJ, Godfrey KR, Smith PJ, Chappell MJ (2004) A mathematical model for the in vitro kinetics of the anti-cancer agent topotecan. Math Biosci 189:185–217. doi:10.1016/j.mbs.2004.01.007, S002555640400063X [pii]

88. Cheung SY, Evans ND, Chappell MJ, Godfrey KR, Smith PJ, Errington RJ (2008) Exploration of the intercellular heterogeneity of topotecan uptake into human breast cancer cells through compartmental modelling. Math Biosci 213:119–134. doi:10.1016/j.mbs.2008. 03.008, S0025-5564(08)00059-X [pii]

89. Chappell MJ, Evans ND, Errington RJ, Khan IA, Campbell L, Ali R, Godfrey KR, and Smith PJ (2008) A coupled drug kinetics-cell cycle model to analyse the response of human cells to intervention by topotecan. Comput Methods Programs Biomed 89:169–178. doi:10. 1016/S0169-2607(07)00265-9, 10.1016/j.cmpb.2007.11.002 [pii]

90. Thanos CG, Bintz B, Emerich DF (2010) Microencapsulated choroid plexus epithelial cell transplants for repair of the brain. Adv Exp Med Biol 670:80–91
91. Uludag H, De Vos P, Tresco PA (2000) Technology of mammalian cell encapsulation. Adv Drug Deliv Rev 42:29–64
92. Workman VL, Dunnett SB, Kille P, Palmer DD (2007) Microfluidic chip-based synthesis of alginate microspheres for encapsulation of immortalized human cells. Biomicrofluidics 1 (1):014105. doi:10.1063/1.2431860
93. Lanza RP, Butler DH, Borland KM, Staruk JE, Faustman DL, Solomon BA, Muller TE, Rupp RG, Maki T, Monaco AP, Chick WL (1991) Xenotransplantation of canine, bovine, and porcine islets in diabetic rats without immunosuppression. Proc Nat Acad Sci USA 88:11100–11104
94. Hollingshead MG, Alley MC, Camalier RF, Abbott BJ, Mayo JG, Malspeis L, Grever MR (1995) In vivo cultivation of tumor cells in hollow fibers. Life Sci 57:131–141
95. Casciari JJ, Hollingshead MG, Alley MC, Mayo JG, Malspeis L, Miyauchi S, Grever MR, Weinstein JN (1994) Growth and chemotherapeutic response of cells in a hollow-fiber in-vitro solid tumor-model. J Nat Cancer Inst 86:1846–1852
96. Roberts I, Baila S, Rice RB, Janssens ME, Nguyen K, Moens N, Ruban L, Hernandez D, Coffey P, Mason C (2012) Scale-up of human embryonic stem cell culture using a hollow fibre bioreactor. Biotechnol Lett 34:2307–2315. doi:10.1007/s10529-012-1033-1
97. Wung N, Acott SM, Tosh D, Ellis MJ (2014) Hollow fibre membrane bioreactors for tissue engineering applications. Biotechnol Lett 36:2357–2366. doi:10.1007/s10529-014-1619-x
98. Bettahalli NM, Groen N, Steg H, Unadkat H, de Boer J, van Blitterswijk CA, Wessling M, Stamatialis D (2014) Development of multilayer constructs for tissue engineering. J Tissue Eng Regen Med 8:106–119. doi:10.1002/term.1504
99. Martin Y, Vermette P (2005) Bioreactors for tissue mass culture: Design, characterization, and recent advances. Biomaterials 26:7481–7503. doi:10.1016/j.biomaterials.2005.05.057
100. Lamers CHJ, Gratama JW, Luider-Vrieling B, Bolhuis RLH, and Bast E (1999) Large-scale production of natural cytokines during activation and expansion of human T lymphocytes in hollow fiber bioreactor cultures. J Immunother 22:299–307
101. Jain E, Kumar A (2008) Upstream processes in antibody production: evaluation of critical parameters. Biotechnol Adv 26:46–72. doi:10.1016/j.biotechadv.2007.09.004
102. Kalbfuss B, Genzel Y, Wolff M, Zimmermann A, Morenweiser R, Reichl U (2007) Harvesting and concentration of human influenza A virus produced in serum-free mammalian cell culture for the production of vaccines. Biotechnol Bioeng 97:73–85. doi:10.1002/bit.21139
103. Schmelzer E, Triolo F, Turner ME, Thompson RL, Zeilinger K, Reid LM, Gridelli B, Gerlach JC (2010) Three-dimensional perfusion bioreactor culture supports differentiation of human fetal liver cells. Tissue Eng Part A 16:2007–2016. doi:10.1089/ten.tea.2009.0569
104. Lu HF, Lim WS, Zhang PC, Chia SM, Yu H, Mao HQ, Leong KW (2005) Galactosylated poly(vinylidene difluoride) hollow fiber bioreactor for hepatocyte culture. Tissue Eng 11:1667–1677
105. De Bartolo L, Salerno S, Curcio E, Piscioneri A, Rende M, Morelli S, Tasselli F, Bader A, Drioli E (2009) Human hepatocyte functions in a crossed hollow fiber membrane bioreactor. Biomaterials 30:2531–2543. doi:10.1016/j.biomaterials.2009.01.011
106. Chen G, Palmer AF (2009) Hemoglobin-based oxygen carrier and convection enhanced oxygen transport in a hollow fiber bioreactor. Biotechnol Bioeng 102:1603–1612. doi:10.1002/bit.22200
107. Westmuckett AD, Lupu C, Roquefeuil S, Krausz T, Kakkar VV, and Lupu F (2000) Fluid flow induces upregulation of synthesis and release of tissue factor pathway inhibitor in vitro. Arterioscler Thromb Vasc Biol 20:2474–2482
108. Godara P, McFarland CD, Nordon RE (2008) Design of bioreactors for mesenchymal stem cell tissue engineering. J Chem Technol Biotechnol 83:408–420
109. Cucullo L, Couraud PO, Weksler B, Romero IA, Hossain M, Rapp E, Janigro D (2008) Immortalized human brain endothelial cells and flow-based vascular modeling: a marriage of

convenience for rational neurovascular studies. J Cereb Blood Flow Metab 28:312–328. doi:10.1038/sj.jcbfm.9600525

110. Pearson NC, Shipley RJ, Waters SL, Oliver JM (2014) Multiphase modelling of the influence of fluid flow and chemical concentration on tissue growth in a hollow fibre membrane bioreactor. Math Med Biol 31:393–430. doi:10.1093/imammb/dqt015

111. Chapman LA, Shipley RJ, Whiteley JP, Ellis MJ, Byrne HM, Waters SL (2014) Optimising cell aggregate expansion in a perfused hollow fibre bioreactor via mathematical modelling. PLoS ONE 9:e105813. doi:10.1371/journal.pone.0105813

112. Li M, Tilles AW, Milwid JM, Hammad M, Lee J, Yarmush ML, Parekkadan B (2012) Phenotypic and functional characterization of human bone marrow stromal cells in hollow-fibre bioreactors. J Tissue Eng Regen Med 6:369–377. doi:10.1002/term.439

113. Lubberstedt M, Muller-Vieira U, Biemel KM, Darnell M, Hoffmann SA, Knospel F, Wonne EC, Knobeloch D, Nussler AK, Gerlach JC, Andersson TB, Zeilinger K (2015) Serum-free culture of primary human hepatocytes in a miniaturized hollow-fibre membrane bioreactor for pharmacological in vitro studies. J Tissue Eng Regen Med 9:1017–1026. doi:10.1002/term.1652

114. Billecke N, Raschzok N, Rohn S, Morgul MH, Schwartlander R, Mogl M, Wollersheim S, Schmitt KR, Sauer IM (2012) An operational concept for long-term cinemicrography of cells in mono- and co-culture under highly controlled conditions—the SlideObserver. J Biotechnol 159:83–89. doi:10.1016/j.jbiotec.2012.01.033

115. Decker S, Hollingshead M, Bonomi CA, Carter JP, Sausville EA (2004) The hollow fibre model in cancer drug screening: the NCI experience. Eur J Cancer 40:821–826

116. Suggitt M, Bibby MC (2005) 50 years of preclinical anticancer drug screening: Empirical to target-driven approaches. Clin Cancer Res 11:971–981

117. Johnson JI, Decker S, Zaharevitz D, Rubinstein LV, Venditti J, Schepartz S, Kalyandrug S, Christian M, Arbuck S, Hollingshead M, Sausville EA (2001) Relationships between drug activity in NCI preclinical in vitro and in vivo models and early clinical trials. Br J Cancer 84:1424–1431

118. Sharma SV, Haber DA, Settleman J (2010) Cell line-based platforms to evaluate the therapeutic efficacy of candidate anticancer agents. Nat Rev Cancer 10:241–253. doi:10.1038/nrc2820

119. Zhang GJ, Chen TB, Bednar B, Connolly BM, Hargreaves R, Sur C, Williams DL (2007) Optical Imaging of tumor cells in hollow fibers: evaluation of the antitumor activities of anticancer drugs and target validation. Neoplasia 9:652–661. doi:10.1593/neo.07421

120. Zhang GJ, Safran M, Wei WY, Sorensen E, Lassota P, Zhelev N, Neuberg DS, Shapiro G, Kaelin WG (2004) Bioluminescent imaging of Cdk2 inhibition in vivo. Nat Med 10:643–648

121. Pearce CJ, Lantvit DD, Shen Q, Jarjoura D, Zhang X, Oberlies NH, Kroll DJ, Wani MC, Orjala J, Soejarto DD, Farnsworth NR, de Blanco EJ, Fuchs JR, Kinghorn AD, Swanson SM (2012) Use of the hollow fiber assay for the discovery of novel anticancer agents from fungi. Methods Mol Biol 944:267–277. doi:10.1007/978-1-62703-122-6_20

122. Mi Q, Pezzuto JM, Farnsworth NR, Wani MC, Kinghorn AD, Swanson SM (2009) Use of the in vivo hollow fiber assay in natural products anticancer drug discovery. J Nat Prod 72:573–580. doi:10.1021/np800767a

123. Shnyder SD (2009) Use of the hollow fibre assay for studies of tumor neovasculature. Methods Mol Biol 467:331–42

124. Shnyder SD, Hasan J, Cooper PA, Pilarinou E, Jubb E, Jayson GC, Bibby MC (2005) Development of a modified hollow fibre assay for studying agents targeting the tumour neovasculature. Anticancer Res 25:1889–1894

125. Temmink OH, Prins HJ, van Gelderop E, Peters GJ (2007) The Hollow Fibre Assay as a model for in vivo pharmacodynamics of fluoropyrimidines in colon cancer cells. Br J Cancer 96:61–66

126. GE Healthcare (2003) G2M cell cycle phase marker assay user manual [cited 2009 December]; GE Healthcare Life Sciences (formerly Amersham Biosciences). Available

from: http://www4.gelifesciences.com/aptrix/upp00919.nsf/Content/0D387E3C4C6D4FFF C1257628001D05DD/$file/25-8010-50UM.pdf

127. Thomas N (2003) Lighting the circle of life: fluorescent sensors for covert surveillance of the cell cycle. Cell cycle 2:545–549 (Georgetown, Tex)

128. Griesdoorn V, Brown MR, Wiltshire M, Smith PJ, Errington RJ (2016) Tracking the cyclin B1-GFP sensor to profile the pattern of mitosis versus mitotic bypass. Methods Mol Biol 1342:279–285. doi:10.1007/978-1-4939-2957-3_17

129. Hassan SB, de la Torre M, Nygren P, Karlsson MO, Larsson R, Jonsson E (2001) A hollow fiber model for in vitro studies of cytotoxic compounds: activity of the cyanoguanidine CHS 828. AntiCancer Drugs 12:33–42

130. Sadar MD, Akopian VA, Beraldi E (2002) Characterization of a new in vivo hollow fiber model for the study of progression of prostate cancer to androgen independence. Mol Cancer Ther 1:629–637

131. Liu D, Queva C, Ready S, Webster K, Zabludoff S (2004) Testing cell cycle regulation effect of a compound using a hollow fibre cell implant. Patent number: WO2004106924

132. Suggitt M, Swaine DJ, Pettit GR, Bibby MC (2004) Characterization of the hollow fiber assay for the determination of microtubule disruption in vivo. Clin Cancer Res 10:6677–6685

133. Bishai WR, Karakousis PC (2006) Hollow fiber technique for in vivo study of cell populations. Patent Number US2006182685

134. Bridges EM, Bibby MC, Burchill SA (2006) The hollow fiber assay for drug responsiveness in the Ewing's sarcoma family of tumors. J Pediatr 149:103–111

135. Wang G, Wang J, Sadar MD (2008) Crosstalk between the androgen receptor and beta-catenin in castrate-resistant prostate cancer. Cancer Res 68:9918–9927. doi:10.1158/0008-5472.can-08-1718

136. Heindryckx F, Colle I, Van Vlierberghe H (2009) Experimental mouse models for hepatocellular carcinoma research. Int J Exp Pathol 90:367–386. doi:10.1111/j.1365-2613.2009.00656.x

137. Hollingshead MG, Bonomi CA, Borgel SD, Carter JP, Shoemaker R, Melillo G, Sausville EA (2004) A potential role for imaging technology in anticancer efficacy evaluations. Eur J Cancer 40:890–898

138. Zhang GJ, Chen TB, Hargreaves R, Sur C, Williams DL (2008) Bioluminescence imaging of hollow fibers in living animals: its application in monitoring molecular pathways. Nat Protoc 3:891–899. doi:10.1038/nprot.2008.52

139. Zhang GJ, Kaelin WG (2005) Bioluminescent imaging of ubiquitin ligase activity: measuring cdk2 activity in vivo through changes in p 27 turnover. In: Ubiquitin and Protein Degradation, Pt B. 2005, Elsevier Academic Press Inc, San Diego, p 530-+

140. Apple SK (2016) Sentinel lymph node in breast cancer: review article from a pathologist's point of view. J Pathol Transl Med 50:83–95. doi:10.4132/jptm.2015.11.23

141. Simonsen TG, Gaustad JV, Rofstad EK (2010) Development of hypoxia in a preclinical model of tumor micrometastases. Int J Radiat Oncol Biol Phys 76:879–888. doi:10.1016/j.ijrobp.2009.09.045

142. Grimes DR, Kelly C, Bloch K, Partridge M (2014) A method for estimating the oxygen consumption rate in multicellular tumour spheroids. J R Soc Interface 11:20131124. doi:10.1098/rsif.2013.1124

143. Chouaib S, Noman MZ, Kosmatopoulos K, Curran MA (2016) Hypoxic stress: obstacles and opportunities for innovative immunotherapy of cancer. Oncogene. doi:10.1038/onc.2016.225

144. Rey S, Schito L, Koritzinsky M, Wouters BG (2016) Molecular targeting of hypoxia in radiotherapy. Adv Drug Deliv Rev. doi:10.1016/j.addr.2016.10.002

145. Ming L, Byrne NM, Camac SN, Mitchell CA, Ward C, Waugh DJ, McKeown SR, Worthington J (2013) Androgen deprivation results in time-dependent hypoxia in LNCaP prostate tumours: informed scheduling of the bioreductive drug AQ4N improves treatment response. Int J Cancer 132:1323–1332. doi:10.1002/ijc.27796

146. Dubois LJ, Niemans R, van Kuijk SJ, Panth KM, Parvathaneni NK, Peeters SG, Zegers CM, Rekers NH, van Gisbergen MW, Biemans R, Lieuwes NG, Spiegelberg L, Yaromina A, Winum JY, Vooijs M, Lambin P (2015) New ways to image and target tumour hypoxia and its molecular responses. Radiother Oncol 116:352–357. doi:10.1016/j.radonc.2015.08.022

147. Ljungkvist AS, Bussink J, Kaanders JH, van der Kogel AJ (2007) Dynamics of tumor hypoxia measured with bioreductive hypoxic cell markers. Radiat Res 167:127–45

148. Tafreshi NK, Lloyd MC, Proemsey JB, Bui MM, Kim J, Gillies RJ, Morse DL (2016) Evaluation of CAIX and CAXII expression in breast cancer at varied O2 levels: CAIX is the superior surrogate imaging biomarker of tumor hypoxia. Mol Imaging Biol 18:219–231. doi:10.1007/s11307-015-0885-x

149. Fleming IN, Manavaki R, Blower PJ, West C, Williams KJ, Harris AL, Domarkas J, Lord S, Baldry C, Gilbert FJ (2015) Imaging tumour hypoxia with positron emission tomography. Br J Cancer 112:238–250. doi:10.1038/bjc.2014.610

150. Tran LB, Bol A, Labar D, Jordan B, Magat J, Mignion L, Gregoire V, Gallez B (2012) Hypoxia imaging with the nitroimidazole 18F-FAZA PET tracer: a comparison with OxyLite, EPR oximetry and 19F-MRI relaxometry. Radiother Oncol 105:29–35. doi:10.1016/j.radonc.2012.04.011

151. Andersen AP, Moreira JMA, Pedersen SF (2014) Interactions of ion transporters and channels with cancer cell metabolism and the tumour microenvironment. Philos Trans Royal Soc B Biol Sci 369:20130098. doi:10.1098/rstb.2013.0098

152. Davies LC, Taylor PR (2015) Tissue-resident macrophages: then and now. Immunology 144:541–548. doi:10.1111/imm.12451

153. Davies LC, Rosas M, Smith PJ, Fraser DJ, Jones SA, Taylor PR (2011) A quantifiable proliferative burst of tissue macrophages restores homeostatic macrophage populations after acute inflammation. Eur J Immunol 41:2155–2164. doi:10.1002/eji.201141817

154. Zhang G, Liu H, Huang J, Chen S, Pan X, Huang H, Wang L (2016) TREM-1low is a novel characteristic for tumor-associated macrophages in lung cancer. Oncotarget 7:40508–40517. doi:10.18632/oncotarget.9639

155. Kim JW, Tchernyshyov I, Semenza GL, Dang CV (2006) HIF-1-mediated expression of pyruvate dehydrogenase kinase: a metabolic switch required for cellular adaptation to hypoxia. Cell Metab 3:177–185. doi:10.1016/j.cmet.2006.02.002

156. Colegio OR, Chu NQ, Szabo AL, Chu T, Rhebergen AM, Jairam V, Cyrus N, Brokowski CE, Eisenbarth SC, Phillips GM, Cline GW, Phillips AJ, Medzhitov R (2014) Functional polarization of tumour-associated macrophages by tumour-derived lactic acid. Nature 513:559–563. doi:10.1038/nature13490

157. Kao SH, Wu KJ, Lee WH (2016) Hypoxia, epithelial-mesenchymal transition, and TET-mediated epigenetic changes. J Clin Med 5. doi:10.3390/jcm5020024

158. Kusuma S, Zhao S, Gerecht S (2012) The extracellular matrix is a novel attribute of endothelial progenitors and of hypoxic mature endothelial cells. Faseb J 26:4925–4936. doi:10.1096/fj.12-209296

159. Ioannou M, Papamichali R, Kouvaras E, Mylonis I, Vageli D, Kerenidou T, Barbanis S, Daponte A, Simos G, Gourgoulianis K, Koukoulis GK (2009) Hypoxia inducible factor-1 alpha and vascular endothelial growth factor in biopsies of small cell lung carcinoma. Lung 187:321–329. doi:10.1007/s00408-009-9169-z

160. Ribatti D (2016) Tumor refractoriness to anti-VEGF therapy. Oncotarget. doi:10.18632/oncotarget.8694

161. Gray LH, Conger AD, Ebert M, Hornsey S, Scott OC (1953) The concentration of oxygen dissolved in tissues at the time of irradiation as a factor in radiotherapy. Br J Radiol 26:638–648. doi:10.1259/0007-1285-26-312-638

162. Thomlinson RH, Gray LH (1955) The histological structure of some human lung cancers and the possible implications for radiotherapy. Br J Cancer 9:539–49

163. Margolin AA (2013) Oncogenic driver mutations: Neither tissue-specific nor independent. Sci Transl Med 5:214ec200
164. Liang D, Miller GH, Tranmer GK (2015) Hypoxia activated prodrugs: factors influencing design and development. Curr Med Chem 22:4313–25
165. Phillips RM (2016) Targeting the hypoxic fraction of tumours using hypoxia-activated prodrugs. Cancer Chemother Pharmacol 77:441–457. doi:10.1007/s00280-015-2920-7
166. Wigerup C, Pahlman S, Bexell D (2016) Therapeutic targeting of hypoxia and hypoxia-inducible factors in cancer. Pharmacol Ther 164:152–169. doi:10.1016/j.pharmthera.2016.04.009
167. Guise CP, Mowday AM, Ashoorzadeh A, Yuan R, Lin WH, Wu DH, Smaill JB, Patterson AV, Ding K (2014) Bioreductive prodrugs as cancer therapeutics: targeting tumor hypoxia. Chin J Cancer 33:80–86. doi:10.5732/cjc.012.10285
168. Mehibel M, Singh S, Chinje EC, Cowen RL, Stratford IJ (2009) Effects of cytokine-induced macrophages on the response of tumor cells to banoxantrone (AQ4N). Mol Cancer Ther 8:1261–1269. doi:10.1158/1535-7163.mct-08-0927
169. Foehrenbacher A, Secomb TW, Wilson WR, Hicks KO (2013) Design of optimized hypoxia-activated prodrugs using pharmacokinetic/pharmacodynamic modeling. Front Oncol 3:314. doi:10.3389/fonc.2013.00314
170. Meng F, Evans JW, Bhupathi D, Banica M, Lan L, Lorente G, Duan JX, Cai X, Mowday AM, Guise CP, Maroz A, Anderson RF, Patterson AV, Stachelek GC, Glazer PM, Matteucci MD, Hart CP (2012) Molecular and cellular pharmacology of the hypoxia-activated prodrug TH-302. Mol Cancer Ther 11:740–751. doi:10.1158/1535-7163.mct-11-0634
171. Albertella MR, Loadman PM, Jones PH, Phillips RM, Rampling R, Burnet N, Alcock C, Anthoney A, Vjaters E, Dunk CR, Harris PA, Wong A, Lalani AS, Twelves CJ (2008) Hypoxia-selective targeting by the bioreductive prodrug AQ4N in patients with solid tumors: results of a phase I study. Clin Cancer Res 14:1096–1104. doi:10.1158/1078-0432.ccr-07-4020
172. Nesbitt H, Byrne NM, Williams N, Ming L, Worthington J, Errington RJ, Patterson LH, Smith P, McKeown SR, McKenna DJ (2016) Targeting hypoxic prostate tumours using the novel hypoxia-activated prodrug OCT1002 inhibits expression of genes associated with malignant progression. Clin Cancer Res. doi:10.1158/1078-0432.ccr-16-1361

Rare Cells: Focus on Detection and Clinical Relevance

Sara De Biasi, Lara Gibellini, Milena Nasi, Marcello Pinti
and Andrea Cossarizza

Abstract The study of rare-cell populations is assuming a growing importance to the advancement of medical diagnostics and therapeutics. In several clinical studies, counting rare cells can provide valuable information on the status of the patient; examples are the search for circulating tumor cells in peripheral blood, tumor stem cells, endothelial cells, hematopoietic progenitor cells and their subpopulations, antigen-specific T-cells, invariant natural killer T cells, and fetal cells in maternal circulation. The study of rare-cell populations is useful not only to understand disease mechanisms, but also to find novel targets. With multiparameter capabilities and a very high analysis rate, flow cytometry is at present the most potent technology to address rare-cell analysis. This chapter will describe the main issues of the pre-analytical phase, including the amount of blood to use, the use of pre-enriched populations, the number of markers to use, and the number of cells to acquire. Moreover, we will discuss the importance of excluding doublets and the

S. De Biasi (✉) · L. Gibellini · M. Nasi
Department of Surgery, Medicine, Dentistry and Morphological Sciences,
University of Modena and Reggio Emilia School of Medicine,
Via Campi 287-41125 Modena, Italy
e-mail: debiasisara@yahoo.it

L. Gibellini
e-mail: lara.gibellini@unimore.it

M. Nasi
e-mail: milena.nasi@unimore.it

M. Pinti
Chair of Pathology and Immunology, Department of Life Sciences,
University of Modena and Reggio Emilia, Via Campi 287-41125 Modena, Italy
e-mail: marcello.pinti@unimore.it

A. Cossarizza
Chair of Pathology and Immunology, Department of Medical and Surgical Sciences
for Children and Adults, University of Modena and Reggio Emilia School of Medicine,
Via Campi 287-41125 Modena, Italy
e-mail: andrea.cossarizza@unimore.it

© Springer Nature Singapore Pte Ltd. 2017
J.P. Robinson and A. Cossarizza (eds.), *Single Cell Analysis*,
Series in BioEngineering, DOI 10.1007/978-981-10-4499-1_2

39

use of a DUMP channel, along with the importance of using optimal methodologies in all phases, including collection of biological samples, adequate controls, and expert use of software and hardware.

Keywords Rare events · Antigen-specific T cells · Invariant natural killer cells · Polyfunctionality · Circulating endothelial cells

1 Introduction

The detection, count, and functional analysis of rare cells can give relevant information either to the basic scientist or to the clinician. The term "rare" typically refers to events with a frequency of 0.01% or less [1, 2]. Applications of rare-cell analysis include the detection in blood of tumors such as metastatic breast cancer cells [3] or neuroblastoma cells infiltrating the bone marrow [4], monitoring of minimal residual disease [5, 6], detection of stem cells and rare HIV-infected cells in peripheral blood [7], identification of antigen (Ag)-specific T cells [8] and invariant natural killer T cells [9, 10], and analysis of mutation frequencies in genetic toxicology [11]. This chapter will discuss the detection of some rare cell populations of interest for immunologists and oncologists, along with some crucial technical issues, along with the requirements for isolating and enumerating such cells in peripheral blood.

2 Ag-Specific T Cells

Ag-specific T cells are clearly fundamental in specific immune responses, and they form the basis of immunological memory. To know their frequency, phenotype, and functional capability can be essential to evaluate the immune status of an individual, to understand the mechanism(s) of an eventual immunopathological event, or eventually to predict immune protection. Considering that the repertoire of T cells can be theoretically unlimited (i.e., on the order of 10^{14}), the frequency of T cells specific for a single Ag in the peripheral blood (the most accessible tissue of the organism) is very low. Indeed, in the absence of acute infections, specific T-cell frequencies are typically much below 1% in either the repertoire of naïve (with a range of 0.2–60 cells/10^6 naïve T cells) or memory cells [12].

Two main approaches can be used to detect Ag-specific T cells (Fig. 1). First, recombinant MHC-peptide multimers recognized by a specific T-cell receptor (TCR) allow one to identify the total pool of T cells with a distinct specificity. The multimerization of peptide-MHC complexes that increases the relative binding avidity and the use of combinatorial color-coded tetramers (enabling the simultaneous use of several tetramers to detect a greater number of different antigen specificities) are the major advantages of this approach [13, 14]. Second, it is

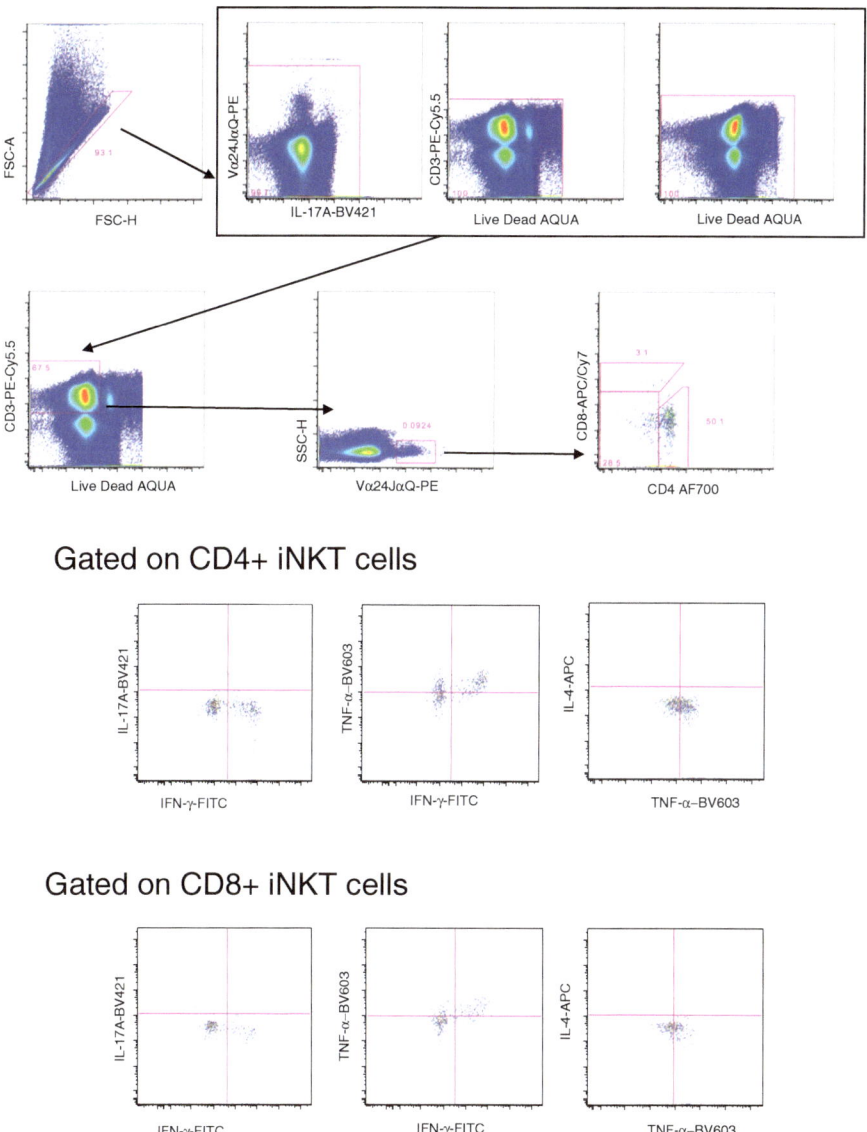

Fig. 1 Gating strategy to analyze the production of cytokines by different types of iNKT cells. Lymphocytes were identified by physical parameters, doublets were removed by using FSC-H versus FSC-A, then iNKT cells were characterized by the expression of CD3 and invariant TCR (TCR Vα24JαQ). In this subset, three populations were identified according to CD4 and CD8 expression. The percentage of CD4+ and CD8+ iNKT cell producing IFN-γ, TNF-α, IL-17A, and IL-4 was quantified. A minimum of 10 million cells per sample were acquired

possible to analyze either the production of cytokines or the expression of activation markers expressed by Ag-specific T cells after specific in vitro stimulation by peptides, intact proteins, or peptide pools that cover the whole sequence of a protein, and thus cover almost all possible T-cell epitopes [15].

Most cytometry-based approaches are limited to the simultaneous screening of up to 100 distinct T-cell specificities in one sample, but recently MHC multimers labeled with individual DNA barcodes have allowed the screening of >1000 peptide specificities in a single sample and detection of low-frequency CD8 T cells specific for virus- or cancer-restricted antigens. This means, for example, that it is possible to detect specific T-cell populations recognizing neoepitopes in tumor-infiltrating lymphocytes or in peripheral blood [16]. Finally, a peptide-MHC dodecamer as a "next-generation" technology has been reported. This alternative technique is able to detect two- to five-fold more antigen-specific T cells in both human and murine CD4+ and CD8+ $\alpha\beta$ T-cell compartments compared to the equivalent tetramers [17].

Interest in the evaluation and quantification of Ag-specific responses derives from the fact that a new era for vaccine development has now arisen, one that generates cellular immune responses instead of the classical antibodies. Thus, the repertoire of specific T cells (in terms of cytokine production, proliferation, and killing capability) could be useful to predict vaccine response [18].

As an example of the analysis of the antiviral capabilities of T cells, the magnitude and duration of anti-smallpox immunity has been studied. Antiviral antibody responses remained stable up to 75 years after vaccination, whereas antiviral T-cell responses declined slowly, with a half-life of 8–15 years [19]. Then, it was shown that the duration of immunity following smallpox infection was remarkably similar to that observed after smallpox vaccination, with antiviral T-cell responses that declined slowly over time and antiviral antibody responses that remained stable for decades after recovery from infection [20].

The evaluation of the production of different cytokines by Ag-specific T cells could mirror not only the response to a given pathogen, but also the effect of different treatments. For example, it has been shown that cytotoxic granule release dominates gag-specific CD4+ T-cell response in different phases of HIV infection [21] and that CD4+ gag-specific T lymphocytes are unaffected by CD4-guided treatment interruption and therapy resumption [8].

Cytomegalovirus (CMV) has an enormous impact on the overall immune profile of each individual. Preserving both CD8+ T-cell memory and a long-term control of CMV requires considerable effort from the host immune system. One hallmark of CMV infection is the maintenance of large populations of CMV-specific memory CD8+ T cells (that can be as high as 10% of peripheral CD8+ T cells)—a phenomenon termed "memory inflation." Recent data suggest that memory inflation is associated with impaired immunity in the elderly [22]. Finally, novel features of *M. tuberculosis* antigen–specific T-cell differentiation have been discovered, which reveals pathways that limit and promote immune control of infection [23].

3   Invariant Natural Killer T (iNKT) Cells

iNKT cell are innate-like lymphocytes characterized by the expression of markers typical of T lymphocytes (CD3, CD4, CD8) and NK markers (CD161, CD56), but they are uniquely identified by the expression of an invariant $V\alpha24J\alpha18$ TCR, and they recognize as cognate antigens self and foreign lipids presented by CD1d [24, 25]. In humans, iNKT cells have a low frequency in peripheral blood, as they represent 0.1–0.001% of T cells. iNKT cells are considered to be a bridge between innate and adaptive immunity because they can rapidly produce cytokines after stimulation and are able to mediate both protective and regulatory immune func- tions. However, on the basis of CD4 and CD8 expression, mature iNKT cells can be divided into functionally distinct subsets, i.e., CD4+CD8−, CD4−CD8−, and CD4−CD8+ [26, 27]. Each subset of iNKT cells can release large amounts of Th1 [interferon (IFN)-γ, tumor necrosis factor (TNF)-α], Th2 [interleukin (IL)-4, IL-12, and IL-13] [28] and Th17 (IL-17, IL-22) cytokines [29]. Different subsets of iNKT cells have different immunological properties: CD4+ iNKT cells release Th1, Th2, and Th17 cytokines, while CD8+ and CD4−CD8− cells exhibit Th1 phenotypes and cytotoxic activity [30] (Fig. 1). Moreover, a regulatory phenotype of iNKT cells has also been described, based upon the production of IL-10 and transforming growth factor–β, and on the expression of classical markers of regulatory T cells such as CD4+, $CD25^{bright}$, and FoxP3+ [31, 32] (Fig. 2).

   iNKT cell defects can predispose an individual to autoimmune diseases because of the failure of immune regulation [33]. For example, NKT cells play a key regulatory role in type 1 diabetes. The absence of NKT cells correlates with exacerbation of type 1 diabetes, whereas an increased frequency and/or activation of NKT cells prevents autoimmunity against beta cells [34]. Similarly, the reduced frequency of iNKT cells in peripheral blood from patients with systemic lupus erythematosus supports the idea of a protective role for these cells in the immunopathology of SLE [35]. In addition, a low number of iNKT cells has been found in blood from patients with rheumatoid arthritis (RA), compared to blood from healthy individuals, and low iNKT cell frequencies were associated with an active form of the disease [36]. Studies on iNKT cells in patients affected by multiple sclerosis (MS) then showed the presence of different defects, but yielded contradictory results, mainly because of non-stringent methods used for the iden- tification of these cells and the limited analysis of the distribution and/or function of their subsets [26]. Recently, we reported that the percentage of iNKT cells from patients affected by different forms of multiple sclerosis are similar, but patients affected by progressive forms are characterized by high level of Th1 and Th17 iNKT cells [9].

   The frequency and activity of iNKT cells was studied in acute and chronic infections, including that by HIV, where iNKT cells can constitute a considerable viral reservoir [37]. In the early phases of HIV infection, CD4+ iNKT cells are rapidly depleted in the gut after primary infection [38], but are preserved in gut-associated lymphoid tissue (GALT) characterized by the presence of a Th2

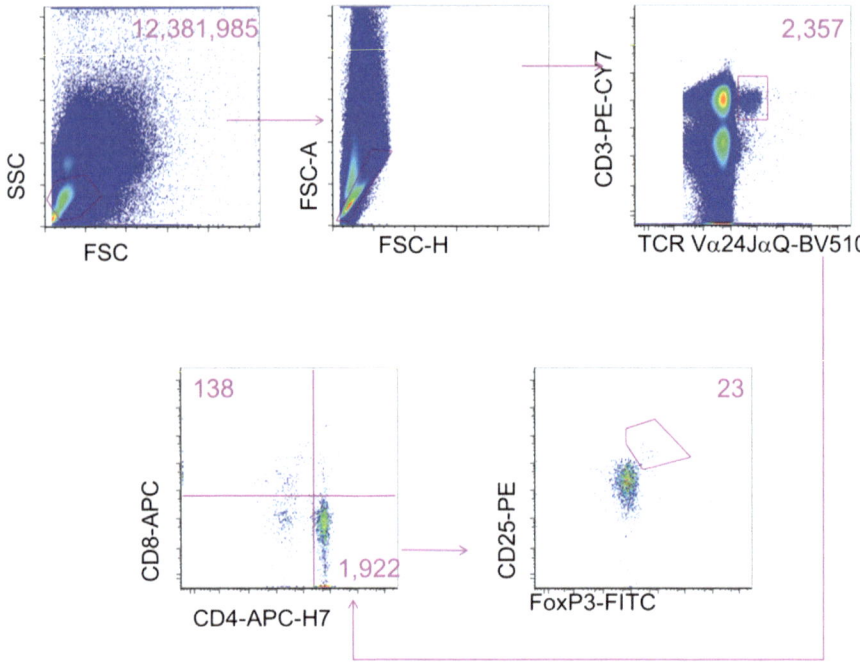

Fig. 2 Gating strategy to analyze iNKT cells with regulatory potential. Lymphocytes were identified by physical parameters, doublets were removed by using FSC-H versus FSC-A, then iNKT cells were characterized by the expression of CD3 and invariant TCR (TCR Vα24JαQ). In this subset, three populations were identified according to CD4 and CD8 expression. Among CD4 + iNKT cell, the expression of CD25 and FoxP3 allowed the identification of iNKT cells with regulatory potential. In pink, the number of cells present in the gate or in the quadrant

iNKT subset [39]. Circulating iNKT cells are functionally impaired, express high levels of PD-1 [40], and can produce IFN-γ [40]. Moreover, they present an inverse correlation between the expression of CD161 and cytokine secretion [41]. In addition, a successful antiretroviral therapy is able to induce a rapid recovery of NKT cells [42], but the reconstitution of the CD4+ iNKT cell subset remains delayed [43]. Recently, we found that in patients with low CD4/CD8 ratio after successful antiretroviral therapy the different subset distributions (CD4+ and CD8+ iNKT cells), the persistent level of activation among iNKT cells, the amount of IL-17-producing CD4+ iNKT cells, and that of IFN-γ-producing CD8+ iNKT cells reflected what happens in the whole T-cell compartment [10].

iNKT cell number and activation have been intensely investigated in patients suffering from different cancers. Contradictory results emerged, dependent on the type of tumor and the methods used to identify these cells [44]. For instance, reduced iNKT cell frequency and function have been observed in patients with hematologic cancers or in those with different solid tumors. In accordance with these observations, reduced iNKT cell frequency was correlated to poor overall survival in patients with acute myeloid leukemia or head and neck squamous cell carcinoma. Interestingly, elevated iNKT cell frequency in some tumors is thought to be a positive prognostic indicator. Concerning activation and functionality, upon activation, iNKT cells can exhibit potent cytotoxic functions to promote the killing of tumor cells such as acute myeloid leukemia [45].

4   Circulating Endothelial Cells and Their Precursors

In 1934, non-hematological cells were found in the blood of cancer patients for the first time [46]. Almost six decades later, in 1991, an endothelial identity was attributed to these cells [47]. Since then, circulating endothelial cells (CECs) have been studied in several pathologic conditions that have in common the presence of vascular injury [48–50].

CECs are mature endothelial cells detaching from the intima monolayer in response to endothelial damage [51]. Endothelial dysfunction can take place during the development and the progression of different cardiovascular disorders. Thus, in order to find an early biomarker using a non-invasive technique, monitoring endothelial activity is assuming an important role in clinical practice [52–54]. Indeed, although CECs are rare in healthy individuals, they could be easily counted in the blood of patients with cardiovascular-related complications, suggesting that they may be taken as indicator of disease severity [55].

CECs are also crucial in the neovascularization of tumors both at primary or at metastatic sites [56–59]. Indeed, it has been found that CEC count increased in cancer patients and could correlate with tumor progression [60, 61], and that their number decreased after surgical removal of the tumor and chemotherapy [60]. Kinetics and in vitro viability of CECs are promising predictors of the response to treatment with anti-angiogenic agents in patients with advanced breast cancer or colorectal cancer [62, 63]. CECs would also reveal tumor growth and disease progression at an early stage, in view of the known fragility of the tumor vasculature [64].

If CECs are believed to be mature endothelial cells that have been released into the circulation from an area of disrupted vessels, circulating endothelial precursor or progenitor cells (EPCs) are bone marrow–derived cells that contribute to vasculogenesis—including tumor-associated vasculogenesis [65]. EPCs, which take part in postnatal vasculogenesis, are recruited from the bone marrow under the stimulation of growth factors and cytokines and reach the sites of neovascularization in both physiological and pathological conditions such as malignancies where they contribute to the "angiogenic switch" and tumor progression. The presence of circulating EPCs in the bloodstream of patients with hematological malignancies has been demonstrated and correlated with a poor prognosis [66].

Despite the number of studies performed, investigation of endothelial circulating cells and their progenitors is technically challenging and conflicting results are frequently reported owing to discrepancies in terms of terminology and protocols used for the detection of these cells, leading to ambiguous conclusions in clinical practice [55]. The huge problem in finding these cells is that a unique marker or a combination of markers that identify circulating endothelial cells and their progenitor has not been yet identified. At present, the most common markers used for this purpose are DNA, CD34, CD45, CD133, CD31, CD146, and CD309. In particular, CECs are defined as events that are DNA+, CD34+, CD45−, CD31 +CD133−, CD309+. EPCs are defined as DNA+, CD34+, CD45^{dim}, CD31+, CD133+, CD309± [67–70].

5 Circulating Tumor Cells

In cancer patients, circulating tumor cells (CTCs) can be found in the peripheral blood at very low concentration, ranging from 1 to 10 cells per 10 mL [71]. CTCs are released into the blood stream from primary tumors or from metastatic deposits, and could be useful for a better understanding not only of the phase of the disease and the efficacy of the treatment, but also of the biology of the metastatic process [72]. Strong evidence for CTCs as prognostic markers was first documented in patients suffering from breast cancer [73], and they are now under investigation in several other types of tumors [70]. Three well-known families of antigens can be present on CTCs, either alone or in different combination, *i.e.*, epithelial, mesenchymal, and/or stemness molecules, and can be considered useful markers for the identification of such cells.

A recent publication has described the presence and heterogeneity of a new CTC population that includes cells positive for epithelial cellular adhesion molecule (EpCAM), for cytokeratins (Cks), and also for the pan-hematopoietic marker CD45 [74]. Thus, these cells have characteristics of both epithelial and hematopoietic elements. Interestingly, it has been pointed out that they are not true cancer cells, but rather are tumor-associated macrophages [74]. So, evaluating the role of and eventual changes in circulating cells of the macrophagic lineage, which are often associated with tumors, play a crucial role in their metabolism [75], and can be identified by anti-CD68 mAbs, is under consideration as possible prognostic factors for cancer patients.

In any case, the choice of markers that are characteristic for CTC subpopulations is a really complicated issue; unfortunately, several publications have focused on just one or two markers, creating a heterogeneous amount of (often contrasting) data. Other studies have established which is the gene profile of CTCs. However, from the point of view of a clinician, the interpretation of genetic data is particularly difficult and not feasible in practice. For this reason, in clinical studies the real utility of CTCs for decisions regarding treatment is still under evaluation.

During the past few years, the use of CTCs as a real-time liquid biopsy has received attention [71]. Liquid biopsy is a new diagnostic concept, i.e., a test done on a sample of blood to look for cancer cells from a tumor that are circulating in the blood or for pieces of tumor-cell DNA that are in the blood. A liquid biopsy may be used to help find cancer cells at an early stage [76], even in the absence of an evident primary tumor.

6 Analysis of Rare Events by Flow Cytometry: The First Step

Flow cytometry is at present the most potent technology to find rare cells, and the so called "next-generation" instruments with very high speed and sensitivity are already allowing an easy detection and analysis of such cells. In order to identify those elements, some practical suggestions have to be kept in mind.

The first step in planning an experiment that involves the estimation of rare cells is to establish the quantity of biological material required. For example, should the endpoint of the study be the evaluation of cytokine production by invariant natural killer T (iNKT) cells after in vitro stimulation, some pre-analytical considerations should be taken into account. In order to define circulating iNKT cells, several markers must be used, including those for recognizing CD3, CD4, CD8, and invariant TCR, as well as those for cell viability. Different cytokines such as TNF-α, IFN-γ, IL-4, and IL-17 could be of interest. So a minimum of nine markers are required. If the study considers patients who are severely immunocompromised, like HIV+ individuals before undergoing antiretroviral therapy, the possible low number of lymphocytes must be taken into account. As a consequence, the amount of blood required to detect a reasonable number of rare cells producing one or more cytokines can be as high as 50 mL, since either resting or stimulated cells have to be analyzed [9, 10].

6.1 Enrichment and Choice of Markers

On the basis of the experimental endpoint(s) (e.g., phenotyping, functional assays), the rare population of interest may be enriched or not, and the number of markers that are needed to unambiguously identify a rare cell population needs to be clearly defined. For example, as discussed above, the accurate quantification of CECs and their progenitors (EPCs) is a matter of debate: which combination of markers is the most adequate, or what is needed for an eventual pre-analytical enrichment (by density gradient, buffy coat, and/or magnetic enrichment). Giving a look to the immune system, quantitative pre-enrichment of target cells via magnetic cell separation, which allows rapid processing of large samples (10^6–10^9 cells) [77, 78] is an excellent approach to increase the relative number of rare antigen-specific T-cell frequencies. The enrichment could be performed by known markers that characterize the rare-cell population, or by using tetramers, or by performing a specific enrichment of cytokine-secreting cells (for example, those that produce IFN-γ or IL-4) [79].

Rare antigen-specific CD4 T cells could also be pre-enriched by using anti-CD154 mAb. This live-cell assay could be applied to detect antigen-specific CD4+ T cells with diverse cytokine profiles. By including fluorescently conjugated CD154-specific antibody during stimulation, the assay is fully compatible with intracellular cytokine staining and can be used for cultures as long as 24 h [80]. Finally, another molecule can be used for enrichment of activated CD4+, CD8+, or CD1d+ T cells, namely CD137 (4-1BB), a member of the TNFR superfamily, which has been shown to be expressed following 16–24 h of stimulation [81]. The combined analysis of CD137 and CD154 following a short-term (6 h) stimulation might be optimal to detect in parallel conventional T cells and regulatory T cells (Tregs) reacting against the same antigen [82] (Fig. 3).

The enrichment can have negative effects if rare cells are lost, or, on the contrary, positive effects if unwanted cells are removed [67, 69, 83]. Another strategy to facilitate the detection of rare cells is to increase the number of antigen-specific T cells by in vitro expansion methods, even if the expansion of a single T cell is affected not only by its functional status (e.g., naïve, memory, anergic), but also by the presence of other reactive or accessory cells. Therefore, it is difficult to obtain the frequency of a given cell population in the original samples from the frequencies obtained after prolonged in vitro culture. Similarly, the phenotype and function of the expanded cells may be significantly altered by culture conditions [9, 12] (Fig. 4).

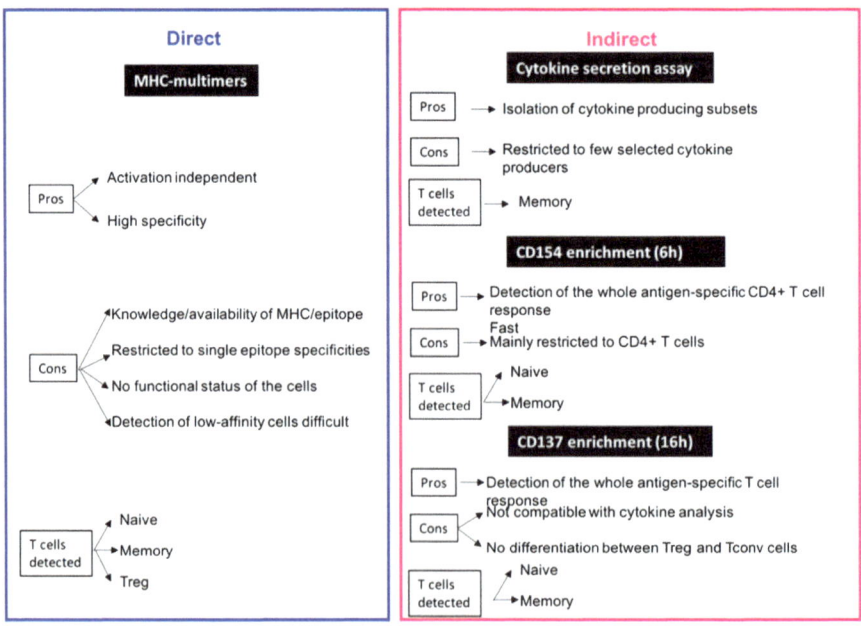

Fig. 3 Enrichment methods for the detection of rare antigen-specific T cells

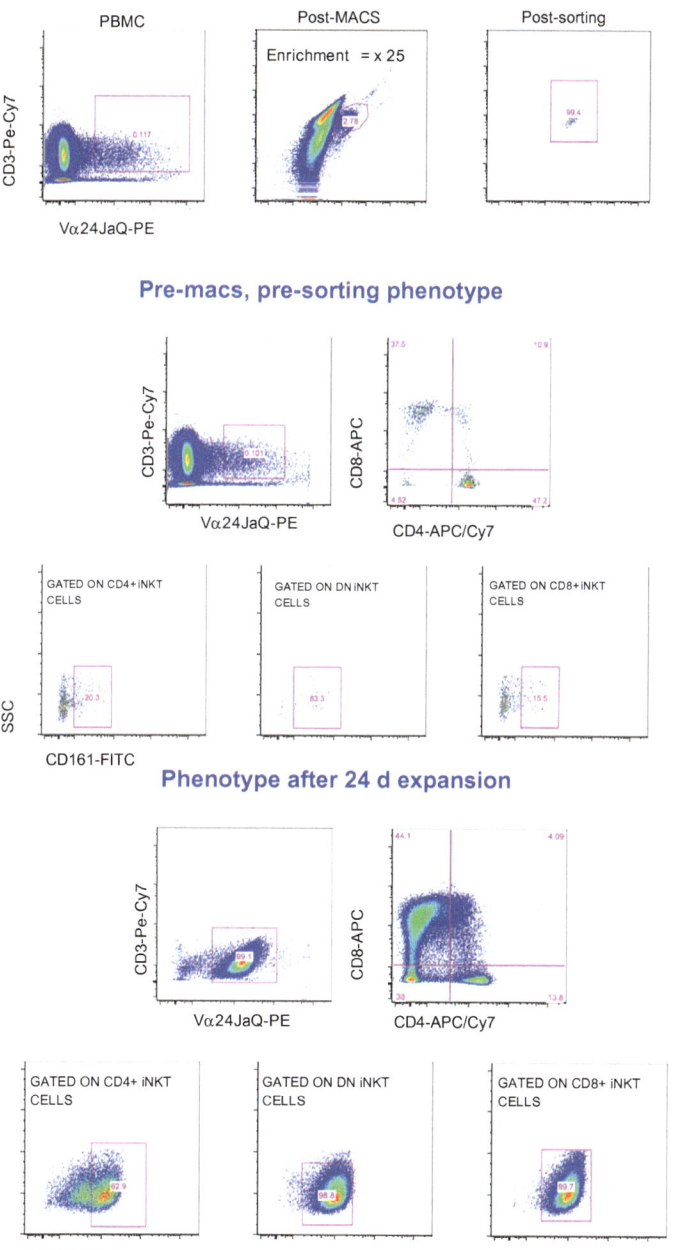

Fig. 4 The phenotype of freshly isolated iNKT cells (upper panel) and that of in vitro expanded iNKT cells may significantly be altered by culture conditions. The middle panel reports the phenotype of these cells just before isolation, while the lower panel shows the phenotype after 24 days of in vitro culture. Note the expression of CD161, well evident in almost all cultured cells

Unfortunately, quite often the lack of well standardized methods for rare-cell detection influences the decision regarding the number of markers and type of antibody and fluorochrome that are necessary for the identification of the population of interest, as well as the rate and speed of acquisition. Depending on the technical characteristics of flow cytometers, which have a varying number of fluorescence channels, sensitivity, and speed, the first thing to decide is always where (in which channel, and by which fluorochrome) the most important marker allowing the identification and characterization of such populations should be detected. For example, in the case of iNKT cells, mAbs recognizing Vα24JQ invariant T-cell receptor allows the unique identification of these cells. Having done that, the panel has to be built following the general rule that the brightest fluorochrome must be used for the marker with the weakest expression. Last but not least, attention should be paid to compensation and acquisition of fluorescence-minus-one (FMO) controls [84–86].

6.2   Number of Acquired Events

Poisson statistics defines the probability that a given number of events will occur in a fixed interval of time/space, assuming that these events would occur with a known average rate and independent of the time elapsed from the previous event [87]. This sort of statistics is crucial for analyzing rare cells, since it can be applied to count randomly distributed cells in a certain volume. If we enumerate a total of N events, of which R meet a certain criterion (i.e., they are positive, P), obviously, the proportion of P events is defined as R/N. The probability of any single event's being positive is between 0 and 1, and this is related to the random manner in which cells are selected for analysis. It is possible to define the variance, Var, as Var $(R) = NP\,(1-P)$. The standard deviation is the square root of Var [88], and the coefficient of variation (CV) is equal to 1/square root of Var [89].

Let's consider a practical situation, the case of human PBMCs stained with a mAb recognizing iNKT cells (e.g., a population that express Vα24βQ TCR and CD3). If the frequency of these cells is 1 in 10,000 (0.01%), P = 0.0001. Good experimental practice suggests keeping the CV below 5%; thus, acquiring one million events gives a CV of 9.99, which is not acceptable, while 10 million events gives a CV of 3.16 (Table 1).

It must be taken into account that sometimes the number of acquired events cannot be high enough to respect this rule, for example because of the amount of blood that researchers can obtain from a patient. We can consider the case in which 1 million peripheral-blood T cells are stimulated with an antigen that activates less than 1 cell in 1000; T-cell activation can be analyzed by polychromatic flow cytometry, which allows the recognition of four functions per cell. Thus, among responding cells, up to 16 populations exist, likely with different frequencies. Clearly each subpopulation contains a few cells that typically must be absent in the

Table 1 The number of acquired events for a cell population with final frequency 0.01%

Acquired events (**N**)	100,000	1,000,000	4,010,000	10,000,000
Positive (**R**)	10	100	401	1,000
Proportion (**P**)	0.0001	0.0001	0.0001	0.0001
Variance (**Var**)	10	100	401	1,000
Standard deviation (SD)	3.16	10.00	20.02	31.62
Coefficient of variation (**CV**) %	31.62	10.00	**4.99**	**3.16**

Note that to obtain a CV of 3.16 it is suggested to acquire 10 million events; acquiring 4,010,000 events gives a CV of 4.99

control sample (that with no stimulation). The number of positive cells is much lower than that suggested by a rigid statistical approach, but it is the general opinion that the events can be considered positive if alternative explanations for the presence of positive events in that channel can be excluded. There must be no noise due to dead cells or fragments, and cell activation must be really due to the antigenic stimulation and not to any in vivo pre-activation of T cells. In principle, there is no reason to fix a threshold for the number of events below which any frequency must be considered "negative" [90]. So, if adequate negative controls are set, "positivity" can be determined after comparison of the measurements (i.e., positive minus negative, namely positive minus the background), using standard statistical tools to compare the frequencies. For example, assuming that from the technical point of view the experiment is well performed, if T cells from "n" unvaccinated controls show no activation after stimulation with the adequate peptides, while T cells from "n" vaccinated individuals do, even extremely low frequencies can be taken as positive. The same logic can be applied in several other cases when negative controls are well chosen.

6.3 Sample Concentration and Flow Rate

Because it is crucial to acquire a high number of events for detection of rare-cell populations, the concentration in the sample and the flow rate are critical parameters, which can typically shorten acquisition time. However, particular attention has to be paid to the fact that increasing the flow rate results in an increase of coincidence, and thus in a higher CV, if flow cytometers use hydrodynamic focusing (the system used at present in most commercially available flow cytometers).

6.4 Thresholds, Gating, and DUMP Channel

Maximizing the signal-to-noise ratio is fundamental to distinguish the signal of the population of interest from the background. Fixing a threshold could help in this, along with the use of a gating strategy that removes from the analysis dead cells (identified by a viability marker, such as amine reactive dyes, sold in different fluorochromes), that excludes doublets/aggregates/debris, and that uses a "DUMP" channel containing antibodies that identify cells of no interest. Furthermore, one has to monitor the parameter "time of acquisition" to remove the event bursts caused by clogs or other possible transient problems during the acquisition.

Of note, two other factors to consider to optimize the sensitivity of an assay are the cleanliness of the instrument and the integrity of the sample. It is important to make sure that the instrument and fluids used are clean and free of particles that could contribute falsely to the rare population.

6.5 Data Analysis

Finally, the analysis of data benefits from powerful hardware (in terms of gigabytes of random access memory, RAM), because depending on how many events and parameters have been acquired, data files tend to be quite huge (if not enormous). To minimize the file size, parameters that are not needed can be turned off, and thresholds can be used.

The high-throughput nature of flow cytometry, combined with the increasing capacity to measure more parameters at once, is generating massive and high-dimensional datasets on a routine daily basis. These data can no longer be adequately analyzed using the classical, mostly manual, analysis techniques and therefore require the development of novel computational techniques, as well as their adoption by the broad community. Computational flow cytometry provides a set of tools to analyze, visualize, and interpret large amounts of cell data in a more automated and unbiased way [91, 92].

In order to inspect data before the analysis, there are several software programs (based on principal component analysis, PCA) that use visualization techniques as alternatives to the traditional two-dimensional dot plots. The first paper on the use of PCA for analyzing flow cytometry data was published in 2007; this approach was applied to eight-color cytofluorimetric analysis on the virgin and memory T-cell compartments in donors of different age (young, middle-aged, and centenarians) [93]. For the first time, it was shown how to use a novel bioinformatic approach to analyze large datasets generated by polychromatic flow cytometry and to obtain the rapid identification of key populations that best characterize a group of subjects.

To date, several tools like SPADE, FlowMap, FlowSOM, viSNE, PhenoGraph, Scaffold map, and DREMI-DREVI are available. These approaches are mainly dimensionality reduction- or clustering-based techniques (reviewed in [92]). Concerning the identification of very rare cell types, there could be some issues related to the fact that they could be mistaken for noise by many clustering algorithms. To identify all relevant populations, it might be necessary to do an exhaustive gating, resulting in strong over-clustering, and then select only those features related to a phenotype. With the traditional clustering algorithms, it is recommended to ensure that only relevant markers are used for clustering. Markers that vary little or that indicate properties not relevant for cell-type identification (for example, activation markers) are best left out, as these will only contribute noise to the similarity calculation. For example, in a recent paper different software programs were compared, i.e., SPADE, t-SNE, and FlowSOM, for the analysis of splenocytes of a wild-type C57BL/6 J mouse. Only very rare cell types, such as natural killer T (NKT) cells (which constitute only 0.3% of the dataset), were not distinguished, whereas relatively rare cell types such as neutrophils (which constitute 0.7% of the dataset) were distinguished by all three methods [92]. Interestingly, such a problem in the identification of rare events is not encountered with SWIFT (a tool present in the software Matlab), which is an automated gating technique specifically developed to identify rare populations [94].

References

1. Donnenberg AD and Donnenberg VS (2007) Rare-event analysis in flow cytometry. Clin Lab Med 27:627–52, viii. doi:10.1016/j.cll.2007.05.013
2. Gross HJ, Verwer B, Houck D, Hoffman RA, Recktenwald D (1995) Model study detecting breast cancer cells in peripheral blood mononuclear cells at frequencies as low as 10(−7). Proc Natl Acad Sci USA 92:537–41
3. Leslie DS, Johnston WW, Daly L, Ring DB, Shpall EJ, Peters WP, Bast RC Jr (1990) Detection of breast carcinoma cells in human bone marrow using fluorescence-activated cell sorting and conventional cytology. Am J Clin Pathol 94:8–13
4. Frantz CN, Ryan DH, Cheung NV, Duerst RE, Wilbur DC (1988) Sensitive detection of rare metastatic human neuroblastoma cells in bone marrow by two-color immunofluorescence and cell sorting. Prog Clin Biol Res 271:249–62
5. Ryan DH, Mitchell SJ, Hennessy LA, Bauer KD, Horan PK, Cohen HJ (1984) Improved detection of rare CALLA-positive cells in peripheral blood using multiparameter flow cytometry. J Immunol Methods 74:115–28
6. Visser JW, De Vries P (1990) Identification and purification of murine hematopoietic stem cells by flow cytometry. Methods Cell Biol 33:451–68
7. Cory JM, Ohlsson-Wilhelm BM, Brock EJ, Sheaffer NA, Steck ME, Eyster ME, Rapp F (1987) Detection of human immunodeficiency virus-infected lymphoid cells at low frequency by flow cytometry. J Immunol Methods 105:71–8
8. Nemes E, Lugli E, Bertoncelli L, Nasi M, Pinti M, Manzini S, Prati F, Manzini L, Del Giovane C, D'Amico R, Cossarizza A, Mussini C (2011) CD4+ T-cell differentiation, regulatory T cells and gag-specific T lymphocytes are unaffected by CD4-guided treatment interruption and therapy resumption. AIDS 25:1443–1453. doi:10.1097/QAD.0b013e328347b5e2
9. De Biasi S, Simone AM, Nasi M, Bianchini E, Ferraro D, Vitetta F, Gibellini L, Pinti M, Del Giovane C, Sola P, Cossarizza A (2016) iNKT cells in secondary progressive multiple sclerosis patients display pro-inflammatory profiles. Frontiers in Immunol 7. doi:10.3389/mmu.2016.00555
10. De Biasi S, Bianchini E, Nasi M, Digaetano M, Gibellini L, Carnevale G, Borghi V, Guaraldi G, Pinti M, Mussini C, Cossarizza A (2016) Th1 and Th17 proinflammatory profile characterizes invariant natural killer T cells in virologically suppressed HIV+ patients with low CD4+/CD8+ ratio. AIDS 30:2599–2610. doi:10.1097/QAD.0000000000001247

11. Jensen RH, Leary JF (1990) Flow cytometry and sorting, 2nd ed. New York
12. Bacher P, Scheffold A (2013) Flow-cytometric analysis of rare antigen-specific T cells. Cytometry A 83:692–701. doi:10.1002/cyto.a.22317
13. Gratama JW, Kern F, Manca F, Roederer M (2008) Measuring antigen-specific immune responses, 2008 update. Cytometry A 73:971–974. doi:10.1002/cyto.a.20655
14. Newell EW, Klein LO, Yu W, Davis MM (2009) Simultaneous detection of many T-cell specificities using combinatorial tetramer staining. Nat Methods 6:497–499. doi:10.1038/nmeth.1344
15. Lamoreaux L, Roederer M, Koup R (2006) Intracellular cytokine optimization and standard operating procedure. Nat Protoc 1:1507–1516. doi:10.1038/nprot.2006.268
16. Bentzen AK, Marquard AM, Lyngaa R, Saini SK, Ramskov S, Donia M, Such L, Furness AJS, McGranahan N, Rosenthal R, Straten PT, Szallasi Z, Svane IM, Swanton C, Quezada SA, Jakobsen SN, Eklund AC, Hadrup SR (2016) Large-scale detection of antigen-specific T cells using peptide-MHC-I multimers labeled with DNA barcodes. Nat Biotechnol 34:1037–1045. doi:10.1038/nbt.3662
17. Huang J, Zeng X, Sigal N, Lund PJ, Su LF, Huang H, Chien YH, Davis MM (2016) Detection, phenotyping, and quantification of antigen-specific T cells using a peptide-MHC dodecamer. Proc Natl Acad Sci USA 113:E1890–E1897. doi:10.1073/pnas.1602488113
18. Casazza JP, Bowman KA, Adzaku S, Smith EC, Enama ME, Bailer RT, Price DA, Gostick E, Gordon IJ, Ambrozak DR, Nason MC, Roederer M, Andrews CA, Maldarelli FM, Wiegand A, Kearney MF, Persaud D, Ziemniak C, Gottardo R, Ledgerwood JE, Graham BS, Koup RA, Team VRCS (2013) Therapeutic vaccination expands and improves the function of the HIV-specific memory T-cell repertoire. J Infect Dis 207:1829–1840. doi:10.1093/infdis/jit098
19. Hammarlund E, Lewis MW, Hansen SG, Strelow LI, Nelson JA, Sexton GJ, Hanifin JM, Slifka MK (2003) Duration of antiviral immunity after smallpox vaccination. Nat Med 9:1131–1137. doi:10.1038/nm917
20. Hammarlund E, Lewis MW, Hanifin JM, Mori M, Koudelka CW, Slifka MK (2010) Antiviral immunity following smallpox virus infection: a case-control study. J Virol 84:12754–12760. doi:10.1128/JVI.01763-10
21. Nemes E, Bertoncelli L, Lugli E, Pinti M, Nasi M, Manzini L, Manzini S, Prati F, Borghi V, Cossarizza A, Mussini C (2010) Cytotoxic granule release dominates gag-specific CD4+ T-cell response in different phases of HIV infection. AIDS 24:947–957. doi:10.1097/QAD.0b013e328337b144
22. Klenerman P, Oxenius A (2016) T cell responses to cytomegalovirus. Nat Rev Immunol 16:367–377. doi:10.1038/nri.2016.38
23. Urdahl KB, Shafiani S, Ernst JD (2011) Initiation and regulation of T-cell responses in tuberculosis. Mucosal Immunol 4:288–293. doi:10.1038/mi.2011.10
24. Brennan PJ, Brigl M, Brenner MB (2013) Invariant natural killer T cells: an innate activation scheme linked to diverse effector functions. Nat Rev Immunol 13:101–117. doi:10.1038/nri3369nri3369 [pii]
25. Godfrey DI, Stankovic S, Baxter AG (2010) Raising the NKT cell family. Nat Immunol 11:197–206. doi:10.1038/ni.1841ni.1841 [pii]
26. Berzins SP, Smyth MJ, Baxter AG (2011) Presumed guilty: natural killer T cell defects and human disease. Nat Rev Immunol 11:131–142. doi:10.1038/nri2904nri2904 [pii]
27. Seino K, Taniguchi M (2005) Functionally distinct NKT cell subsets and subtypes. J Exp Med 202:1623–1626. doi:10.1084/jem.20051600 [pii]
28. Osada T, Morse MA, Lyerly HK, Clay TM (2005) Ex vivo expanded human CD4+ regulatory NKT cells suppress expansion of tumour antigen-specific CTLs. Int Immunol 17:1143–1155. doi:10.1093/intimm/dxh292
29. Goto M, Murakawa M, Kadoshima-Yamaoka K, Tanaka Y, Nagahira K, Fukuda Y, Nishimura T (2009) Murine NKT cells produce Th17 cytokine interleukin-22. Cell Immunol 254:81–84. doi:10.1016/j.cellimm.2008.10.002

30. O'Reilly V, Zeng SG, Bricard G, Atzberger A, Hogan AE, Jackson J, Feighery C, Porcelli SA, Doherty DG (2011) Distinct and overlapping effector functions of expanded human CD4+, CD8α+ and CD4−CD8α− invariant natural killer T cells. PLoS ONE 6: e28648. doi:10.1371/journal.pone.0028648

31. Sag D, Krause P, Hedrick CC, Kronenberg M, Wingender G (2014) IL-10-producing NKT10 cells are a distinct regulatory invariant NKT cell subset. J Clin Invest 124:3725–3740. doi:10.1172/JCI72308

32. Monteiro M, Almeida CF, Caridade M, Ribot JC, Duarte J, Agua-Doce A, Wollenberg I, Silva-Santos B, Graca L (2010) Identification of regulatory Foxp3+ invariant NKT cells induced by TGF-beta. J Immunol 185:2157–2163. doi:10.4049/jimmunol.1000359

33. Yamamura T, Sakuishi K, Illes Z, Miyake S (2007) Understanding the behavior of invariant NKT cells in autoimmune diseases. J Neuroimmunol 191:8–15. doi:10.1016/j.jneuroim.2007.09.014

34. Novak J, Griseri T, Beaudoin L, Lehuen A (2007) Regulation of type 1 diabetes by NKT cells. Int Rev Immunol 26:49–72. doi:10.1080/08830180601070229

35. Gabriel L, Morley BJ, Rogers NJ (2009) The role of iNKT cells in the immunopathology of systemic lupus erythematosus. Ann N Y Acad Sci 1173:435–441. doi:10.1111/j.1749-6632.2009.04743.x

36. Parietti V, Chifflot H, Sibilia J, Muller S, Monneaux F (2010) Rituximab treatment overcomes reduction of regulatory iNKT cells in patients with rheumatoid arthritis. Clin Immunol 134:331–339. doi:10.1016/j.clim.2009.11.007

37. Motsinger A, Haas DW, Stanic AK, Van Kaer L, Joyce S, Unutmaz D (2002) CD1d-restricted human natural killer T cells are highly susceptible to human immunodeficiency virus 1 infection. J Exp Med 195:869–79

38. Ibarrondo FJ, Wilson SB, Hultin LE, Shih R, Hausner MA, Hultin PM, Anton PA, Jamieson BD, Yang OO (2013) Preferential depletion of gut CD4-expressing iNKT cells contributes to systemic immune activation in HIV-1 infection. Mucosal Immunol 6:591–600. doi:10.1038/mi.2012.101

39. Paquin-Proulx D, Ching C, Vujkovic-Cvijin I, Fadrosh D, Loh L, Huang Y, Somsouk M, Lynch SV, Hunt PW, Nixon DF, SenGupta D (2016) Bacteroides are associated with GALT iNKT cell function and reduction of microbial translocation in HIV-1 infection. Mucosal Immunol. doi:10.1038/mi.2016.34

40. Moll M, Kuylenstierna C, Gonzalez VD, Andersson SK, Bosnjak L, Sonnerborg A, Quigley MF, Sandberg JK (2009) Severe functional impairment and elevated PD-1 expression in CD1d-restricted NKT cells retained during chronic HIV-1 infection. Eur J Immunol 39:902–911. doi:10.1002/eji.200838780

41. Snyder-Cappione JE, Loo CP, Carvalho KI, Kuylenstierna C, Deeks SG, Hecht FM, Rosenberg MG, Sandberg JK, Kallas EG, Nixon DF (2009) Lower cytokine secretion ex vivo by natural killer T cells in HIV-infected individuals is associated with higher CD161 expression. AIDS 23:1965–1970. doi:10.1097/QAD.0b013e32832b5134

42. van der Vliet HJ, van Vonderen MG, Molling JW, Bontkes HJ, Reijm M, Reiss P, van Agtmael MA, Danner SA, van den Eertwegh AJ, von Blomberg BM, Scheper RJ (2006) Cutting edge: rapid recovery of NKT cells upon institution of highly active antiretroviral therapy for HIV-1 infection. J Immunol 177:5775–8

43. Yang OO, Wilson SB, Hultin LE, Detels R, Hultin PM, Ibarrondo FJ, Jamieson BD (2007) Delayed reconstitution of CD4+ iNKT cells after effective HIV type 1 therapy. AIDS Res Hum Retroviruses 23:913–922. doi:10.1089/aid.2006.0253

44. Wilson SB, Delovitch TL (2003) Janus-like role of regulatory iNKT cells in autoimmune disease and tumour immunity. Nat Rev Immunol 3:211–222. doi:10.1038/nri1028

45. McEwen-Smith RM, Salio M, Cerundolo V (2015) The regulatory role of invariant NKT cells in tumour immunity. Cancer Immunol Res 3:425–435. doi:10.1158/2326-6066.CIR-15-0062

46. Pool EH, Dunlop GR (1934) Cancer cells in the blood stream. Am J Cancer 21:99–103

47. Sbarbati R, de Boer M, Marzilli M, Scarlattini M, Rossi G, van Mourik JA (1991) Immunologic detection of endothelial cells in human whole blood. Blood 77:764–9

48. Solovey A, Lin Y, Browne P, Choong S, Wayner E, Hebbel RP (1997) Circulating activated endothelial cells in sickle cell anemia. N Engl J Med 337:1584–1590. doi:10.1056/NEJM199711273372203

49. Sowemimo-Coker SO, Meiselman HJ, Francis RB Jr (1989) Increased circulating endothelial cells in sickle cell crisis. Am J Hematol 31:263–5

50. Lefevre P, George F, Durand JM, Sampol J (1993) Detection of circulating endothelial cells in thrombotic thrombocytopenic purpura. Thromb Haemost 69:522

51. Hill JM, Zalos G, Halcox JP, Schenke WH, Waclawiw MA, Quyyumi AA, Finkel T (2003) Circulating endothelial progenitor cells, vascular function, and cardiovascular risk. N Engl J Med 348:593–600. doi:10.1056/NEJMoa022287

52. Werner N, Kosiol S, Schiegl T, Ahlers P, Walenta K, Link A, Bohm M, Nickenig G (2005) Circulating endothelial progenitor cells and cardiovascular outcomes. N Engl J Med 353:999–1007. doi:10.1056/NEJMoa043814

53. Bakogiannis C, Tousoulis D, Androulakis E, Briasoulis A, Papageorgiou N, Vogiatzi G, Kampoli AM, Charakida M, Siasos G, Latsios G, Antoniades C, Stefanadis C (2012) Circulating endothelial progenitor cells as biomarkers for prediction of cardiovascular outcomes. Curr Med Chem 19:2597–604

54. De Biasi S, Cerri S, Bianchini E, Gibellini L, Persiani E, Montanari G, Luppi F, Carbonelli CM, Zucchi L, Bocchino M, Zamparelli AS, Vancheri C, Sgalla G, Richeldi L, Cossarizza A (2015) Levels of circulating endothelial cells are low in idiopathic pulmonary fibrosis and are further reduced by anti-fibrotic treatments. BMC Med 13:277. doi:10.1186/s12916-015-0515-0

55. Boraldi F, Bartolomeo A, De Biasi S, Orlando S, Costa S, Cossarizza A, Quaglino D (2016) Innovative flow cytometry allows accurate identification of rare circulating cells involved in endothelial dysfunction. PLoS One 11:e0160153. doi:10.1371/journal.pone.0160153

56. Furstenberger G, von Moos R, Senn HJ, Boneberg EM (2005) Real-time PCR of CD146 mRNA in peripheral blood enables the relative quantification of circulating endothelial cells and is an indicator of angiogenesis. Br J Cancer 93:793–798. doi:10.1038/sj.bjc.6602782

57. Goon PK, Lip GY, Boos CJ, Stonelake PS, Blann AD (2006) Circulating endothelial cells, endothelial progenitor cells, and endothelial microparticles in cancer. Neoplasia 8:79–88. doi:10.1593/neo.05592

58. Narazaki M, Tosato G (2006) Tumour cell populations differ in angiogenic activity: a model system for spontaneous angiogenic switch can tell us why. J Natl Cancer Inst 98:294–295. doi:10.1093/jnci/djj099

59. Weis SM, Cheresh DA (2011) Tumour angiogenesis: molecular pathways and therapeutic targets. Nat Med 17:1359–1370. doi:10.1038/nm.2537

60. Mancuso P, Burlini A, Pruneri G, Goldhirsch A, Martinelli G, Bertolini F (2001) Resting and activated endothelial cells are increased in the peripheral blood of cancer patients. Blood 97:3658–61

61. Beerepoot LV, Mehra N, Vermaat JS, Zonnenberg BA, Gebbink MF, and Voest EE (2004) Increased levels of viable circulating endothelial cells are an indicator of progressive disease in cancer patients. Ann Oncol 15:139–45

62. Mancuso P, Colleoni M, Calleri A, Orlando L, Maisonneuve P, Pruneri G, Agliano A, Goldhirsch A, Shaked Y, Kerbel RS, Bertolini F (2006) Circulating endothelial-cell kinetics and viability predict survival in breast cancer patients receiving metronomic chemotherapy. Blood 108:452–459. doi:10.1182/blood-2005-11-4570

63. Malka D, Boige V, Jacques N, Vimond N, Adenis A, Boucher E, Pierga JY, Conroy T, Chauffert B, Francois E, Guichard P, Galais MP, Cvitkovic F, Ducreux M, Farace F (2012) Clinical value of circulating endothelial cell levels in metastatic colorectal cancer patients treated with first-line chemotherapy and bevacizumab. Ann Oncol 23:919–927. doi:10.1093/annonc/mdr365

64. Strijbos MH, Rao C, Schmitz PI, Kraan J, Lamers CH, Sleijfer S, Terstappen LW, Gratama JW (2008) Correlation between circulating endothelial cell counts and plasma thrombomodulin levels as markers for endothelial damage. Thromb Haemost 100:642–7

65. DuBois SG, Stempak D, Wu B, Mokhtari RB, Nayar R, Janeway KA, Goldsby R, Grier HE, Baruchel S (2012) Circulating endothelial cells and circulating endothelial precursor cells in patients with osteosarcoma. Pediatr Blood Cancer 58:181–184. doi:10.1002/pbc.23046

66. Reale A, Melaccio A, Lamanuzzi A, Saltarella I, Dammacco F, Vacca A, Ria R (2016) Functional and biological role of endothelial precursor cells in tumour progression: a new potential therapeutic target in haematological malignancies. Stem Cells Int 2016:7954580. doi:10.1155/2016/7954580

67. Duda DG, Cohen KS, Scadden DT, Jain RK (2007) A protocol for phenotypic detection and enumeration of circulating endothelial cells and circulating progenitor cells in human blood. Nat Protoc 2:805–810. doi:10.1038/nprot.2007.111

68. Van Craenenbroeck EM, Conraads VM, Van Bockstaele DR, Haine SE, Vermeulen K, Van Tendeloo VF, Vrints CJ, Hoymans VY (2008) Quantification of circulating endothelial progenitor cells: a methodological comparison of six flow cytometric approaches. J Immunol Methods 332:31–40. doi:10.1016/j.jim.2007.12.006

69. Estes ML, Mund JA, Ingram DA, Case J (2010) Identification of endothelial cells and progenitor cell subsets in human peripheral blood. Curr Protoc Cytom, Chapter 9:Unit 9–33,1–11. doi:10.1002/0471142956.cy0933s52

70. Danova M, Comolli G, Manzoni M, Torchio M, Mazzini G (2016) Flow cytometric analysis of circulating endothelial cells and endothelial progenitors for clinical purposes in oncology: a critical evaluation. Mol Clin Oncol 4:909–917. doi:10.3892/mco.2016.823

71. Alix-Panabieres C, Pantel K (2014) Challenges in circulating tumour cell research. Nat Rev Cancer 14:623–631. doi:10.1038/nrc3820

72. Pantel K, Speicher MR (2016) The biology of circulating tumour cells. Oncogene 35:1216–1224. doi:10.1038/onc.2015.192

73. Mu Z, Benali-Furet N, Uzan G, Znaty A, Ye Z, Paolillo C, Wang C, Austin L, Rossi G, Fortina P, Yang H, Cristofanilli M (2016) Detection and characterization of circulating tumour associated cells in metastatic breast cancer. Int J Mol Sci 17. doi:10.3390/ijms17101665

74. Barriere G, Fici P, Gallerani G, Fabbri F, Zoli W, Rigaud M (2014) Circulating tumour cells and epithelial, mesenchymal and stemness markers: characterization of cell subpopulations. Ann Transl Med 2:109. doi:10.3978/j.issn.2305-5839.2014.10.04

75. Mantovani A, Locati M (2016) Macrophage metabolism shapes angiogenesis in tumours. Cell Metab 24:887–888. doi:10.1016/j.cmet.2016.11.007

76. Alix-Panabieres C, Pantel K (2013) Real-time liquid biopsy: circulating tumour cells versus circulating tumour DNA. Ann Transl Med 1:18. doi:10.3978/j.issn.2305-5839.2013.06.02

77. Miltenyi S, Muller W, Weichel W, Radbruch A (1990) High gradient magnetic cell separation with MACS. Cytometry 11:231–238. doi:10.1002/cyto.990110203

78. Radbruch A, Recktenwald D (1995) Detection and isolation of rare cells. Curr Opin Immunol 7:270–3

79. Campbell JD, Foerster A, Lasmanowicz V, Niemoller M, Scheffold A, Fahrendorff M, Rauser G, Assenmacher M, Richter A (2011) Rapid detection, enrichment and propagation of specific T cell subsets based on cytokine secretion. Clin Exp Immunol 163:1–10. doi:10.1111/j.1365-2249.2010.04261.x

80. Chattopadhyay PK, Yu J, Roederer M (2005) A live-cell assay to detect antigen-specific CD4+ T cells with diverse cytokine profiles. Nat Med 11:1113–1117. doi:10.1038/nm1293

81. Wolfl M, Kuball J, Ho WY, Nguyen H, Manley TJ, Bleakley M, Greenberg PD (2007) Activation-induced expression of CD137 permits detection, isolation, and expansion of the full repertoire of CD8+ T cells responding to antigen without requiring knowledge of epitope specificities. Blood 110:201–210. doi:10.1182/blood-2006-11-056168

82. Bacher P, Kniemeyer O, Schonbrunn A, Sawitzki B, Assenmacher M, Rietschel E, Steinbach A, Cornely OA, Brakhage AA, Thiel A, Scheffold A (2014) Antigen-specific expansion of human regulatory T cells as a major tolerance mechanism against mucosal fungi. Mucosal Immunol 7:916–928. doi:10.1038/mi.2013.107

83. Mancuso P, Antoniotti P, Quarna J, Calleri A, Rabascio C, Tacchetti C, Braidotti P, Wu HK, Zurita AJ, Saronni L, Cheng JB, Shalinsky DR, Heymach JV, Bertolini F (2009) Validation of a standardized method for enumerating circulating endothelial cells and progenitors: flow cytometry and molecular and ultrastructural analyses. Clin Cancer Res 15:267–273. doi:10.1158/1078-0432.CCR-08-0432

84. Roederer M (2001) Spectral compensation for flow cytometry: visualization artifacts, limitations, and caveats. Cytometry 45:194–205

85. Roederer M (2002) Compensation in flow cytometry. Curr Protoc Cytom, Chapter 1:Unit 1–14. doi:10.1002/0471142956.cy0114s22

86. Perfetto SP, Ambrozak D, Nguyen R, Chattopadhyay P, Roederer M (2006) Quality assurance for polychromatic flow cytometry. Nat Protoc 1:1522–1530. doi:10.1038/nprot.2006.250

87. Cox C, Reeder JE, Robinson RD, Suppes SB, Wheeless LL (1988) Comparison of frequency distributions in flow cytometry. Cytometry 9:291–298. doi:10.1002/cyto.990090404

88. Lin Y, Weisdorf DJ, Solovey A, Hebbel RP (2000) Origins of circulating endothelial cells and endothelial outgrowth from blood. J Clin Invest 105:71–77. doi:10.1172/JCI8071

89. Haight FA (1967) Handbook of the Poisson distribution. Wiley, New York

90. Roederer M (2008) How many events is enough? are you positive? Cytometry A 73:384–385. doi:10.1002/cyto.a.20549

91. Lugli E, Roederer M, Cossarizza A (2010) Data analysis in flow cytometry: the future just started. Cytometry A 77:705–713. doi:10.1002/cyto.a.20901

92. Saeys Y, Van Gassen S, Lambrecht BN (2016) Computational flow cytometry: helping to make sense of high-dimensional immunology data. Nat Rev Immunol 16:449–462. doi:10.1038/nri.2016.56

93. Lugli E, Pinti M, Nasi M, Troiano L, Ferraresi R, Mussi C, Salvioli G, Patsekin V, Robinson JP, Durante C, Cocchi M, Cossarizza A (2007) Subject classification obtained by cluster analysis and principal component analysis applied to flow cytometric data. Cytometry Part A 71a:334–344. doi:10.1002/cyto.a.20387

94. Naim I, Datta S, Rebhahn J, Cavenaugh JS, Mosmann TR, Sharma G (2014) SWIFT-scalable clustering for automated identification of rare cell populations in large, high-dimensional flow cytometry datasets, part 1: algorithm design. Cytometry A 85:408–421. doi:10.1002/cyto.a.22446

"E All'ottavo Giorno, Dio Creò La Citometria … and on the 8th Day, God Created Cytometry"

J. Paul Robinson

Abstract Flow cytometry has been available for about 50 years, but the core technology has changed very little. This constancy achieves two things: it makes the technology stronger as it has long-term development, but it also means that many applications may be a little stale. This chapter is an attempt to outline how the technology evolved into its current form and to identify the key changes that occurred along the way. While not every aspect of the evolution is documented herein, it is an attempt at framing the discoveries within the context of flow cytometry–specific demands. Flow cytometry evolved slowly for the first 15 years and then rapidly developed into one of the most significant cellular analysis technologies available to cell biologists.

Keywords Cytometry history · Cytometry invention · Ink-jet printing · Cell sorting · Multicolor phenotyping · Monoclonal antibodies

1 Introduction

Depending upon when you start the clock, the field of flow cytometry is either 50, 60, or 400 years old. Regardless of age, the history of how the field developed into the most mature technology for single-cell analysis is a fascinating story bringing together a diverse group of technologies that have successfully created a toolset critical for today's research and clinical diagnostic environments. Many technologies purport to focus on single cells, but only cytometry has the capacity to perform

J.P. Robinson (✉)
Purdue University Cytometry Laboratories, Department of Biomedical Engineering,
Purdue University, West Lafayette, IN, USA
e-mail: wombat@purdue.edu

and

Purdue University Cytometry Laboratories, Department of Basic Medical Sciences,
Weldon School of Biomedical Engineering, Purdue University,
West Lafayette, IN, USA

© Springer Nature Singapore Pte Ltd. 2017
J.P. Robinson and A. Cossarizza (eds.), *Single Cell Analysis*,
Series in BioEngineering, DOI 10.1007/978-981-10-4499-1_3

59

both qualitative and quantitative analysis and rapidly differentiate a large number of subsets into unique phenotypic classifiers.

There have been several periods of cytometry development during which important developments in other fields significantly enhanced cell-analysis technologies. While this is not unusual, the crossover of engineering discoveries into cytometry is often crucial; many of the events discussed below have their origins in fundamental engineering advances. For example, the Coulter counter was developed by an engineer, not a biologist, the cell sorter was developed by two engineers focusing on two quite separate problems, the laser was introduced into a flow cytometer well before it was available in virtually any other commercial product, high-throughput cytometry followed discoveries in robotic engineering, and more recently the cell-analysis technology of Coulter's 50–60-year old technology was re-discovered as materials engineers were able to manufacture microfluidic devices using silicon chip–manufacturing technologies. While these advances were driving the field of flow cytometry, similar events drove advances in the imaging field.

1.1   Before the Beginning

Before the beginning of cytometry, mankind had eyes. The human eye is a remarkable organ, as it functions like a camera—sending light impulses to the brain for the translation of light into structural detail. If there is sufficient light, the human brain can reconstruct the light scattered from objects, giving us the ability to differentiate with amazing definition between tiny dots of perhaps 100 microns up to giant mountains. Whatever we see, however, we are dependent on the arrival of photons at the right wavelength. The human eye is most sensitive at about 540 nm—which is pretty much green! We really cannot see much below about 365 nm (near-UV) and we cannot see at all above about 750 nm (near-IR). So our eyes are pretty important in the cytometry domain. Zacharias Jansen (1585–1632) (probably invented the first microscope), Robert Hooke (1635–1703) (author of *Micrographia*, which has recently been digitally reproduced [1]), and Anton van Leeuwenhoek (1632–1723) (first identified many tissues) all worked entirely with a single detection technology—the human eye. What they saw was amplified only by a few lenses—or in van Leeuwenhoek's case a single lens. Each of these scientists over 350 years ago was observing single cells; they were cytometrists of the first kind. Their observations were entirely spatio-temporal; everything was done in real time. They created drawings while observing the specimen under the microscope. Their ability to communicate the details of scientific experiments was almost entirely based on their ability to transform an observation into art. This is most evident in reading Hooke's *Micrographia*, which is filled with the most incredible details and beautiful drawings [2]. Hooke did not have the advantage of digital cameras or graphics engines and drew by hand all the illustrations in his book. One of the most famous shows a drawing of cork, where he describes the structure of what he termed "cells." Some years ago, our lab decided to replicate Hooke's

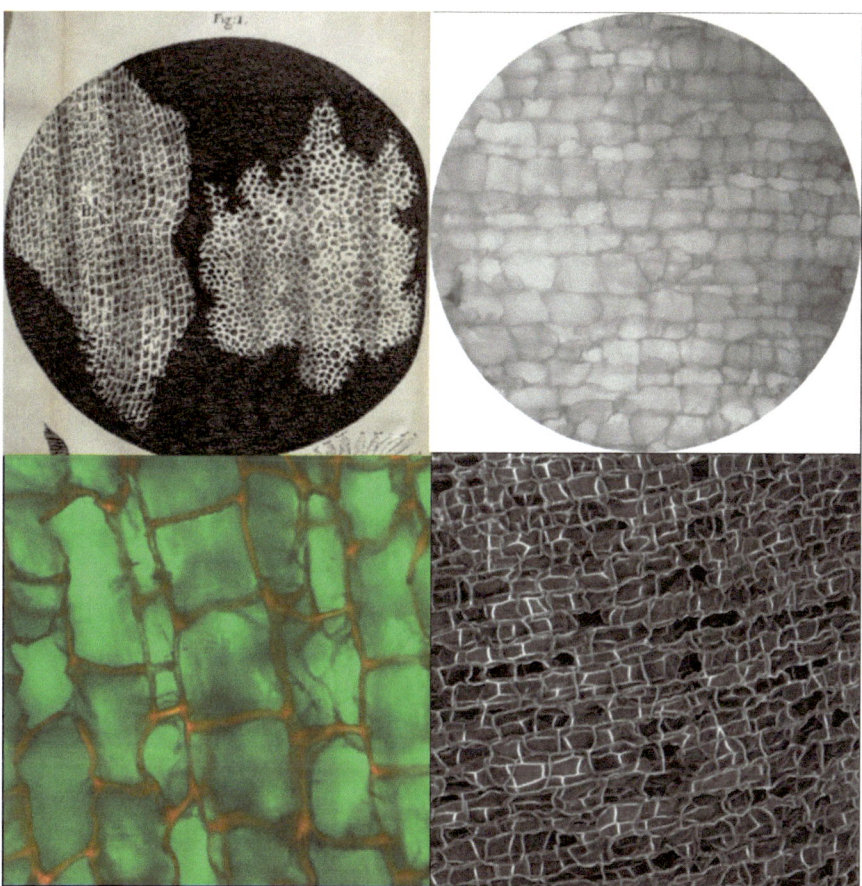

Fig. 1 One of the core deliveries of the world of microscopic discovery was the nature of the cell. Robert Hooke made his famous drawing of cork in 1665. Top *left*: The original image from the 1665 edition of *Micrographia*. It is accompanied by pictures of identical cork tissue imaged on a confocal microscope in our laboratory (photos copyright by J.P. Robinson—used by permission) (Images created by Jennie Sturgis)

original detailed microscopy of cork. This is reproduced in Fig. 1 together with images of a similar slice of cork imaged by a modern confocal microscope. We created some amazing images; if you look at an electronic reconstruction, you can view the cork structure based on its 3D structure from confocal microscopy. I suspect that everyone who has done microscopy knows the story of Hooke's reason for the name "cell" (which you find on page 113 of *Micrographia*, for those interested).

1.2   In the Beginning

In the 200 years between the beginning of the 18th century and the end of the 19th century much of our understanding of microscopy was established. Interestingly, there was no significant improvement in magnification, even though key developments included understanding how the resolution of systems was impacted by understanding aberration, defining how to achieve good optical resolution by matching refractive indices, and determining appropriate numerical aperture. Between 1700 and 1900 thousands of microscopes were manufactured; eventually individuals like Zeiss and Abbe drove production into more defined standards. It is arguable that the most significant analytical technology for scientists was the microscope. It was mostly transportable, did not require vast laboratory resources, and was usable by anyone. For virtually all developments in cytology, pathology, and histology, the microscope was the tool of choice. Our entire cellular background has been built around microscopy.

2   And then There Was Electronics

While microscopy established the fundamentals of cellular systems, defining intracellular structures and their functions required an expansion of spatial imaging to optical detection using photon detection systems. Prior to this, cellular evaluation could be achieved by use of specialty stains. Much of Paul Ehrlich's work[1] was based on manipulating a variety of chemistries, which led to the definition of many pathways for identifying cellular components and structural characteristics. In 1924 Feulgen [3] demonstrated that DNA was present in both animal and plant cell nuclei by using a stoichiometric procedure for staining DNA involving a derivatizing dye (fuchsin) and the formation of a Schiff base. In 1934 Moldavan [4] identified a technology that was to drive single-cell analysis for many years, but it was during and after World War II that Gucker [5, 6] designed a PMT-based system (Fig. 2) using a Ford headlight as a light source and was able to measure small particles in an air-based medium. One can argue that this technology was one of the fundamental drivers for particle analysis.

Caspersson developed his technologies using UV-absorption measurements of DNA and cytoplasm for a grasshopper metaphase chromosome [7]. By measuring densitometric traces across a region of the chromosome, Caspersson was able to use extinction values for chromosome and cytoplasts plotted against wavelength. He expanded this concept decades later when he was able to measure frequency distributions of DNA content in normal, malignant, premalignant, and atypical cell populations from the human uterine cervix [8]. Others used similar approaches; Robert Mellors built a microfluorometic scanner for differential detection of cells by

---

[1]1908 Nobel Prize in Physiology or Medicine.

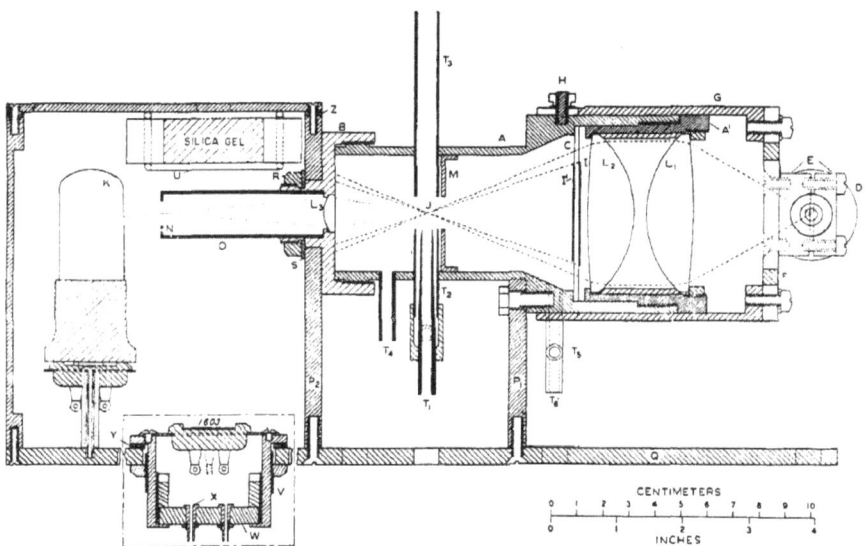

Fig. 2 Gucker's smoke-cell unit, composed of a light source and a PMT for measuring small particles (Illustration from Gucker and O'Konski [6])

1951. His technology utilized a CRT and a scanning image technology that was leading-edge at that time.

However, it was during the 1940s that one of the most significant discoveries for the future cytometry world occurred: the first successful application of fluorescence labeling. In 1941, Coons developed the fluorescence antibody technique; antipneumococcal antibodies labeled with anthracene allowed detection of both the organism and the antibody in tissue by means of UV-excited blue fluorescence [9]. From the manuscript: "Moreover, when Type II and III organisms were dried on different parts of the same slide, exposed to the conjugate for 30 min, washed in saline and distilled water, and mounted in glycerol, individual Type III organisms could be seen with the fluorescence microscope...." This dramatic experiment was the first example of what would end up as the most common technique in the world of cytometry—antibody conjugation using a variety of stains. By 1950 Coons and Kaplan had defined a more user-friendly technique for conjugating antibodies, conjugating fluorescein with isocyanate, which gave a far better blue-green fluorescent signal that was, most significantly, further from tissue autofluorescence [10].

In the strictly imaging world, the work of Papanicolaou established a relationship between observations of cellular morphology and staging of cancers [11]. Papanicolaou defined the state of carcinoma of the uterus based on vaginal smear morphology. This resulted in what is referred to as the Pap test, still the basis of what is performed to this day. This process was ideal for the technology of the day, essentially a microscope, and it allowed accurate staging diagnostics of uterine cancer from only a stained smear.

Two other quite important discoveries can be considered critical to our understanding of cellular systems and drove several new approaches to cell detection. First, Torbjorn Caspersson showed that "nucleic acids, far from being waste products, were necessary prerequisites for the protein synthesis in the cell and that they actively participated in those processes" (published in *Naturwissenschaften*) [12]. In order to expand upon this theory, Caspersson developed a UV spectrometer–based instrument that was one of the early cell-analysis instruments of importance in our field [13]. The second important discovery was that of Oswald T. Avery, whose fundamental work on the role of DNA as the carrier of genetic information [14] generated a significant amount of study on nucleic acid, which within a few years yielded a Nobel prize for the discovery of its structure [15].

2.1   Evolving Instrumentation

Among the many contributors to the field, perhaps Wallace Coulter had the most longest-lasting impact. There are a number of versions of Wallace Coulter's entry into the world of cell analysis. Some are true, some are interesting but probably not so accurate—and some scientists will argue this forever! Regardless of who says they know what actually happened nearly 70 years ago, we do know a few facts. First, Wallace Coulter was a great inventor. Second, he invented the Coulter counter. Third, he transformed clinical diagnostics. The how, where, when, and why would be wonderful to know, but we can certainly infer some of these. We know it took him several years, from about 1949 to about 1956, to perfect the technology [16]. His work was based on the measurement of impedance changes as cells traversed from one side of a very small orifice into another chamber; this early instrument is shown in Fig. 3. Coulter's instrument had a profound effect on many fields, not just hematology. The bottom line is that no matter how we think Wallace Coulter achieved his success, what he did changed the way *point-of-care* diagnostics operated. This was not the only time that Coulter impacted cell analysis, as we will see when we discuss the state of the art in the mid 1960s and early 1970s. Coulter expanded both his instrument line and his company, and created a competitive environment in a very fast moving field of hematology.

While Wallace Coulter was playing with impedance-based systems, P. J. Crosland-Taylor was developing the concept of the sheath-flow principle [17]. His initial paper, published as a letter to *Nature*, describes the sheath-flow principle that has survived for the past 60-plus years as the basis for almost every flow cytometer. Crosland-Taylor developed a blood-cell counter based on a lamp and PMT-based detection as cells passed through the beam, but the technology [18] was not as commercially successful as Coulter's impedance-based systems, which ruled the hematology world for the next few decades.

The Coulter counter dramatically changed the clinical picture of diagnostics. Instead of the hours required to do a manual cell count and differential, the Coulter counter provided data in minutes. By integrating accurate dilution techniques,

Fig. 3 Wallace Coulter's first commercial version of the Coulter counter (photo copyright J. P. Robinson; used by permission)

Coulter was able to demonstrate a simple technology that provided solid, reproducible results. The technology was successful in many fields—chemical engineering, solid-particle analysis (for example, the paint industry), soil analysis, and of course medical diagnostics. From about 1958 to 1961 Coulter sold over 1500 Coulter counters[2] and had a strong user base. Advertisements were commonly found in the scientific literature; a classic example is show in Fig. 4.

By 1962 Coulter was advertising applications for measuring contamination in clean-room air, jet fuel, hydraulic fluids, and a variety of solvents. However, the field of biology was without doubt the area that Coulter and others considered the most significant. One group that began using the Coulter counter was at the Los Alamos National Laboratories. A huge effort had been mounted after the end of WWII to better understand the effects of radiation; one particular pathologist by the name of C.C. Lushbaugh had begun to use the Coulter counter in his studies of anemia and in particular red blood cells [19].

At this point occurred one of the most significant factors in the making of flow cytometry as we know it today. Had C.C. Lushbaugh understood the technology

---

[2]Advertisement in Anal. Chem., 1961, 33 (4), p. 28A Publication Date: April 1, 1961 (Article) doi:10.1021/ac60172a719.

Fig. 4 Coulter's advertisement in Science in 1962 indicated that the Model A was selling for $3550. By 1962 Coulter had sold 1500 instruments

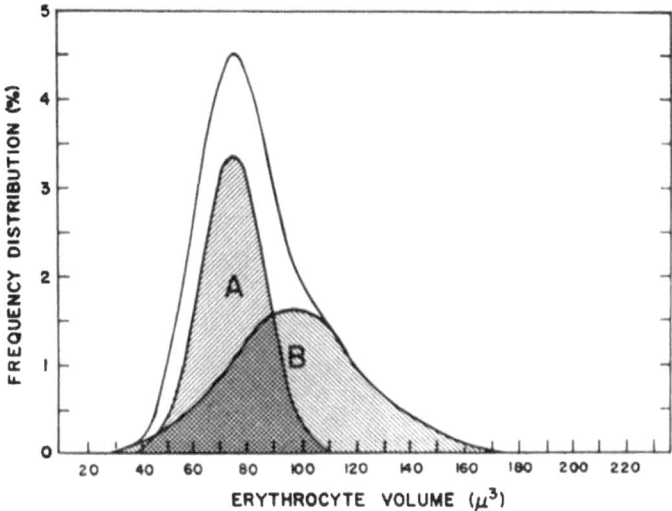

Fig. 5 Lushbaugh's famous figure claiming two populations of RBCs, from which Mack Fulwyler identified the misuse of the Coulter counter electronics

and used it correctly, his studies on anemia would likely have gone in a different direction. One then wonders how the field of flow cytometry might have evolved, if at all. The bottom line is, though, that Lushbaugh ended up misusing the Coulter counter to evaluate red blood cells (RBCs). He published a paper in 1963 describing two populations (Fig. 5) of RBCs, claiming that this was determined based on some electronic modifications that Los Alamos had made to the Coulter counter; his paper set in play a series of motions that changed the face of cytometry.

These two populations of RBCs described in his 1963 paper [20] identified a concern about the use of the Coulter technology. Fortunately, in the same division certain people with engineering backgrounds found that Lushbaugh's claims were not supported by engineering principles. So in 1964 an engineer by the name of Mack Fulwyler challenged Lushbaugh's contention that the Coulter counter data could be interpreted as showing different populations. Fulwyler hypothesized that the dual populations were artefactual and based on an abuse of the technology (personal communication). Fulwyler proposed to test his hypothesis by separating the so-called "two populations" into pure populations of RBCs and comparing them again on a Coulter counter. To achieve this, Fulwyler sought assistance from an unusual source—Richard Sweet, a scientist working on a military-funded project at the University of California, San Francisco. Fulwyler had heard a talk by Sweet and surmised that he might be able to use Sweet's electrostatic droplet-separation technology [21] to separate RBCs. He thought that if he could place RBCs into single droplets, analyze those droplets using the same criteria as Lushbaugh, and then re-analyze those two pure populations, he would have the answer.

It took Fulwyler less than 12 months to design, build, and test the first electrostatic cell sorter. He published his findings in *Science* in 1965 [22]. A more

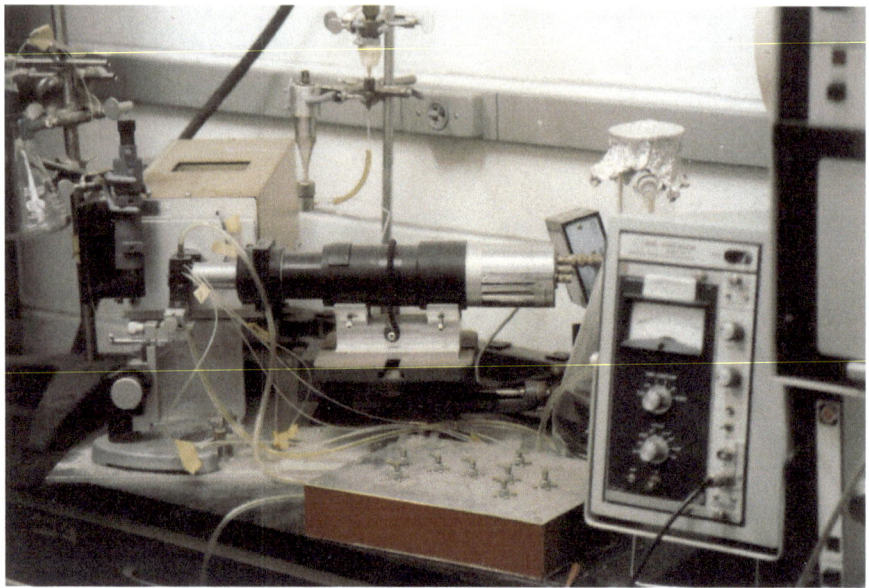

Fig. 6 Part of Fulwyler's cell sorter showing the impedance sorting chamber and a number of other components (photo copyright J.P. Robinson; used by permission)

detailed account of how this invention happened is found elsewhere [23] but remains one of the most important discoveries for the field of cytometry. A part of the original instrument design is shown in Fig. 6.

Within a year or so, Len and Lee Herzenberg were onto this technology, and in a partnership that was to last for almost 50 years, the Herzenbergs guided flow cytometry into a mature and successful field that became absolutely fundamental to the world of immunology (Fig. 7).

2.2   The Technology-Biology Interface

One more important development between 1965 and 1967 involved a scientist studying beta-ᴅ-galactosidase in *E. coli*. Boris Rotman had been interested in this area for many years and worked in the laboratory of Joshua Lederberg (Nobel Prize in Medicine, 1958, at the age of 33). It is unclear how Rotman found out about Fulwyler's cell sorter, but when Rotman was offered his first faculty position at Brown University in 1967, he contracted with Fulwyler to build for him a replica of Fulwyler's 1965 cell sorter (which still exists today and is the oldest flow cell sorter in existence). This instrument, however, also included a PMT for fluorescence analysis since Rotman intended on sorting fluorescent *E. coli*. Rotman paid $5000

Fig. 7 Len and Lee Herzenberg with one of the early cell sorters developed in their laboratory at Stanford (photo copyright Lee Herzenberg, used by permission)

for the instrument[3] and it operated in his laboratory for 40 years before being rescued by the author. It is unclear if Rotman ever published any papers using the technology, but he was the first person other than Mack Fulwyler to have a cell sorter in his laboratory.

While C.C. Lushbaugh was experimenting with the Coulter counter there was a lot of activity to develop cell-separation and analysis technologies on the east coast, mainly Boston. Lou Kamentsky had been experimenting with single-cell analysis for some years. He had been an engineer at the IBM research laboratories, where one of his main projects was apparently to develop character-recognition technology for the purposes of converting telephone books to electronic documents (Kamentsky personal communication). His early patents focused on character recognition [24, 25] and were assigned to Bell Laboratories. He expanded this technology to design and build a number of instruments focused on UV spectroscopy around 260 nm [26, 27]. The instruments were sophisticated, multiparameter, and most significantly, controlled by rather huge IBM computer systems to which Kamentsky had almost unique access. One of the two instruments (Fig. 8) that Kamentsky built was shipped to Len Herzenberg in 1967 (Kamentsky personal communication). By 1969 Kamentsky had left IBM, created his own company, Biophysics Systems, and developed first a cytometer with a red laser for light scatter only (Fig. 9) and eventually one with a PMT for fluorescence measurement. At this same time, Wolfgang Göhde in Münster, Germany, was experimenting with

---

[3]http://siarchives.si.edu/collections/siris_arc_217722.

Fig. 8 The 4-parameter cell sorter developed by Lou Kamentsky about 1962–1963. Two of these instruments were constructed; one was shipped to the Herzenberg laboratory in 1967 (photo courtesy of Lou Kamentsky)

Fig. 9 Kamentsky's Biophysics Cytograph. This early-model analyzer measured only light scatter from a red laser. Eventually, a blue laser was added to give the instrument a fluorescence capability and the instrument was renamed Cytofluorograph (photo Copyright J.P. Robinson; used by permission)

cytometry using a microscope base; he demonstrated that cytometry was a perfect technology for measuring fluorescence [28]—and opened a new dimension in cytometry with rapid commercialization of the technology.

There are some fascinating stories about how the field of flow cytometry grew and insufficient space to detail them in this chapter. There was a tug-of-war between Wallace Coulter and Becton Dickinson to take over this exciting technology. As it happened, Coulter gained control of Fulwyler and his invention (which by the way was publicly available as it was developed in the National Laboratory) by creating a company at Los Alamos called Particle Technologies. Meanwhile, Becton Dickinson under the guidance of Bernie Shaw created Becton Dickinson Immunodiagnostic Systems (BDIS) and partnered with the Herzenbergs to develop independent technologies. These independent agreements by competing companies were signed off within 30 days of each other after many months of negotiations.[4] It is interesting to take a look at how industry viewed the potential for cytometry back in the 1970s. Unlike today, when cytometry is a billion-dollar enterprise, back then it was an unknown technology that appeared to have no real applications. BD at the time had internal arguments as to whether the technology was of any use at all. In one memo, a senior manager even questioned why a laser would be necessary for a cell-analysis system. So little was known about the potential that it was left to some key scientists such as the Herzenbergs, Kamentsky, Fulwyler, and Shapiro to foster different directions of discovery. The result was two competing companies building what became known as cell sorters. Initially, only sorters were of interest, as the idea of analysis was apparently of primary concern to no one except those interested in clinical diagnostics, mainly hematology analyzers. At this time, there were no monoclonal antibodies, and no understanding of the diversity of white blood-cell populations, with the exception of small and large lymphocytes, monocytes, and a variety of polymorphonuclear cells. Indeed, in the late 1960s the knowledge of immune cells was minimal.

2.3 Immunology Changes Technology

The advent of monoclonal antibodies was the most significant defining factor in the success of flow cytometry as a technology. When Köhler and Milstein came up with the idea of monoclonal antibodies [29], one would imagine that this would be of immediate interest to the community. But first it had to be published, and it is legend that the paper submitted to *Nature* was not considered significant enough to be published as a full paper—thus it lives forever as a letter to *Nature* [30]. Wouldn't it be interesting to read the reviewer's comments now? It did not take long for the scientific community to understand the impact of Köhler and Milstein's

---

[4]The author has a package of letters, memos, and agreements from this period of time that outline these events.

genius; Köhler and Milstein received the Nobel Prize in Medicine in 1984. In his review of the discovery of monoclonal antibodies (mAbs) [30] Milstein makes it very clear that the resultant naming of the term "hybridoma" came from Len Herzenberg during his sabbatical in Milstein's lab at Cambridge [30], so it is not surprising that when Len and Lee Herzenberg returned to Stanford University, they began a 50-year exploration of immunological diversity and drove the field firmly down the immunological pathway, which was a most successful venture. As a side note, in the above article Milstein also discussed the reasons that the hybridoma technology was not patented. To find out why not, I encourage you to read the paper. The fact is that the failure to patent may have been the most fortuitous event in medical history, as literally hundreds of thousands of discoveries in science resulted from the unbridled creation of mAbs for decades. Milstein was the guest of honor at the XII International Meeting of the Society for Analytical Cytology held at Cambridge University, Cambridge, England, on Aug. 9–15, 1987; an ebullient Milstein is shown with raised hands in Fig. 10. James Watson, the local host of the meeting, told me that Milstein fell off his bike while riding home from the banquet and broke his arm. Such is life for a Noble Laureate!

The Herzenbergs continued to expand the knowledge base in flow cytometry as they developed their instrument group and published a seminal paper in *Science*

Fig. 10 César Milstein (standing, right) at the SAC (ISAC) meeting in Cambridge University, in 1987. Center is Howard Shapiro (as usual with guitar), and left, Gunter Valet (photo Copyright J. P. Robinson; used by permission)

describing the sorting of FDA-labeled CHO cells using a mercury arc–lamp light source and a PMT detector [31]. This classic paper describes what was state-of-art technology based on their evaluation of Sweet's work [21], Fulwyler's sorter [22], and Kamentsky's rapid cell spectrometer, which Herzenberg had the opportunity to test [32].

In the United States, the technology of flow cytometry was beginning to be appreciated. Herzenberg's 1969 paper [31] had drawn some serious attention and together with the impact of Fulwyler's and Kamentsky's papers attracted corporate interest. As noted above, two agreements by the two competing groups, Stanford with BD and Fulwyler with Coulter, were signed in the last quarter of 1971, and competition now began in earnest. In New Mexico, Mack Fulwyler and Wallace Coulter set up a new company called Particle Technologies (PTI). Coulter owned this company outright, but Fulwyler was the president. They not only made beads for calibration, but also continued the tradition of Fulwyler (who had by this time competed his Ph.D.) of building cell sorters and analyzers. Fulwyler had a strong vision of bringing advanced multiparameter instruments to the biology marketplace. He was joined by a team of engineers, one of whom was Bob Auer. Meanwhile at Stanford, things were moving quickly. Richard Sweet had joined the Herzenberg team and BD had signed an agreement to build several cell sorters. Herzenberg pushed ahead to replace the arc-lamp light source in his sorter with an argon laser. He published this in a paper with Sweet as a co-author in 1972 [33]. This instrument essentially defined the design and operation of flow cytometers for the next 45 years. It had a single PMT for fluorescence. In an ironic twist, since the only source of fluorescent calibration beads at the time were those developed by Mack Fulwlyer, these were used as a calibration source by the Herzenberg lab.

Meanwhile, at the competing camp in New Mexico, Fulwyler was keen to expand his repertoire and build a large multiparameter machine. The PTI team started to build the TPS-1 (two-parameter sorter), which was designed to combine Coulter volume with fluorescence for sorting. This project was going rather slowly and when the lead engineer left, Bob Auer was put in charge of the project. At the same time, Fulwyler knew that competition in the field would be driven by large, expensive, multiparameter instruments, in the manner of those under construction in the Herzenberg lab. In fact, the PTI team had begun a project called the *Super Dooper Sorter* (SDS-1), a huge instrument capable of two colors of fluorescence, forward scatter, and Coulter volume and operated by a PDP 8 computer console with graphics displays and a Tektronics 4010 display. The computer and display occupied three six-foot-high instrument racks. Data and protocols were stored on Tridata cartridge tapes (Robert Auer, personal communication). This instrument was built for a special program at the NIH; there is a fascinating letter from Dr. James Corll of PTI to Mr. Sherman Oxendine of the NIH procurement branch dated May 24, 1974, in which Dr. Corll attempts to assist the NIH in determining what a cell-sorting device might be called. The technology could not be easily described and the letter starts:

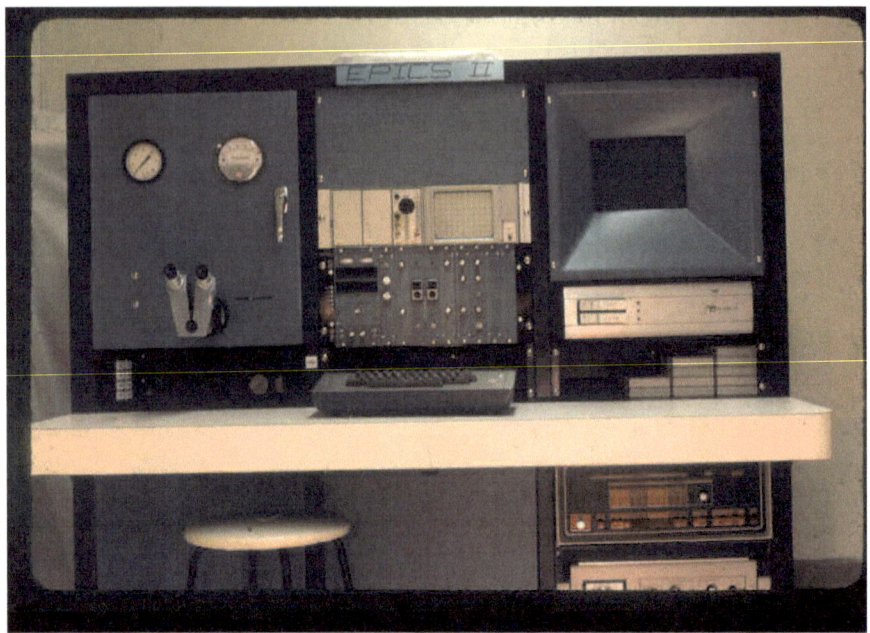

Fig. 11 The SDS-1 sorter developed by Fulwyler and shipped to the NIH in 1974. This instrument had two fluorescence PMTs and a light scatter sensor (0–12 degrees). Its official name was the EPICS II (photo courtesy of Bob Auer used by permission)

> I can empathize with your quandary of how to describe the device Chet Herman requires. These devices are very complex and of such recent development that the terminology—the very words, names, and function descriptions—associated with such devices is still not universally understood or accepted.[5]

The instrument was apparently identified by the NIH, who purchased it, and it was delivered that same year. It consisted of a 4-parameter cell sorter with two fluorescence detectors (see Fig. 11). The irony of this huge success for Fulwyler is that it drew significant attention from Wallace Coulter, mainly because of its huge size, very high power demand (water-cooled argon laser), and complexity of operation. What occurred next was calamitous and destructive for the PTI team. There is room for only a short summary here. Coulter and his chief engineer Walter Hogg visited PTI on their way to the Engineering Foundation Meeting at Asilomar in 1973 and the rest is history. While the team had demonstrated high-quality, advanced systems for multiparameter cell sorting, Coulter was convinced that anything that did not fit into the back of a station wagon would never sell and he stopped the development of large systems in early 1975. Within weeks Mack Fulwyler had resigned from the PTI team; because he had a 12-month non-compete contract he went to Germany and spent a year on sabbatical in the Jovin laboratory in the Max Planck Institute. Fulwyler also announced that on his return he

---

[5]Letter from James A. Corll to Mr. Sherman Oxendine, NIH Procurement Branch, 24 May, 1974.

would become a member of Wallace's direct competition, Becton Dickinson Immunocytometry Systems (BDIS). By March 1975 the entire PTI team was moved into Coulter's operation at Hialeah, Florida, where they continued to build the much smaller TPS-1 sorter. The EPICS division of Coulter was established with Bob Auer as the head.

It did not take Wallace Coulter very long to realize he had made a tactical error. Large instruments for biological research were a major blind spot for him. He focused so much on personal demonstrations of equipment that he could not imagine how big instruments could be demonstrated, and therefore he felt they would have no future. Within a couple of years, however, Bob Auer had convinced him that this attitude was behind the times. The very fact that Fulwyler had turned to BD, who were producing their Fluorescence Activated Cell Sorter (FACS) instrument line very successfully, was probable a significant factor in Coulter's about-face. We will never know precisely what changed his mind, but the delay in establishing a big-instrument line was a costly mistake for Coulter Corporation, and one may well argue that they never again managed to compete successfully with BD on large instruments.

There may be interesting lessons to be drawn from this series of events over 40 years ago. Companies that focus on the end-product of technology, i.e., are content to sell only technology that is already well understood and mature, will end up losing a large segment of the marketplace. It is critical for high-technology companies to be pushing the envelope and developing the next generation of technology at the highest level if they want to manage technology at the benchtop level. It is also interesting to note that all the companies that began to build flow cytometers were driven by engineers who understood the notion that being a leading-edge developer was necessary to maintain the lower-level business of small benchtop analyzers. Today, many companies are run by accountants, who do not understand this principle. The writing is on the wall when a company sees a high-technology area as simply a sales and marketing opportunity and not a technology development opportunity. Companies that forget the fact that lower-cost technology is a subset of leading-edge research and development programs and drop these major programs will soon find themselves without customers for any instruments. Keeping the leading edge of technology is a critical-path component to getting smaller, lower-cost analytical instruments into labs.

Two other aspects of the early days of cytometry deserve to be included in this "creation" story. Both are related to the development of standards. First is the bead standards themselves; the other is the decision to create a file standard for the field. Both these decisions were paths critical to the success of cytometry as a major technology in science.

2.4 The Creation of Standards and Education

As noted earlier, Mack Fulwyler was heavily involved in making standard beads while at Los Alamos National Laboratory. Particle Technologies Inc. was initially formed to expand the idea of flow cytometry (although it was not called that then) and calibration beads were not a fundamental requirement. However, as flow cytometry become a more standard procedure with several companies manufacturing multiparameter flow cytometers, calibration beads became a necessary component. One of the most significant contributions was made by Abe Schwartz, who recognized the need for better design of calibration beads, and in particular using the same dyes as were being used for biological measurements. Schwartz designed a bead that contained the dye within the bead [34] in 1984, when it was clear that improved calibration was desirable. His patent on the manufacture of beads established some of the basics of modern bead chemistry. Poncelet also developed a bead for what he termed a quantitative indirect immunofluorescence assay (QIFI) [35]; this was a rather complex procedure for flow cytometry labs, requiring radioisotope labeling. Schwartz recognized a desire to properly calibrate biological systems and focused for some years on developing standards, obtaining a number of patents that were important for creating a commercial success in standards [34, 36]. His establishment of molecules of equivalent fluorescence (MESF) was an important contribution [37] allowing for the determination of equivalence between a bead and the number of fluorescent molecules on the cell surface.

At the same time, a number of individuals started to evaluate the nature of data files and reporting of experimental data. With many individual instruments creating different file structures and data outputs, the question was what would constitute a reasonable reporting model. The first such report was by a group of scientists who had been working with DNA cytometry [38]. At the same time, Murphy proposed a universal file structure for listmode files [39] that was adopted and became the basis for all future flow cytometry data. The creation of a file standard was a very important decision by the field. The organization responsible for flow cytometry at the time, the Society for Analytical Cytology (SAC) (later renamed the International Society for Analytical Cytology (ISAC); more recently Analytical Cytology was changed to Advancement of Cytometry) continued to drive the adoption of newer standards based on the initial FCS 1.0, FCS 2.0 [40], FCS 3.0 [41], and FCS 3.1 [42]. One of the more important results of the early focus on file standards has been the continuation of the model whereby key individuals in the scientific field establish the standards and not industry. This is important since it to some extent forces industry to abide by the standards (well mostly anyway) so that all instruments produce a data file that closely follows the standard.

3   The Impact of Software

Instrument manufacturers concerned themselves with hardware accompanied by a minimum of software. Those who experienced the development of instruments from the 1980s to about 2000 were very well aware of the lack of quality analytical software that came with flow cytometers. A senior engineer of one instrument corporation once maintained to the author that it was not the instrument manufacturer's responsibility to develop software! Of course, in the early days, users had no choice but to use the software provided, unless they could write their own; many did so, but what they produced was usually not suitable for general distribution. Much of the early software developed for flow cytometry in general was focused on DNA analysis and functional studies, as these were major interest for flow cytometry at the time [43–45]. Until the mid 1990s, the number of parameters collected by most flow cytometers was not high, ranging from four to six; 3-color immunofluorescence was introduced only in 1984 by Loken [46] but not used routinely for another six or eight years. Three colors could be reasonably represented in relatively simple software. However, as the number of parameters increased, so do did interest in advanced analysis tools.

3.1 Advancing Technology—Digital Flow Cytometry

For decades, flow cytometry was almost entirely an analog technology. There are several reasons for this, but the primary one is the core speed of analog and the lack of speed for digital systems. Eventually, several groups demonstrated the advantages of digital electronics; Galbraith's group showed the power of digitizing signals, which allowed the application of very advanced algorithms in software [47]. Others, such as Leary, also utilized digitization of signals and implementation of digital logic to perform complex processes [48]. The introduction of the FACSDiva option[6] by Becton Dickenson around 2001 drove a frenzy of change within the instrumentation community. The battle was between the speed of the processors and the design and manufacture time of a large, expensive instrument—a continuing problem for all technologies, of course. This led to an entire industry of third-party software companies who produced advanced software of very high quality. Indeed, flow cytometry software is now almost entirely driven by third-party sources.

4 Cytometry as an Education Tool

Cytometry represents complexity in many areas, including the technology itself, the design and function of assays, the chemistry of fluorochromes, and the difficulty of data analysis. Without doubt the earliest and possibly the most important educational tool was the famous *Practical Flow Cytometry* written by Howard Shapiro [49, 50]. Shapiro first developed his book to explain the principles of cytometry to colleagues and to assist people to build small instruments.[7] Eventually, he created his own manuscript using a dot-matrix printer. The result was the first edition of his

---

[6]http://www.bdbiosciences.com/ds//is/others/23-6566.pdf.
[7]Howard M. Shapiro, personal communication.

book, in which Shapiro dedicated many pages to the design of the "Cytomutt," a small, custom-built instrument that introduced many people to the field of cytometry. In subsequent versions of the book, he advanced the physics, optics, and electronics as well as applications for the entire field. There is little doubt that Shapiro's book is the single most effective educational tool in the field. It has introduced many neophytes to the field, given cytometry-based scientists detailed documentation on many applications, and provided difficult-to-find information on just about every aspect of the field.

4.1   The Internet Is Here

We probably all know the story of the invention by Tim Berners-Lee of the common internet we use every day [51–53]. But many of us forget that for the great majority of individuals there was actually a time where we did not use email. The original internet was initially designed by the US military to operate as a secondary network to the telephone system, which was considered vulnerable to enemy attack. How things change, and remain the same! ARPAnet was based on computer-to-computer connection and allowed interactions independent of the telephone system.[8]

4.2   The Purdue Email Discussion List

In 1988 the Purdue University Cytometry Laboratories (PUCL) was established at Purdue University at a time when computers were literally just becoming readily available. We set up links to the University computer network and Steve Kelley in our lab managed data archiving and computer operation for both the shared facility and our R&D laboratory. Steve introduced us to email. In the early days of email, I used to check my mail at least once per month, in case something important was there! One day Steve asked me if other people in the field of cytometry might have email addresses. I investigated and found very few. The list started with about five or ten people with email addresses whom we could locate, and who by the way I called on the phone to ask if they actually used their email. At first the discussions were few and far between, but as people were added, the list because one of the most important electronic tools for communication in the field.[9] By 2000 there were almost 2000 members, an increase of 200 per year, and by 2016 there were 4500 members. The archive for the early years (pre-1992) was lost, as we did not realize that the list would extend for over 25 years. It is possible that this list may be one of

---

[8]http://www.history.com/topics/inventions/invention-of-the-internet accessed Nov 10, 2016.

[9]The Cytometry Listserve address is http://www.cyto.purdue.edu/hmarchiv/index.htm.

the longest lasting discussion lists in science—at 27 continuous years it certainly has some basis for a record!

4.3   The World Wide Web

As technology developed, a network of computers became connected to each other around the world and files could be exchanged. By 1991, however, Berners-Lee had a slightly different problem, to create a technology that could rapidly exchange electronic information, and one of the first tasks implemented was to create an internal electronic telephone directory. It is interesting to consider that this was not the only major discovery to be made in relationship to creating telephone directories. As noted earlier from Lou Kamentsky, he originally worked for Bell Telephone and one of his major tasks was to create an electronic version of a telephone book![10] Berners-Lee's efforts in creating a model for instant communication was an important step in bringing education to the masses. In 1993, the PUCL lab established www.cyto.purdue.edu; the address has not changed. We initially created lists of cytometry labs worldwide; search engines were either non-existent or useless. Eventually, of course, everyone built lab websites that have become a powerful communication and teaching tool in the field of cytometry. By 1994 we were able to put the Cytometry Listserve onto the web so that everyone had access to it, as opposed to only those who used its email links.

5   Conclusion

We just passed the 50th anniversary of the founding of the traditional field of flow cytometry [54]. The heady days of the early inventors has been transformed into both small and large companies designing and building a breadth of instruments, assays, and reagents for the field of flow cytometry. New approaches to old problems have created new opportunities for discovery. For example, imaging flow cytometry has become a new part of the field with diverging approaches, all of which have merit [55–58]. All these advances address slightly different problems by creating technologies that have advantages for either structural or chemical differences in cells or particles. Technologies are being adapted to deal with small particles such as exosomes, which are well below the diffraction limit and therefore cannot be properly analyzed using current flow cytometers, most of which were designed for blood cell–sized particles. This will not be an easy task and will require many iterations of instruments. Finally, with the advances in automation and the introduction of very-high-content flow cytometers such as the CyTOF

---

[10]Lou Kamentsky, personal communication.

[59, 60] more advanced approaches to high-content analysis have been required [61], which brings informatics-driven technologies well within the confines of a field that started over a dispute about two types of red blood cells.

References

1. Hooke R (1665) Micrographia. Octavo, Republished in 1998 as CDrom edition. 1891788027
2. Robert H (1665) Micrographica. Royal Society, London
3. Feulgen R, Rossenback H (1924) Mikroskopisch-chemischer Nachweis einer Nucleinsäure vom Typus der Thymonucleinsäure und die darauf beruhende elektive Färbungvon Zellkernen in mikroskopischen Präparaten. Hoppe Seyler ZPhysiolChem 135
4. Moldavan A (1934) Photo-electric technique for the counting of microscopical cells. Science 80:188–189
5. Gucker FT Jr, O'Konski CT, Pickard HB, Pitts JN Jr (1947) A photoelectronic counter for colloidal particles. J Am Chem Soc 69:2422–2431
6. Gucker FT Jr, O'Konski CT (1949) Electronic methods of counting aerosol particles. Chem Rev 44:373–388
7. Caspersson TO (1936) Uber den chemischen Aufbau der Strukturen des Zellkernes. SkandArchPhysiol 73
8. Caspersson TO (1964) Quantitative cytochemical studies on normal, malignant, premalignant and atypical cell populations from the human uterine cervix. Acta Cytologica 8:45
9. Coons AH, Creech HJ, Jones RN (1941) Immunological properties of an antibody containing a fluorescent group. Proc Soc Exp Biol Med 47:200–202
10. Coons AH, Kaplan MH (1950) Localization of antigen in tissue cells. II. Improvements in a method for the detection of antigen by means of fluorescent antibody. J Exp Med 91:1–13
11. Papanicolaou GN, Traut R (1941) The diagnostic value of vaginal smears in carcinoma of the uterus. Am J Obstet Gynecol 42:193–206
12. Caspersson TO (1941) Ribonucleic acids and the synthesis of cellular proteins. Naturwiss 28:33
13. Caspersson TO (1950) Cell growth and cell function. A cytochemical study. W.W. Norton & Company, New York
14. Avery OT, McCarty M, MacLeod M (1944) Studies on the chemical nature of the substance inducing transformation of pneumococcal types. Induction of transformation by a desoxyribonucleic acid fraction isolated from Pneumococcus type III. J Exp Med 79:137–158
15. Watson JD, Crick FHC (1953) Molecular structure of nucleic acids; a structure for deoxyribose nucleic acid. Nature 171:737–738
16. Coulter WH (1953) Means for counting particles suspended in a fluid.
17. Crosland-Taylor PJ (1953) A device for counting small particles suspended in fluid through a tube. Nature 171:37–38
18. Crosland-Taylor PJ, Stewart JW, Haggis G (1958) An electronic blood-cell-counting machine. Blood 13:398
19. Lushbaugh CC, Basmann NJ, Glascock B (1962) Electronic measurements of cellular Volumes II. Frequency distribution of erythrocyte volumes. Blood 20:241–248
20. Lushbaugh CC, Hale DB (1963) Electronic measurement of cellular Volume VII. Biologic evidence for two volumetrically distinct subpopulations of red blood cells, in annual report biological and medical research group (H-4) of the health division, July 1962 through June 1963. Los Alamos Scientific Laboratory of the University of California, p 270–278
21. Sweet RG (1965) High frequency recording with electrostatically deflected ink jets. Rev Sci Instru 36:131–136

22. Fulwyler MJ (1965) Electronic separation of biological cells by volume. Science 150:910–911
23. Fulwyler MJ (2005) Mack Fulwyler in his own words. Interview by Robinson JP, Kondratas RA. CytometryPart A. J Int Soc Anal Cytol 67:61–67
24. Kamentsky LA (1958) Spatially oriented character recognition. Bell Telephone Labs
25. Kamentsky LA, Liu CN (1963) Computer-automated design of multifont print recognition logic. J Res Dev 7:2–13
26. Kamentsky LA, Derman H, Melamed MR (1963) Ultraviolet absorption in epidermoid cancer cells. Science 142:1580–1583
27. Kamentsky LA, Melamed MR, Derman H (1965) Spectrophotometer: new instrument for ultrarapid cell analysis. Science 150:630–631
28. Dittrich W, Gohde W (1969) Impulse fluorometry of single cells in suspension. Z Naturforsch B 24:360–361
29. Kîhler G, Milstein C (1975) Continuous cultures of fused cells secreting antibody of predefined specificity. Nature 256:495–497
30. Milstein C (1999) The hybridoma revolution: an offshoot of basic research. Bioessays 21:966–973
31. Hulett HR, Bonner WA, Barrett J, Herzenberg LA (1969) Cell sorting: automated separation of mammalian cells as a function of intracellular fluorescence. Science 166:747–749
32. Kamentsky LA, Melamed MR (1967) Spectrophotometric cell sorter. Science 156:1364–1365
33. Bonner WA, Hulett HR, Sweet RG, Herzenberg LA (1972) Fluorescence activated cell sorting. Review of Scientific Instruments 43:404–409
34. Schwartz A (1988) Calibration method for flow cytometry using fluorescent microbeads and synthesis thereof, U.P. Office, Editor. USA
35. Poncelet P, Carayon P (1985) Cytofluorometric quantification of cell-surface antigens by indirect immunofluorescence using monoclonal antibodies. J Immunol Methods 85:65–74
36. Schwartz A (1990) A method for calibrating a flow cytometer of fluorescence microscope for quantitating binding antibodies on a selected sample, and microbead calibration kit thereof, U.P. Office, Editor, Carribean Microparticle Corporation, USA
37. Schwartz A, Gaigalas AK, Wang L, Marti GE, Vogt RF, Fernandez-Repollet E (2004) Formalization of the MESF unit of fluorescence intensity. Cytometry B Clin Cytom 57:1–6
38. Hiddemann W, Schumann J, Andreeff M, Barlogie B, Herman CJ, Leif RC, Mayall BH, Murphy RF, Sandberg AA (1984) Convention on nomenclature for DNA cytometry. Committee on nomenclature, Society for Analytical Cytology. Cancer Genet Cytogenet 13:181–183
39. Murphy RF, Chused TM (1984) A proposal for a flow cytometric data file standard. Cytometry 5:553–555
40. Dean PN, Bagwell CB, Lindmo T, Murphy RF, Salzman GC (1990) Data file standard for flow cytometry. Cytometry 11:323–332. doi:10.1002/(SICI)1097-0320(19970601)28:2<118: AID-CYTO3>3.0.CO;2-B
41. Spidlen J, Gentleman RC, Haaland PD, Langille M, Le Meur N, Ochs MF, Schmitt C, Smith CA, Treister AS, Brinkman RR (2006) Data standards for flow cytometry. OMICS 10:209–214
42. Spidlen J, Moore W, Parks D, Goldberg M, Bray C, Bierre P, Gorombey P, Hyun B, Hubbard M, Lange S, Lefebvre R, Leif R, Novo D, Ostruszka L, Treister A, Wood J, Murphy RF, Roederer M, Sudar D, Zigon R, Brinkman RR (2010) Data file standard for flow cytometry, version FCS 3.1. CytometryPart A. J Int Soc Anal Cytol 77:97–100
43. Dytch HE, Bibbo M, Puls JH, Bartels PH, Wied GL (1986) Software design for an inexpensive, practical, microcomputer-based DNA cytometry system. Anal Quant Cytol Histol 8:8–18
44. Keij JF, Griffioen AW, The TH, Rijkers GT (1989) INCA: Software for Consort 30 analysis of flow cytometric calcium determinations. Cytometry 10:814–817
45. Redelman D (1991) MS-DOS software to analyze flow cytometry data. Cytometry Suppl 5:136–136

46. Loken MR, Lanier LL (1984) Three-color immunofluorescence analysis of Leu antigens on human peripheral blood using two lasers on a fluorescence-activated cell sorter. Cytometry 5:151–158. doi:10.1002/cyto.990050209
47. Godavarti M, Rodriguez JJ, Yopp TA, Lambert GM, Galbraith DW (1996) Automated particle classification based on digital acquisition and analysis of flow cytometric pulse waveforms. Cytometry 24:330–339
48. Leary JF (2001) High-speed cell sorting. Curr Protoc Cytom Chapter 1:Unit 1–7. doi:10.1002/0471142956.cy0107s01
49. Shapiro HM (1985) Practical flow cytometry. Alan R. Liss, New York
50. Shapiro HM (1988) Practical flow cytometry, vol 2. Alan R. Liss Inc., New York
51. Berners-Lee T, Secret A, Manning G (1993) The virtual library. s.n., S.l
52. Berners-Lee T, Fischetti M (1999) Weaving the Web: the original design and ultimate destiny of the world wide web by its inventor. 1st edn. HarperSanFrancisco, San Francisco. xi, p 226
53. Berners-Lee T (2006) A framework for web science. Now, Boston. ix, p 134
54. Robinson JP, Roederer M (2015) History of science. Flow cytometry strikes gold. Science 350:739–740. doi:10.1126/science.aad6770
55. Yamamoto M, Robinson JP (2014) Scanning flow cytometer
56. Jacobs KM, Lu JQ, Hu XH (2009) Development of a diffraction imaging flow cytometer. Opt Lett 34:2985–2987
57. McFarlin BK, Williams RR, Venable AS, Dwyer KC, Haviland DL (2013) Image-based cytometry reveals three distinct subsets of activated granulocytes based on phagocytosis and oxidative burst. Cytometry A 83:745–751. doi:10.1002/cyto.a.22330
58. Zhang C, Huang K-C, Rajwa B, Li J, Yang S, Lin H, C-s Liao, Eakins G, Kuang S, Patsekin V, Robinson JP, Cheng J-X (2017) Stimulated Raman scattering flow cytometry for label-free single-particle analysis. Optica 4:103–109. doi:10.1364/OPTICA.4.000103
59. Bendall SC, Nolan GP, Roederer M, Chattopadhyay PK (2012) A deep profiler's guide to cytometry. Trends Immunol 33:323–332. doi:10.1016/j.it.2012.02.010
60. Bodenmiller B, Zunder ER, Finck R, Chen TJ, Savig ES, Bruggner RV, Simonds EF, Bendall SC, Sachs K, Krutzik PO, Nolan GP (2012) Multiplexed mass cytometry profiling of cellular states perturbed by small-molecule regulators. Nat Biotechnol 30:858–867. doi:10.1038/nbt.2317
61. Qiu P, Simonds EF, Bendall SC, Gibbs KD Jr, Bruggner RV, Linderman MD, Sachs K, Nolan GP, Plevritis SK (2011) Extracting a cellular hierarchy from high-dimensional cytometry data with SPADE. Nature Biotechnology

Cytomics of Oxidative Stress: Probes and Problems

José-Enrique O'Connor, Guadalupe Herrera,
Francisco Sala-de-Oyanguren, Beatriz Jávega
and Alicia Martínez-Romero

Abstract Oxidative stress has been implicated in cellular senescence and aging, as well as in the onset and progression of many diverse genetic and acquired diseases and conditions. However, reactive oxygen (ROS) and nitrogen (RNS) species initiating oxidative stress also serve important regulatory roles, mediated by intercellular and intracellular signaling, adaptation to endogenous and exogenous stress, and destruction of invading pathogens. Fluorescence-based analysis of oxidative stress and related processes is an important cytomic application; almost 4000 papers were published between 1989 and 2016. To ascertain the specific role of ROS and RNS in oxidative stress studies by cytomic methodologies, it is essential to detect and characterize these species accurately. Unfortunately, the detection and quantitation of individual intracellular ROS and RNS remains a challenge, but different, complementary cytometric strategies directed toward other endpoints of oxidative stress may also be considered. In this chapter we present and briefly discuss the limitations and perspectives of such approaches.

Keywords Cytomics · Fluorescence · Flow cytometry · Image cytometry · Reactive oxygen species · Reactive nitrogen species · Antioxidants

J.-E. O'Connor · B. Jávega · A. Martínez-Romero
Laboratory of Cytomics, Joint Research Unit CIPF-UVEG, The University
of Valencia and Principe Felipe Research Center, Valencia, Spain

G. Herrera
Cytometry Service, Central Research Unit (UCIM), Incliva Foundation, The University
of Valencia and University Clinical Hospital, Valencia, Spain

A. Martínez-Romero
Cytomics Technological Service, Principe Felipe Research Center, Valencia, Spain

J.-E. O'Connor (✉)
Department of Biochemistry and Molecular Biology, Faculty of Medicine, University
of Valencia, Av Blasco Ibáñez, 15, 46010 Valencia, Spain
e-mail: jose.e.oconnor@uv.es

F. Sala-de-Oyanguren
Ludwig Institute for Cancer Research, Département D'Oncologie Fondamentale, Faculté de
Biologie et Médecine, Université de Lausanne, Épalinges, Switzerland

© Springer Nature Singapore Pte Ltd. 2017
J.P. Robinson and A. Cossarizza (eds.), *Single Cell Analysis*,
Series in BioEngineering, DOI 10.1007/978-981-10-4499-1_4

83

1 Introduction to Reactive Oxygen (ROS) and Nitrogen (RNS) Species

Life on Earth has evolved by creating organisms that need oxygen to live. Most living beings depend on oxygen to obtain large amounts of metabolic energy from the oxidation of biomolecules [1, 2]. Paradoxically, the oxygen functions essential to living things depend on a chemical property dangerous to them: the structure of the oxygen molecule (O_2) has two unpaired electrons, and O_2 can accept individual electrons to generate highly unstable and highly reactive molecular forms known as reactive oxygen species (ROS). The term ROS may be applied to a variety of molecules not derived from O_2 alone and includes both free radicals and species derived from free radicals [3, 4].

Main ROS include singlet-oxygen radical, superoxide anion radical, hydrogen peroxide (H_2O_2), hydroxyl radical, hypochlorous acid (HOCl), lipid peroxides (ROOH), and ozone (O_3) [3, 4]. There are also free radicals and reactive nitrogen-containing molecules, the reactive nitrogen species (RNS), including nitric oxide (NO) and peroxynitrite [5, 6]. Because they also contain oxygen and their generation is connected to ROS generation, they are often considered as ROS. Thus, ROS and RNS are not single entities but represent instead a broad range of chemically distinct reactive species with diverse biological reactivities.

The generation of ROS and RNS has been implicated in cellular senescence and aging [7–9], as well as in the onset and progression of genetic [10, 11] and acquired diseases and conditions, including, but not limited to, inflammatory conditions [12–14], cardiovascular diseases [15–18] and thrombosis [19], cancer [20–23] and anticancer chemotherapy [24], HIV progression [25, 26], neurodegenerative diseases [27–29], and metabolic disorders [30]. However, ROS and RNS also serve important regulatory roles, mediated by intercellular and intracellular signaling [31–34], and cell function–modifying processes involved both in the destruction of invading pathogens [35] and in the fine tuning of cellular adaptation to endogenous and exogenous stress [36–38]. Phagocytes use ROS and NOS as a powerful antimicrobial weapon (oxidative burst), and in low concentrations, ROS and NOS serve also as second messengers of signal transduction.

2 The Physiological Side of ROS and NOS

2.1 Sources of ROS and RNS

The generation of ROS and RNS can be endogenous, associated with oxidative processes (such as the mitochondrial chain of electronic transport, NADPH oxidase, xanthine oxidase and various flavoproteins) [39] or exogenous, derived from inflammatory pathologies, exposure to xenobiotics, ionizing radiation, etc. [40]

Most ROS and RNS arise physiologically in specific subcellular compartments. The intracellular location of ROS and RNS is of great importance, as microenvironment will affect both the intrinsic functions of the reactive species and the population of molecules and structures that will be eventually affected by their interaction with ROS and RNS. In higher organisms, the major generation of ROS takes place in the mitochondria, during the tetravalent reduction of O_2 occurring in the electron-transport chain associated with oxidative phosphorylation. In prokaryotic cells, this mechanism takes place on the plasma membrane. This process, directed to the production of ATP, gives rise to H_2O as the final product, via a sequence of univalent reductions that generate ROS [5]. Other organelles with localized ROS generation include the phagosomes, where ROS and RNS are focused on pathogen killing, and the peroxisomes, where many catabolic oxidation reactions are confined [39, 40].

Nitric oxide (NO) is synthesized from L-arginine in a reaction catalyzed by NO synthase [32, 33]. NO reacts readily with superoxide to form the peroxynitrite anion, a RNS with strong oxidant properties [34]; activated macrophages and neutrophils produce NO and superoxide, and thus peroxynitrite, at similar rates.

2.2   ROS in Phagocytosis

The stimulated production of ROS and RNS by phagocytic cells is known as the respiratory burst, because of the increased consumption of O_2 by these cells during phagocytosis necessary for the bactericidal action of phagocytes [41]. This process is initiated by NADPH oxidase, a multicomponent membrane-bound enzyme complex that generates superoxide anion radicals, which in turn give rise to further ROS. Similarly, activated nitric oxide synthase 2 catalyzes the production of nitric oxide radicals, which leads to the formation of reactive nitrogen intermediates. ROS and RNS can interact to form further reactive species. While each of these antimicrobial systems operates independently, they are synergistic in destroying invading pathogens [42].

2.3   ROS and RNS in Cellular Signaling

Most cytoplasmic proteins contain free SH groups, which may undergo oxidation/reduction cycles. In coordination with antioxidant proteins and molecules, ROS may turn functional proteins on and off by redox cycling [43]. A large number of such proteins are involved in signal transduction or in the regulation of gene expression in eukaryotes and prokaryotes [44].

ROS signaling is involved in cell survival and adaptation to stress. Signaling through mitogen-activated protein kinases (MAPKs) leads to the generation of H_2O_2 from several enzymes, including NADPH oxidases [45]; production of H_2O_2 at nanomolar levels is required for proliferation in response to growth factors [46]. In synchronized cells ROS increase along the cell cycle, peaking at the G2/M phase [47]; it has been suggested that small increases of H_2O_2 result in increased reentry into the cell cycle, while sustained high levels of H_2O_2 lead to cell arrest and apoptosis [36]. Apoptosis induced by prolonged activation of c-Jun N-terminal kinase (JNK) has been shown to be caused by exposure to ROS [48, 49]. Conversely, autophagy is also triggered as an adaptive response, among other stressors, to intracellular ROS [38].

ROS have also recently been related to signaling in platelets [50]. ROS produced after platelet stimulation with collagen are responsible for a series of platelet-activating events owing to oxidative inactivation of SHP-2, which promotes tyrosine phosphorylation–based signal transduction.

NO plays a critical role as a molecular mediator of a variety of physiological processes, including blood-pressure regulation and neurotransmission [32, 33]. NO that diffuses into smooth muscle cells binds to the heme group of guanylate cyclase. Peroxynitrite, a RNS considered as an inflammatory mediator in various cardio-vascular pathologies, has more recently been recognized as a modulator of signal-transduction pathways owing to its ability to nitrate tyrosine residues, thereby influencing responses dependent on tyrosine phosphorylation [34].

3 Oxidative Stress: Definition, Causes, and Consequences

3.1 Definition of Oxidative Stress

Despite the powerful and complex antioxidant machinery of higher organisms, when the capacity of these protective mechanisms is overcome by the intensity or duration of oxidative processes, a situation occurs called oxidative stress, which is defined as an alteration in the equilibrium between ROS production and antioxidant defenses, producing oxidative damage [39, 40].

Oxidative stress can result from two separate but not exclusive processes. On the one hand is the decrease in the levels or the activity of enzymes of the antioxidant defense by mutation or destruction of the active center, induced by the ROS themselves [40]. Deficiencies in the dietary supply of soluble antioxidants can also cause oxidative stress. On the other hand, increased production of ROS, exposure of cells or organisms to elevated levels of exogenous ROS or their metabolic precursors, and even excessive induction of protective (immunological, detoxifying) systems that produce ROS can lead to the situation of oxidative stress [5].

3.2 Causes of Oxidative Stress

To prevent the harmful effects of the in vivo production of ROS and RNS, evolution has provided prokaryotes and higher organisms with complex and effective antioxidant systems that include enzymatic antioxidant mechanisms and antioxidant molecules, broadly understood as those molecules that protect a biological target against oxidative stress [5, 39, 40, 51].

The first line of antioxidant enzymes is the superoxide dismutase family of enzymes (SOD) that catalyze the dismutation of superoxide to H_2O_2. Catalase converts H_2O_2 to water and O_2, and thus completes the detoxification initiated by SOD. Glutathione (GSH) peroxidase includes a group of Se-containing enzymes that also catalyze the decomposition of H_2O_2, as well as of organic peroxides. In the GSH peroxidase process, GSH is consumed by oxidation, so that GSH reductase is required to transform oxidized GSH (GSSG) into GSH [40].

The non-enzymatic antioxidants are a large group of molecules that exert various protective antioxidant mechanisms, and include molecules that react with ROS, such as GSH, tocopherol and β-carotene, or proteins such as transferrin and ceruloplasmin, capable of chelating transition metals. GSH is the most important intracellular defense against the toxic effects of ROS. Vitamin C or ascorbic acid is a water-soluble molecule capable of reducing ROS, while vitamin E (α-tocopherol) is a lipid-soluble molecule that has been suggested as playing a similar role in membranes [40].

3.3 Consequences of Oxidative Stress

Oxidative damage to cells and tissues produced by ROS is associated with free-radical chain reactions with all kinds of biomolecules, such as carbohydrates, lipids, proteins, and DNA.

3.3.1 Protein Damage

Protein oxidation plays an important role in many of the effects of oxidative and nitrosative stress. The modifications produced may be irreversible, such as carbonylation of lysine (Lys) and arginine (Arg), formation of di-tyrosine bonds, protein-protein bonds, and nitration of Tyr and tryptophan [52]. These changes generally result in loss of permanent function of damaged proteins. In complex enzymes, free-radical interaction can also be damaging at the level of the prosthetic group, leading to functional inactivation [53].

3.3.2   Lipid Peroxidation

Lipid peroxidation is a process that occurs in three phases: initiation, propagation, and termination. The initiation phase involves the reaction of free radicals with cellular lipids, generating peroxyl radicals. In the propagation phase, the reaction of these newly formed peroxyl radicals with intact lipid triggers a chain reaction that may be terminated by the action of antioxidants. It is a phenomenon detrimental to the cell, since changes in the physico-chemical properties of the membrane, as its fluidity, as well as the inactivity of transporters and membrane enzymes can occur [53, 54].

3.3.3   Oxidative Lesions to DNA

More than 100 different free radical–induced DNA modifications have been described, either in nitrogenated bases or in deoxyribose. The hydroxyl ion seems to be the main cause of these lesions, an effect that is facilitated by the polyanionic character of the phosphodiester bond, since it is attracted by metals such as Fe^{2+}, favoring the Fenton reaction. NO and its derivatives have the capacity to induce lesions in the DNA, mainly through the deamination of bases, although other processes are involved [55].

Modifications produced by ROS in the bases may be mutagenic, leading to incorrect pairing of the bases, or cytotoxic if an arrest of replication occurs. Of special relevance are oxidation reactions with purines that generate different products with high mutagenic capacity, such as 8-oxo-7, 8-dihydroguanine (8oxoG), widely used as an important marker of oxidative DNA damage [56, 57]. Cross-linking reactions between nucleotides of the same chain may be cytotoxic or mutagenic, the best known case being the formation of pyrimidine dimers.

4   Strategies and Reagents for Cytomic Analysis of Oxidative Stress

Given the participation of ROS and RNS in physiological and pathological issues, active search for biomarkers of oxidative stress has become relevant to many biomedical fields [58], and many different methods are applied to assess the redox state of the body or of specific tissues and cells [59–65]. Fluorescence-based analysis of oxidative stress and related processes is an extended application of flow cytometry [66, 67]; more than 3700 papers on this topic have been published between 1989 and 2016, according to PubMed Central (Fig. 1). Imaging approaches, including confocal microscopy [68], high-content analysis by automated microscopy [69], and the more recent imaging flow cytometry [70, 71] allow one in addition to visualize and quantify topographical issues of intracellular ROS production and action (Fig. 2).

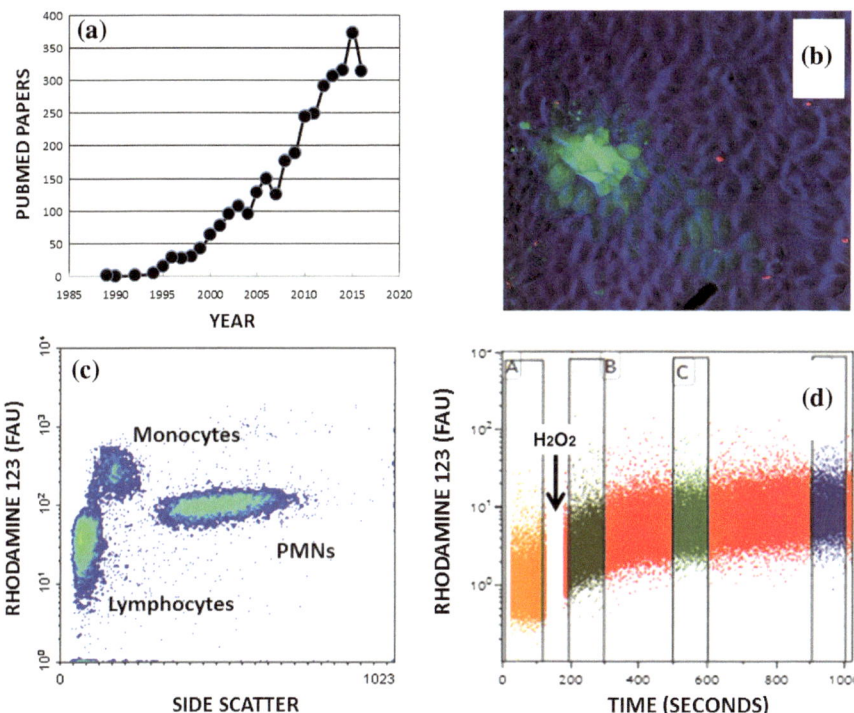

Fig. 1 Examples of relevant cytometric techniques for the study of ROS, RNS and oxidative stress. **a** PubMed-indexed papers containing the general terms "Cytometry" AND "oxidative stress" published between 1985 and 2016. Panels b-d show three complementary cytometric applications having in common the same fluorogenic substrate, dihydrorhodamine 123 (DHR123), which detects mainly H_2O_2 and peroxynitrite, as explained in the paper. **b** Example of analysis by fluorescence microscopy of ROS generation. A confluent monolayer of MDCK cells was stained with DHRH123 and treated with 25 μM $CdCl_2$ for 1 h, before being photographed with a standard fluorescence microscope. General experimental conditions are similar to those described in [181]. **c** Example of a whole-blood assay by flow cytometry of ROS/RNS generation by resting leukocytes. A sample of whole blood in heparin from a healthy volunteer was stained with CD45-PC5 antibody, to exclude erythrocytes from the analysis, and with DHRH123. Leukocyte populations are distinguished by their side-scatter properties. The fluorescence intensity of rhodamine 123, the oxidation product of DHR123, shows that monocytes at rest generate more ROS/RNS than do neutrophils and lymphocytes. General experimental details can be found in [111]. **d** Example of a whole-blood assay by real-time cytometry of ROS/RNS generation by resting leukocytes. Experimental conditions were similar to those in panel c, but data acquisition was started before H_2O_2 was added to the sample stained with CD45, CD14, and DHRH123, already running in the flow cytometer. The generation of ROS/RNS in monocytes was followed by means of a kinetic plot of rhodamine 123 fluorescence intensity versus time. Analytical regions allow quantification of the rate of fluorescence variation. FAU: Fluorescence Arbitrary Units

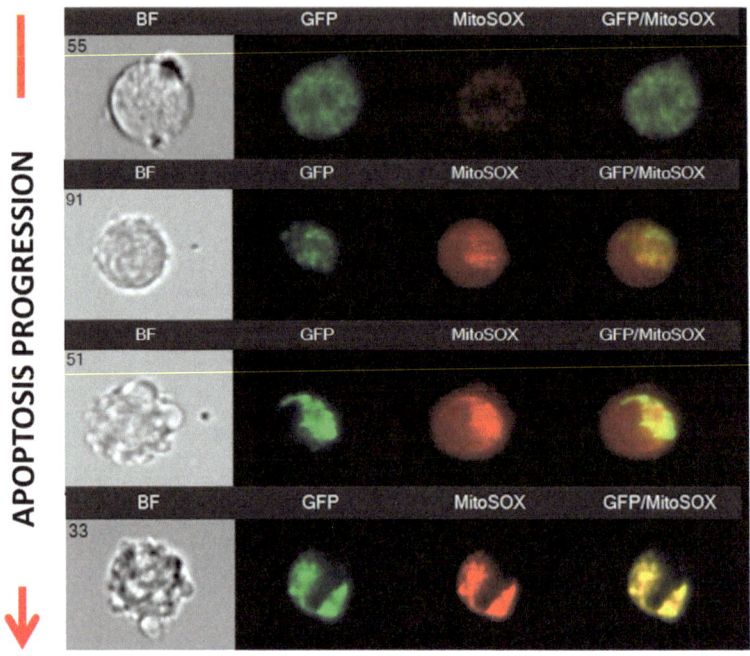

Fig. 2 Imaging flow cytometry applied to the analysis of ROS generation and oxidative stress in vitro. Colocalization by imaging flow cytometry of mitochondria and the superoxide radical–sensitive probe MitoSox Red dye during apoptosis. The human liposarcoma SW872/GFP cell line obtained by transfection with a retroviral vector expresses the GFP-tagged mitochondrial LON protease [182]. Apoptosis was induced by treatment with camptothecin and at appropriate times cells were trypsinized and stained with the mitochondrial superoxide sensor MitoSox Red dye. Cell suspensions were analyzed in an Amnis ImageStream flow cytometer that collects multispectral images of single cells at high speed. In each row of images, channels show from left to right, bright field illumination, expression of GFP-Lon protease (i.e., mitochondrial compartment), MitoSox Red dye fluorescence (i.e., generation of superoxide radical), and the merged image wherefrom colocalization may be observed and quantified. The series of images from top to bottom show that the morphological progression of apoptosis (brightfield channel) is accompanied by generation of superoxide that is associated with the mitochondrial compartment

However, to ascertain the specific role of ROS and RNS in oxidative stress studies by cytomic methodologies, it is essential to detect and characterize these species accurately. Unfortunately, the detection and quantitation of individual intracellular ROS and RNS remains a challenge [60–65], but different, complementary cytometric strategies aimed to other endpoints of oxidative stress may also be considered. In this chapter we present and discuss briefly the limitations and perspectives of such approaches.

4.1 Cytomic Strategies in the Analysis of Oxidative Stress

The complex processes involved in the generation of ROS and RNS, their control by the antioxidant system, and the physiological or pathological consequences of their action may be approached by cytomic analysis at different levels or stages, using complementary methodologies based upon fluorescence, in multiple cell types and clinical situations or experimental models. Thus, the most common cytomic strategies to the study of oxidative stress include:

(a) Direct detection of ROS and RNS, the initiators of the oxidative stress process. This task is complex owing to the low concentration, short half-life, and extensive interactions of ROS and RNS, as well as by intrinsic limitations of both probes and experimental conditions.
(b) Detection of more stable products of ROS and RNS reaction with cell components or with exogenous probes, including the analysis of lipid peroxidation and oxidative damage to DNA.
(c) Assessment of antioxidant defenses, mostly GSH and SH-containing proteins. This indirect approach to oxidative stress may be limited by issues related to the complexity of the antioxidant defense by itself and to the specificity of enzymes required for fluorescent reporting of the process.

4.2 Detection of ROS and RNS Using Fluorogenic Substrates

The use of fluorescent probes and fluorogenic substrates (Fig. 3) appears a simple and easy approach for the detection and quantification of ROS production in cellular systems. However, there are many limitations and artifacts in this methodology. In this section we mention the principal fluorescent probes and fluorogenic substrates used in cytometric analysis of ROS and RNS. Their main limitations and potential sources of artifacts will be considered further along in this chapter.

4.2.1 2',7'-Dichlorodihydrofluorescein Diacetate (H_2DCF-DA) and Related Probes

The cell-permeant H_2DCF-DA is one of the most commonly fluorogenic substrates used in studies related to ROS and RNS generation [62–64]. Upon cleavage of the acetate groups by intracellular esterases, the intracellular oxidation of 2',7'-dichlorodihydrofluorescein (DCFH) produces 2',7'-dichlorofluorescein (DCF), a fluorescent compound (λ excitation = 498 nm; λ emission = 522 nm). Initially, H_2DCF was widely accepted as a specific indicator for H_2O_2 [72] but as discussed later in this chapter, H_2DCF is oxidized by other ROS, such as hydroxyl and

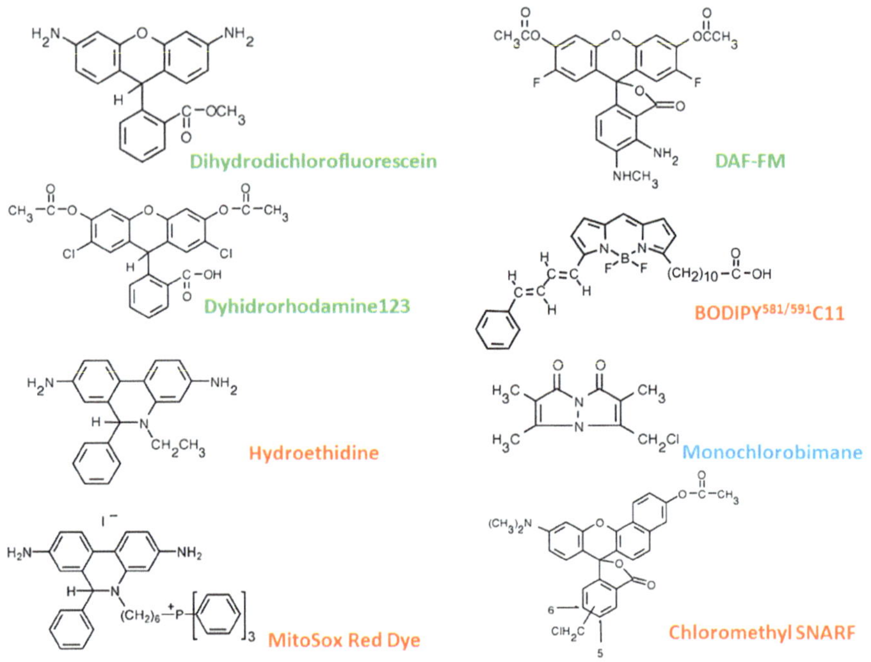

Fig. 3 Some examples of fluorescent probes and fluorogenic substrates frequently used in the cytometric analysis of ROS, RNS and oxidative stress. The properties, applications, and limitations of these reagents are described in the corresponding sections of the chapter. The letter colors of reagent names indicate the spectral region of their fluorescence emission, according to [80]

peroxyl radicals, and also by RNS like peroxynitrite [62–64]. On the other hand, it seems well established that H_2DCF is not oxidized by superoxide anion, hypochlorous acid, or NO [62]. With these caveats, H_2DCF has been successfully used for studies of oxidative burst in phagocytes [73, 74] and to follow the generation of prooxidants in many cell models [75–79].

Intracellular oxidation of H_2DCF tends to be accompanied by leakage of the product, DCF. To enhance retention of the fluorescent product, several analogs with improved retention have been designed, such as carboxylated H_2DCFDA (carboxy-H_2DCFDA), which has two negative charges at physiological pH, and its di-(acetoxymethyl ester) [80]. The halogenated derivatives 5-(and 6-)chloromethyl-2',7'-dichlorodihydrofluorescein diacetate, acetyl ester (CM-H_2DCFDA), and 5-(and 6-)carboxy-2',7'-difluorodihydrofluorescein diacetate (carboxy-H_2DFFDA) exhibit much better retention in live cells; they have been used for following oxidative burst in inflammatory and infectious processes and have been applied to different experimental settings related to oxidative stress [80].

4.2.2 Dihydrorhodamine 123 (DHR123)

DHR123 is a non-fluorescent molecule that upon oxidation generates rhodamine 123, a fluorescent cationic and lipophilic probe (λ excitation = 505 nm; λ emission = 529 nm) [62, 80]. The lipophilicity of DHR123 facilitates its diffusion across cell membranes. Upon oxidation of DHR123 to the fluorescent rhodamine 123, one of the two equivalent amino groups tautomerizes into an imino group, effectively trapping rhodamine 123 within cells [81]. Like H_2DCF, DHR123 is oxidized by H_2O_2 in the presence of peroxidases, but this probe has low specificity, since it can also be oxidized by other reactive oxidants, namely peroxynitrite, Fe^{2+}, Fe^{3+} in the presence of ascorbate or EDTA, cytochrome c, or HOCl [62, 81]. DHR123 is not directly oxidizable by H_2O_2 alone, nor by superoxide anion or by the system xanthine/xanthine oxidase [62, 81].

4.2.3 New Fluorescent Probes for H_2O_2 Detection

New chemoselective fluorescent indicators are being developed to provide improved selectivity for H_2O_2 over other ROS [82]. A very promising approach is based on the selective H_2O_2-mediated transformation of arylboronates to phenols [83]. Arylboronates are linked to fluorogenic moieties, such that reaction with H_2O_2 generates a fluorescent product.

Arylboronate probes include peroxyfluor-2 (PF2), peroxy yellow 1 (PY1), peroxy orange 1 (PO1), peroxyfluor-6 acetoxymethyl ester (PF6-AM), and mitochondria peroxy yellow 1 (MitoPY1) [82–84]. This family of probes can detect physiological changes in endogenous H_2O_2 levels, allowing various combinations for multicolor imaging experiments. The addition of acetoxymethyl ester groups gives rise to the dye peroxyfluor-6 acetoxymethyl ester (PF6-AM), which increases cellular retention and further increases sensitivity to H_2O_2 [84, 85]. In addition, the recently developed Ratio Peroxyfluor 1 (RPF)-1 provides a ratiometric change of two fluorescent signals upon reaction with H_2O_2, allowing normalization of fluorescence ratio to probe concentration [86].

Combining the boronate–based probe design with appropriate functional groups has resulted in organelle-specific probes that can measure H_2O_2 levels with spatial resolution [87]. In particular, several mitochondria-targeted probes have been generated, including MitoPY1 and SHP-Mito [84, 88, 89] for mitochondrial targeting and Nuclear Peroxy Emerald (NucPE), for nuclear targeting [90].

4.2.4 Hydroethidine or Dihydroethidium (HE)

HE is widely used as a fluorogenic substrate for detecting superoxide anion [62–64, 80, 91]. HE is membrane-permeant and cytosolic HE exhibits blue fluorescence, but once oxidized by superoxide, it generates 2-hydroxy-ethidium (E^+), a fluorescent compound (λ excitation = 520 nm; λ emission = 610 nm). E^+ is retained in the nucleus, intercalating with the DNA, a fact that increases its fluorescence [62].

HE has been repeatedly used in studies related to the oxidative burst in leukocytes [92, 93] and to inflammation [94–97]. HE has been used also for mitochondrial superoxide detection [80, 98, 99] although the more recently developed MitoSOX Red indicator provides more specific mitochondrial localization, as discussed later in this section [80, 100]. Moreover, since mitochondria play a fundamental role in apoptosis, which can be triggered by ROS and RNS, through mitochondrial membrane permeabilization and release of proapoptotic factors, HE and Mito-Sox have been also used to detect changes in mitochondrial superoxide generation associated with the induction and execution of apoptosis [98, 99].

HE may have important limitations when used for analysis of intracellular superoxide. It has been shown that cytochrome c is able to oxidize HE, an aspect that might be important in situations where the detected superoxide is mainly of mitochondrial origin or in conditions leading to apoptosis, where cytochrome c is released to cytosol [97]. Owing to the interconnection between oxidative stress and the apoptotic processes, it will be difficult, in these situations, to assume that HE oxidation to E^+ depends only on superoxide. Furthermore, HE can also be oxidized by a variety of reactive species, including peroxynitrite. Thus, HE should be considered as an indicator of ROS and RNS production [62–64, 97].

4.2.5 MitoSOX Red Mitochondrial Superoxide Indicator (MitoSox Red)

MitoSOX Red, a cationic derivative of HE, was introduced for selective detection of superoxide in the mitochondria of live cells [64, 80, 101]. MitoSOX Red contains a cationic triphenylphosphonium substituent that selectively targets this cell-permeant probe to actively respiring mitochondria, where it accumulates as a function of mitochondrial membrane potential and exhibits fluorescence upon oxidation and subsequent binding to mitochondrial DNA [80]. MitoSOX Red has been used for detection of mitochondrial superoxide production in a wide variety of cell types and conditions [80, 100, 101].

Oxidation of MitoSOX Red by superoxide results in hydroxylation of the ethidium moiety at the 2-position, to yield a 2-hydroxyethidium substituent. Therefore, the fluorescence spectral properties of oxidized MitoSox Red are identical to those of HE [80]. On the other hand, since the chemical reactivity of MitoSOX Red with superoxide is similar to the reactivity of HE with superoxide, all the limitations of HE apply to MitoSOX Red as well [64, 102].

4.2.6 CellROX® Reagents as General Probes for ROS

The CellROX® reagents are a series of proprietary reagents from Life Technologies-Thermofisher. These cell-permeant dyes are weakly fluorescent in the reduced state and exhibit photostable fluorescence upon oxidation by ROS [103–105].

CellROX® green becomes fluorescent only with subsequent binding to DNA, limiting its presence to the nucleus or mitochondria. This compound has an excitation wavelength of 485 nm and an emission wavelength of 520 nm. This reagent can be formaldehyde-fixed and its signal survives detergent treatment, allowing it to be it multiplexed with other compatible dyes and antibodies. CellROX® Orange and CellROX® Deep Red do not require DNA binding for fluorescence and are localized in the cytoplasm. CellROX® orange has an excitation wavelength of 545 and an emission of 565, while CellROX® Deep Red has an excitation peak of 640 nm and an emission peak of 665 nm [104].

4.2.7 4,5-Diaminofluorescein Diacetate (DAF-2 DA)

The NO radical is short-lived and physiological concentrations are very low. NO is readily oxidized to the nitrosonium cation (NO^+), which is moderately stable in aqueous solutions but highly reactive with nucleophiles or other nitrogen oxides. Under aerobic conditions, these reactive nitrogen oxides, but not nitric oxide itself, can be bound by aromatic 1,2-diamines to form fluorescent benzotriazoles [80, 106].

DAF-2 was the first fluorogenic probe for NO [106]. DAF-2 DA is a membrane-permeant substrate that can be hydrolyzed to DAF-2 and trapped within the cell [107]. The fluorescent chemical transformation of DAF is based on the reactivity of the aromatic vicinal diamines with NO in the presence of molecular oxygen. DAF-2, which shows low fluorescence, reacts with NO-derived NO^+ to produce the highly fluorescent triazolofluorescein (DAF-2T). The fluorescence quantum efficiency increases more than 100 times after the modification of DAF-2 by NO^+. DAF-2 DA has been used to detect intracellular NO by fluorescence microscopy and flow cytometry with high sensitivity [108].

4.2.8 4-Amino-5-Methylamino-2',7'-Difluorofluorescein Diacetate (DAF-FM DA)

DAF-FM DA is a cell-permeant diacetate derivative with properties similar to DAF-2 [80]. DAF-FM DA is cleaved by esterases to generate intracellular DAF-FM, which is then oxidized by NO to a triazole product much more fluorescent. Indeed, the fluorescence quantum yield of DAF-FM increases about 160-fold after reacting with NO [80].

DAF-FM has been used in many studies related to NO generation under a variety of experimental conditions [80, 109, 110], including the kinetic analysis of NO generation and consumption in whole-blood monocytes using real-time cytometry [111].

4.2.9 Dihydrorhodamine 123 (DHR123) for Detecting Peroxinitrite

Although DHR123 was described initially as a fluorogenic susbstrate for H_2O_2 [64], currently it is the most frequently used probe for measuring peroxynitrite [62–64], based on the oxidative conversion of DHR123 to its corresponding two-electron oxidized fluorescent product, rhodamine 123, (λ excitation = 505 nm; λ emission = 529 nm) mediated by an intermediate DHR123 radical [64]. The oxidation of DHR123 by peroxynitrite is not induced directly by this ROS, but is mediated by intermediate oxidants formed from the rapid and spontaneous decomposition of peroxynitrite [64, 112].

4.3 Detection of Lipid Peroxidation

Peroxyl radicals are formed by the decomposition of various peroxides and hydroperoxides, including lipid hydroperoxides. The hydroperoxyl radical is also the protonated form of superoxide, and approximately 0.3% of the superoxide in the cytosol is present as this protonated radical [80].

4.3.1 *cis*-Parinaric Acid

cis-Parinaric acid is a fluorescent 18-carbon polyunsaturated fatty acid, containing four conjugated double bonds in positions 9, 11, 13, and 15 [62, 80]. *cis*-Parinaric acid can be metabolically integrated into membrane phospholipids of cultured cells, where its conformation and mobility are comparable to endogenous phospholipids. Moreover, its fluorescent and peroxidative properties are combined in the conjugated system of unsaturated carbon–carbon bonds. The fluorescence of *cis*-parinaric acid (λ excitation = 320 nm; λ emission = 432 nm) is lost upon oxidation [62, 80]

This probe has been repeatedly used to measure lipid peroxidation in a multiplicity of cell systems and conditions [113, 114]. However, there are some problems associated with the use of *cis*-parinaric acid in living cells, such as its absorption in the UV region, a wavelength still absent in most routine cytometers, and where most test compounds may absorb. In addition, *cis*-parinaric acid is most sensitive to air and undergoes photodimerization under illumination, which results in loss of fluorescence and overestimation of the extent of lipid peroxidation [80].

4.3.2 4,4-Difluoro-5-(4-Phenyl-1,3-Butadienyl)- 4-Bora-3a,4a-Diaza-S-Indacene-3-Undecanoic Acid (BODIPY[581/591]C11) and Related BODIPY Probes

BODIPY[581/591] C11 is a fluorescent probe (λ excitation = 510 nm; λ emission = 595 nm) used for evaluating lipid peroxidation and antioxidant efficacy in different experimental models [62, 80, 115]. BODIPY$^{581/591}$ C11 has a long-chain unsaturated fatty acid (C11) of non-polar character, which makes this probe liposoluble, while the conjugated double bonds in the fluorophore make it susceptible to oxidation by peroxyl radicals [62, 80]. BODIPY $^{581/591}$ C11 undergoes a shift from red to green fluorescence emission upon oxidation [80]. This oxidation-dependent emission shift enables fluorescence ratiometric analysis of free radical–mediated oxidation in the lipophilic domain of the membranes. The primary target for ROS is the diene interconnection, leading to the formation of three different oxidation products that are responsible for the shift from red to green fluorescence [62].

BODIPY$^{581/591}$ C11 is sensitive to multiple oxidizing species. It has been demonstrated that this probe is oxidized by peroxyl, hydroxyl radicals, and peroxynitrite, while being insensible to H_2O_2, singlet oxygen, superoxide, NO radical, transition metals, and hydroperoxides in the absence of transition metals [116].

Lipid peroxidation has been detected in cell membranes using BODIPY$^{581/591}$ C11 [117–119] and other similar BODIPY derivatives, such as BODIPY$^{493/503}$ [120], BODIPY FL EDA (a water-soluble dye) [80], or BODIPY FL hexadecanoic acid [80].

4.3.3   Lipophilic Fluorescein Derivatives

The probe 5-(N-dodecanoyl) aminofluorescein (C11-Fluor), a lipophilic derivative of fluorescein, has been used in flow cytometry for determining membrane-lipid peroxidation [60, 121]. This probe remains associated with cellular membranes in a stable and irreversible way. Other lipophilic derivatives of fluorescein include 5-hexadecanoylaminofluorescein   (C16-Fluor),   5-octadecanoyl-aminofluorescein (C18-Fluor),   and   di-hexadecanoyl-glycerophosphoethanolamine   (Fluor-DHPE) [122].

4.4   Detection of Metabolic Derivatives of Peroxidized Lipids

4.4.1   Immunofluorescent Detection of 4-Hydroxy-2-Nonenal (4-HNE)

As a final consequence of the peroxidation process, a variety of aldehydes may be formed. 4-HNE is an unsaturated aldehyde arising from peroxidation of ω-6 unsaturated fatty acids. 4-HNE has been found to be a reliable biomarker of lipid peroxidation, as it is highly reactive towards free SH groups of proteins, producing thioether adducts that further undergo cyclization to form hemiacetals. HNE induces heat-shock protein, inhibits cellular proliferation, and is highly cytotoxic and genotoxic to cells [123, 124].

Monoclonal antibodies recognizing adducts of 4-HNE with histidine, lysine, and cysteine in proteins are now commercially available [125]. These antibodies have been conjugated with distinct fluorochromes and can be used for in situ detection of advanced stages of lipid peroxidation in different cell types with high specificity [126].

4.5 Immunofluorescent Detection of Oxidized Bases in DNA

The oxidized DNA base 8-oxodeoxyguanine (8-oxoDG) is a major form of oxidative DNA damage derived from the attack by hydroxyl radical on guanine at the C8-position, resulting in a C8-OH-adduct radical. Thus, 8-oxoDG is formed during free radical damage to DNA and is a sensitive and specific indicator of DNA oxidation [56, 57].

8-oxoDG can be quantified with the OxyDNA assay, based on the specific binding of a monoclonal antibody conjugated with FITC to 8-oxoDG moieties in the DNA of fixed and permeabilized cells [127]. This assay has been used to detect oxidative genotoxicity in vitro [128], including environmental studies [129]. Of particular interest, the OxyDNA assay has been used in a number of fertility studies related to oxidative stress during cryopreservation of sperm cells [130] and the relation of oxidative DNA damage to fertility in humans [131–133] and animals [134].

4.6 Assessment of Antioxidant Defenses: GSH and Thiol Groups

Cellular thiols, especially GSH, act as nucleophiles and can protect against toxicity, mutagenicity, or transformation by ionizing radiation and many carcinogens [40]. The availability of many thiol-reactive fluorescent probes allowed development since the early 1980s of cytometric assays for GSH [135, 136] and free thiol groups [137] in living cells. Currently, the analysis of intracellular levels of GSH and activity of GSH S-transferase (GST) is a relevant application of functional cytometry in oxidative stress and drug resistance [138], as the more than 1800 papers indexed in PubMed between 1981 and 2016 attest. Cytometric assays for GSH and intracellular SH groups have been critically reviewed on several occasions [139–142].

The probes most used for cytometric analysis of GSH and GST have been the UV-excited, cell-permeant bimanes, particularly monobromobimane (mBrB) and the more selective monochlorobimane (mClB). Both probes are essentially nonfluorescent until conjugated to GSH [138–141]. o-Phthaldialdehyde, another UV reagent, reacts with both the thiol and the amine functions of GSH, yielding a cyclic derivative with excitation and emission maxima shifted from those of its protein adducts, improving the specificity of GSH detection [138–141].

ThiolTracker Violet is up to 10-fold brighter than the bimanes, when excited at 405 nm, yielding emission at 525 nm. An advantage of this cell-permeant probe is that it resists formaldehyde fixation and detergent extraction, allowing analysis of fixed cells [138, 142].

GSH can be determined using visible light–excitable probes, including 5-chloromethylfluorescein diacetate (CellTracker Green CMFDA), and chloromethyl SNARF-1 acetate. Both probes form adducts with intracellular thiols that are well retained by viable cells. CellTracker Green CMFDA is brighter than MClB and is highly specific for GSH over free SH groups [138]. The GSH adduct of chloromethyl SNARF-1 emits beyond 630 nm, allowing multicolor protocols and reducing the impact of cellular autofluorescence.

5 Problems and Limitations in the Determination of ROS and RNS

As commented above, detection of ROS and RNS, the initiators of the oxidative stress process, is a complex task owing to the low concentration, short half-life, and extensive interactions of ROS and RNS, as well as by intrinsic limitations of both probes and experimental conditions. Such limitations and potential sources of artifacts make quantitative measurements of intracellular generation of ROS and RNS a difficult challenge and require careful design of the experiments and cautious interpretation of data.

5.1 Short Half-Life and Intracellular Location of ROS and RNS

Because of their reactivity, most ROS and RNS are short-lived molecules. For example, the half-life of hydroxyl radical within a cell is only about 10^{-9} s, compared to about 1 ms for H_2O_2 [39]. This means that hydroxyl radical will react at or very near its origin, whereas H_2O_2 can diffuse away from its source [39].

The variability in ROS half-life and the complexity of the microenvironments where they are produced and consumed make ROS and RNS quantification almost impossible in cellular systems [63, 64]. While ROS and RNS of low reactivity may accumulate with time, the more reactive ROS will reach a steady state in which the rate of their generation will be equal to the rate of disappearance, the rate of disappearance being the sum of reaction rates of this ROS with various components of the system, plus the rate of self-reaction, plus the rate of reaction with the fluorescent probe.

To attenuate this complication, flow cytometric techniques based on real-time measurements [143, 144] and imaging cytometry of intracellular location of ROS can be used [111], as exemplified in Figs. 1 and 2.

5.2 Complex Interactions Among and Between ROS, RNS, and Fluorescent Probes

A clear example of ROS interplay is mitochondrial respiration, where superoxide anion, H_2O_2, and hydroxyl radical are sequentially produced by a series of partial reductions. Incorporation of an electron into O_2 gives rise to superoxide anion, which is a poorly reactive radical but which can oxidize thiols and ascorbic acid [5, 10, 40]. Superoxide gives rise to H_2O_2 by spontaneous reaction or by the action of superoxide dismutase. H_2O_2, in turn, can react with different organic compounds to produce peroxyl radicals that will eventually release hydroxyl radicals during their metabolism. Moreover, by way of the Fenton reaction, hydroxyl radicals are produced when H_2O_2 and a transition metal, such as Fe^{2+}, react together, yielding Fe^{3+} that consumes superoxide for recycling Fe^{2+}. In the Haber-Weiss reaction, superoxide and H_2O_2 react together to produce hydroxyl radicals [5, 10].

The interaction of ROS with nitrogen derivatives can generate RNS. NO, a gas that is synthesized from L-arginine in many cell types by various isoforms of the enzyme NO synthetase, is a weak reductor and reacts with O_2 to form NO_2, but reacts much faster with superoxide to produce peroxynitrite (ONOO-), a powerful oxidant [33, 34, 111].

5.3 Influence of the Probes on the Experimental System

All reduced fluorogenic substrates are subject to auto-oxidation, which usually produces singlet oxygen, superoxide, and by its dismutation, H_2O_2. If the auto-oxidation rate is significant, it may result in artifactual detection of ROS and higher background, a problem especially important for probes such as HE or MitoSox Red [62–65].

The concentration of the probe is also relevant, as it may affect the stoichiometry of the process under study. For instance, the stoichiometry of the reaction between HE and superoxide depends on the ratio of superoxide flux and HE concentration. Owing to HE-catalyzed superoxide dismutation, the efficiency of HE oxidation decreases at higher rates of superoxide generation, and high HE concentrations might lead to fluorescence increase independent of superoxide [62].

Fluorescent probes at high concentration may perturb cells and be toxic. For example, when irradiated with UVA, H_2DCF auto-oxidizes and photo-sensitizes cells [63]. In addition, probes may affect the activity of ROS-producing enzymes.

For instance, H_2DCF can be a source of electrons for the oxidation of arachidonic acid by prostaglandin H synthase [63], while dihydrocalcein was reported to inhibit the activity of mitochondrial complex I [64].

5.4 Experimental Artifacts

Artifactual generation of ROS may result from photochemical reactions of components of culture media [60, 145]. The presence of ROS has been detected even in natural environments, such as seawater [146, 147]. Xenobiotics and endogenous compounds such as catechols, dopamine, hydralazine, and molecules with SH groups may also produce significant ROS upon interaction with media [60].

On the other hand, binding to macromolecules in the medium may lead to quenching of fluorescent probes. For example, quenching of DCF fluorescence has been reported by binding to native or glyoxal-modified human serum albumin [60, 148].

5.5 Cell Integrity and Functional Competence and Intracellular Localization of Probes

As previously commented in this chapter, ROS and RNS are usually produced and act in discrete intracellular locations. This situation is successfully approached by chemical modifications in the probes that allow them to cross the plasma and, eventually, the organelle membranes to be targeted to specific intracellular environments [62–65]. However, artifacts may arise when these assumptions are not realized. For instance, dihydrocalcein accumulates in mitochondria, in contrast to H_2DCF, which usually localizes in the cytoplasm [149], but preferential localization of H_2DCF in the mitochondria of rat cardial myocytes has been reported [150].

A much more common problem involves extracellular leakage of fluorogenic probes or their oxidation products. Passive probe leakage will always be present, to an undetermined extent, in necrotic or apoptotic cells, owing to enhanced plasma membrane permeability, leading to artifacts or erroneous interpretation of results (Fig. 4).

The presence of active multidrug transporters in the plasma membrane of cells may result in probe extrusion and underestimation of oxidative stress [151], as multidrug-resistant cells with elevated level of expression of some transporters can appear to produce less ROS. Substances such as rhodamine 123 and ethidium are good substrates for P glycoprotein, while substances such as fluorescein and dihydrofluorescein are substrates for MRP1 [152]. Dihydrocalcein has been preferred to H_2DCF because its oxidation product calcein is believed to not leak out of cells; however, calcein is also a good substrate for MRP1 and MRP2 transporters [60].

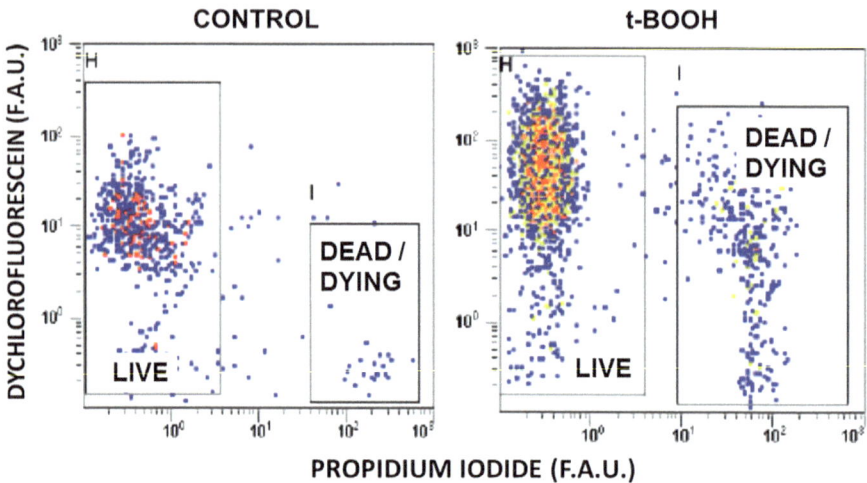

Fig. 4 Example of passive H$_2$DCF leakage from necrotic or apoptotic cells due to enhanced plasma membrane permeability. Human kidney adenocarcinoma A.704 cells were resuspended from monolayer culture by tripsinization and treated for 1 h with 150 μM t-butyl hydroperoxide (t-BOOH), a prooxidant model compound, or with DMSO as vehicle (control). Cells were stained for 30 min with 5 μg/mL H$_2$DCF-DA; 2.5 μg/mL propidium iodide was added immediately before analysis with a standard flow cytometer. Treatment with t-BOOH increases in live cells the fluorescence of DCF, the product of H$_2$DCF-DA oxidation. Dead or dying cells (positive for propidium iodide), exhibit decreased intracellular DCF fluorescence. FAU: Fluorescence Arbitrary Units

5.6   Intrinsic Limitations of Fluorogenic Substrates and Probes

5.6.1   Probes Used for Detection of H$_2$O$_2$ and Organic Peroxides

H$_2$DCF-DA is possibly the probe most widely used for detecting intracellular H$_2$O$_2$ and oxidative stress. Traditionally, H$_2$DCF-DA and DHR123 are believed to be oxidized by H$_2$O$_2$ and organic peroxides, and have been used for assaying peroxides [72, 80]. However, these probes do not react directly with H$_2$O$_2$ in the absence of peroxidases [63, 64], and the fluorescence of DCF or rhodamine 123 is not a direct measure of H$_2$O$_2$. Even if H$_2$DCF oxidation also occurs by action of H$_2$O$_2$ or O$_2$ in the presence of Fe^{2+}, the hydroxyl radical is the species responsible for such oxidation [62].

Since the oxidation of H$_2$DCF and DHR123 by H$_2$O$_2$ under physiological conditions requires peroxidase-dependent systems, enzyme activity may become a limiting factor; thus measurement of probe oxidation might be rather considered a measure of peroxidase activity. However, H$_2$DCF and DHR123 can be oxidized not only by the peroxidases, but also by other related enzymes, such as xanthine oxidase, superoxide dismutase, and cytochrome c [62].

H$_2$DCF and DHR123 are not oxidized by NO or superoxide to any significant extent, but they are very efficiently oxidized by peroxynitrite via the radicals generated during peroxynitrite decomposition [153, 154].

DCF may undergo photoreduction by visible light or by UVA radiation [155]. This mechanism may generate a semiquinone radical from DCF that produces superoxide by reaction with O$_2$. Sequentially, the dismutation of superoxide generates H$_2$O$_2$, which leads to an artificial increase of H$_2$DCF oxidation and to amplification of DCF fluorescence.

Mito PY-1 and other aromatic boronate derivatives have been proposed for analysis of intramitochondrial generation of H$_2$O$_2$ [82–84]. However, aromatic boronates also react nearly stoichiometrically with peroxynitrite a million times faster than they do with H$_2$O$_2$ [156]. Because of this reactivity, it is critical to perform proper controls when using a boronate-based fluorescent probe, such as expression of catalase, or using a peroxynitrite-specific probe.

5.6.2 Probes Used for Detection of Superoxide

Measurement of intracellular and mitochondrial superoxide using HE and Mito-SOX Red is also a widely used strategy for studying oxidative stress [62–64]. The red fluorescence of the two-electron oxidation product of HE, ethidium (E$^+$), is usually considered proof of intracellular superoxide formation. However, it has been demonstrated that E$^+$ is not formed from the direct oxidation of HE by superoxide [157, 158]. Instead, 2-hydroxyethidium (2-OH-E$^+$), a different product with similar fluorescence characteristics, is the reaction product of HE with superoxide [102]. E$^+$ and other dimeric products, but not 2-OH-E$^+$, are generated during the reaction between HE and other oxidants such as peroxynitrite, hydroxyl, H$_2$O$_2$, and peroxidase intermediates. Thus, 2-OH-E$^+$ is only a qualitative indicator of intracellular superoxide [64, 102].

The chemistry of Mito-SOX with superoxide is similar to that of HE and the same caveats apply [64]. Because of its positive charges, Mito-SOX reacts slightly faster with superoxide compared to HE [101]. Mito-SOX reacts with superoxide and forms a red fluorescent product, 2-hydroxymitoethidium (2-OH-Mito-E$^+$), and not Mito-E$^+$. 2-OH-Mito-E$^+$, the specific product of superoxide with Mito-SOX, and Mito-E$^+$, the nonspecific product of Mito-SOX, have overlapping fluorescence spectra. Thus, the red fluorescence formed from Mito-SOX localized in mitochondria is not a reliable indicator of mitochondrial formation of superoxide, as it might arise also from an oxidation product of Mito-SOX induced by one-electron oxidants (such as cytochrome c, peroxidase, and H$_2$O$_2$) [64, 101, 102, 157, 158].

HE is oxidized directly by ferricytochrome c [91] and by other heme proteins. Oxidation of the probe by cytochromes c, c1, b$_{562}$, b$_{566}$, and aa3 is oxygen-independent, whereas oxidation by methemoglobin and metmyoglobin is strictly oxygen-dependent, with products consisting of a mixture of species resulting from 1- to 4-electron abstraction from HE. Although they are different from the superoxide oxidation product, their excitation/emission peaks are close to those generated by superoxide [60, 159].

5.6.3 Probes Used for Detection of NO and Peroxynitrite

Diaminofluoresceins were initially reported to be specific for NO, but DAF-2 reacts mainly with peroxynitrite rather than with nitric oxide [60, 160]. DAF-FM also reacts with peroxynitrite, but under conditions of physiological concentrations of NO and peroxynitrite it is fairly specific for NO.

Oxidants also interfere with the reaction of DAF-2 with NO [161], while reducing compounds such as catecholamines, ascorbate, dithiothreitol, mercaptoethanol, and glutathione attenuate the fluorescence of the reaction product [162]. Peroxidases in the presence of H_2O_2 oxidize DAF-2 to a relatively stable nonfluorescent intermediate that reacts directly with NOS, thus increasing fluorescence yield. Therefore, intracellular oxidation of DAF-2 may result in increase of DAF-2 fluorescence, erroneously indicating increased NO production [163].

DHR123 is the most frequently used probe for measuring peroxynitrite [62–64], but oxidation to rhodamine 123 is actually mediated by the radicals •NO_2 and •OH formed from the rapid and spontaneous decomposition of peroxynitrite, and is not induced directly by peroxynitrite itself. In addition, the intermediate radical, DHR•, formed from the one-electron oxidation of DHR123, also reacts rapidly with O_2 and Fe^{2+} [112, 164], triggering a redox cycling mechanism leading to artifactual amplification of the fluorescence signal intensity. Thus, DHR123 can be used only as a nonspecific indicator of intracellular peroxynitrite and HO radical formation [64].

5.6.4 Probes Used for Detection of Lipid Peroxides

The presence of four double bonds in *cis*-parinaric acid makes this probe very susceptible to oxidation if not rigorously protected from air [62, 80]. During experiments, *cis*-parinaric samples should be handled under inert gas and the solutions prepared with degassed buffers and solvents. *cis*-Parinaric acid is also photolabile and undergoes photodimerization when exposed to intense illumination, resulting in loss of fluorescence [80].

BODIPY[581/591] C11 is photosensitive, degrading under high-intensity illumination conditions, such as those typical of laser confocal microscopy [165]. In addition, BODIPY[581/591] C11 is more sensitive to oxidation than are endogenous lipids, and therefore tends to overestimate oxidative damage and underestimate antioxidant protection effects [80].

5.6.5   Probes Used for the Determination of GSH

Several fluorescent probes are used to determine intracellular GSH, but all of them may have limitations for quantitative studies. In many cases, the fluorescent reagents designed to measure GSH may react with other free or protein-bound intracellular thiols [137, 140, 166]. An important aspect in the use of GSH reagents is the large interspecies and tissue variability of cellular GSH content and the presence of GST isozymes, which may complicate enzyme-based measurements under saturating substrate conditions [166]. For instance, mClB, which is highly selective for GSH in rodents, should not be applied with quantitation purposes to human cells because of its low affinity for human GST [140].

5.7   Controls in the Cytometric Analysis of ROS, RNS, and Oxidative Stress

According to the limitations and caveats presented above, including appropriate positive and negative controls is very important when performing cytometric experiments or analyses related to ROS, RNS, and oxidative stress. When possible, direct visualization of intracellular ROS and RNS generation by co-localization techniques is highly recommended [70] (Fig. 2); detailed discussion of possible controls in such studies is beyond the scope of this chapter, as the biochemical complexity of experimental oxidants and antioxidants parallels that of their bio-logical counterparts [40, 167].

In general, the most frequent controls are positive controls, molecules or com-plex systems that directly or indirectly increase the intracellular level of ROS or RNS or mimic the cellular effects of oxidative stress. Prooxidants are chemicals that induce oxidative stress, either by generating reactive oxygen species or by inhibiting antioxidant systems [40, 167]. To mimic mitochondrial H_2O_2 production, cells can be treated with the complex I respiratory-chain inhibitor rotenone [84]. Peroxyl radicals, including alkylperoxyl and hydroperoxyl radicals, can be gener-ated from compounds such as 2,2'-azobis(2-amidinopropane) and from hydroper-oxides such as t-butyl hydroperoxide or cumene hydroperoxide [84]. The hydroxyl radical can be generated from superoxide donors (e.g., plumbagin or menadione) [111] or by exogenous H_2O_2 in a Fenton reaction catalyzed by Fe^{2+} or other transition metal, as well as by the effect of ionizing radiation [80]. Superoxide can be most effectively produced by the hypoxanthine/xanthine oxidase–generating system [168]. Many xenobiotics, including anticancer agents such as anthracyclines and cis-platin [169], and natural redox-active toxins, like pyocyanin, [170] generate ROS and can be used as positive controls.

Intracellular levels of ROS can also be increased by attenuating or inhibiting antioxidant defenses. A convenient strategy involves depletion of intracellular GSH stores by inhibiting GSH biosynthesis or by accelerating GSH oxidation [171].

Inhibitors of antioxidant enzymes, such as superoxide dismutase [172] and catalase [173], have also been used to increase intracellular ROS and induce oxidative stress.

The intracellular content of RNS can be increased by using NO donors, a heterogeneous group of chemicals (including ester nitrates, furoxans, benzofuroxans, NONOates, S-nitrosothiols, and metal complexes) that cross the cell membrane and generate intracellular NO [111, 174, 175] or peroxynitrite [176].

In addition, negative controls may be designed to reduce the levels of ROS or RNS or attenuate their biological effects. If possible, controls should be specific with respect to which particular reactive species or enzyme system is involved, but in most cases, controls do not attempt that degree of specificity [63–65]. Antioxidants can be categorized as enzymatic and nonenzymatic [167]. Enzymatic antioxidants work by transforming oxidative products to H_2O_2 and then to H_2O, in a sequential process. Cell-permeable forms of antioxidant enzymes, such as polyethyleneglycol–superoxide dismutase [177] can be also used to decrease specifically intracelllular ROS. Non-enzymatic antioxidants work by interrupting free radical–initiated chain reactions. Such antioxidants can be classified depending on whether they are hydrophilic (e.g., ascorbic acid, N-acetyl cysteine, GSH-esters) or lipophilic (e.g., α-tocopherol and Trolox) [40]. In general, water-soluble antioxidants react with oxidants in the cytosol while lipid-soluble antioxidants protect cell membranes from lipid peroxidation [40]. In addition, chelators of transition metals [178] also exert antioxidant effects, based upon the attenuation of Fenton-type reactions [40, 167].

Regarding the use of chemical antioxidants as negative controls, it should be kept in mind that reducing agents may become prooxidants. For instance, ascorbate has antioxidant activity when it reduces oxidizing substances such as H_2O_2, but it can also reduce metal ions, leading to the generation of free radicals through the Fenton reaction [40, 167]. When considering the specificity of antioxidants, all organic compounds react with hydroxyl radicals with rate constants approaching the diffusion limitation. Thus, in solution, no compound really has any more significant hydroxyl radical–scavenging activity than other compounds (proteins, lipids, nucleic acids, amino acids, numerous metabolites, etc.) already present in any biological system [40]. On the contrary, α-tocopherol, owing to its specific uptake into membranes and relatively rapid kinetics of reaction with lipid hydroperoxyl radicals compared with their propagation reaction, may be an effective chain breaker in lipid peroxidation [40].

In recent years, novel approaches to design positive and negative controls in studies of oxidative stress have involved genetically modified organisms. For instance, Guo et al. [85] used an enzymatic method to generate cytoplasmic H_2O_2 in astrocytes. Primary astrocytes were transduced with adenoviruses containing the cDNA for cytoplasmic D-amino acid oxidase (DAAO). DAAO oxidatively deaminates D-amino acids using FAD as the electron acceptor. At the same time, DAAO uses O_2 to oxidize FAD, thus generating H_2O_2 in a dose-dependent manner relative to the concentration of D-alanine.

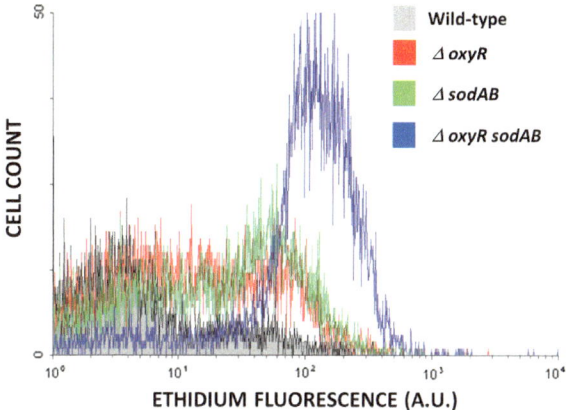

Fig. 5 Example of enhanced sensitivity to oxidative stress in oxyR-deficient strains of *Escherichia coli* B WP2. The WP2 strain of *Escherichia coli* B is characterized by increased membrane permeability to low-molecular compounds, including fluorescent probes. In this genetic background, inactivation of key genes involved in the sensing of ROS (oxyR) and/or the antioxidant defense (sodA, sodB) provoked increased accumulation of intracellular ROS as compared to the wild-type WP2 strain when incubated for 30 min with 10 μM plumbagin, a superoxide donor. Bacteria were stained with hydroethidine and analyzed by flow cytometry, in experimental conditions similar to those described in [180, 181]. Δ oxyR: strain deficient in the oxyR operon; Δ sodAB: strains deficient in the sodA and sodB genes, codifying for superoxide dismutases. Δ oxyRsodAB: triple mutant strain. A. U.: Fluorescence Arbitrary Units

To provide biosensors of oxidative stress our own group [179] has developed a collection of genetically modified strains of *Escherichia coli* B WP2, based on the inactivation of the oxyR operon, a main sensor of oxidative stress [44]. *Escherichia coli* B WP2 strains possess an altered cell-wall lipopolysaccharide that results in increased membrane permeability; we have previously shown that flow cytometric analysis of WP2 strains is a convenient alternative for cytometric assays of bacterial function [180]. Such oxyR-deficient bacterial strains show enhanced sensitivity to oxidative stress and increased accumulation of intracellular ROS when examined by flow cytometry using fluorogenic susbstrates (Fig. 5).

Acknowledgements The authors wish to acknowledge the financial support of the University of Valencia (UVEG) through grants UV-INV-AE15-349700 (Convocatòria Accions Especials 2015) and MOGDETECT (Programa VLC-BIOMED).

References

1. Fridovich I (1998) Oxygen toxicity: a radical explanation. J Exp Biol 210:1203–1209
2. Fridovich I (1999) Fundamental aspects of reactive oxygen species, or what's the matter with oxygen? Ann N Y Acad Sci 893:13–18 PubMed PMID: 10672226

3. Clancy D, Birdsall J (2013) Flies, worms and the free radical theory of ageing. Ageing Res Rev 12:404–412. doi:10.1016/j.arr.2012.03.011
4. Forman HJ, Augusto O, Brigelius-Flohe R, Dennery PA, Kalyanaraman B, Ischiropoulos H, Mann GE, Radi R, Roberts LJ 2nd, Viña J, Davies KJ (2015) Even free radicals should follow some rules: a guide to free radical research terminology and methodology. Free Radic Biol Med 78:233–235. doi:10.1016/j.freeradbiomed.2014.10.504
5. Di Meo S, Reed TT, Venditti P, Victor VM (2016) Role of ROS and RNS sources in physiological and pathological conditions. Oxid Med Cell Longev 2016:1245049. doi:10.1155/2016/1245049
6. Speckmann B, Steinbrenner H, Grune T, Klotz LO (2016) Peroxynitrite: from interception to signaling. Arch Biochem Biophys 595:153–160. doi:10.1016/j.abb.2015.06.022
7. Imlay JA (2003) Pathways of oxidative damage. Annu Rev Microbiol 57:395–418
8. Viña J, Borrás C, Miquel J (2007) Theories of ageing. IUBMB Life 59:249–254 PubMed PMID: 17505961
9. Forman HJ (2016) Redox signaling: an evolution from free radicals to aging. Free Radic Biol Med 97:398–407. doi:10.1016/j.freeradbiomed.2016.07.003
10. Hayashi G, Cortopassi G (2015) Oxidative stress in inherited mitochondrial diseases. Free Radic Biol Med 88:10–17. doi:10.1016/j.freeradbiomed.2015.05.039
11. Moulin M, Ferreiro A (2016) Muscle redox disturbances and oxidative stress as pathomechanisms and therapeutic targets in early-onset myopathies. Semin Cell Dev Biol. pii: S1084–9521(16) 30240-3. doi:10.1016/j.semcdb.2016.08.003
12. Beltrán B, Nos P, Dasí F, Iborra M, Bastida G, Martínez M, O'Connor JE, Sáez G, Moret I, Ponce J (2010) Mitochondrial dysfunction, persistent oxidative damage, and catalase inhibition in immune cells of naïve and treated Crohn's disease. Inflamm Bowel Dis 16:76–86. doi:10.1002/ibd.21027
13. Battacharyya A, Chattopadhyay R, Mitra S, Crowe SE (2014) Oxidative stress: an essential factor in the pathogenesis of gastrointestinal mucosal diseases. Physiol Rev 94:329–354. doi:10.1152/physrev.00040.2012
14. Li S, Tan HY, Wang N, Zhang ZJ, Lao L, Wong CW, Feng Y (2015) The role of oxidative stress and antioxidants in liver diseases. Int J Mol Sci 16:26087–26124. doi:10.3390/ijms161125942
15. Siti HN, Kamisah Y, Kamsiah J (2015) The role of oxidative stress, antioxidants and vascular inflammation in cardiovascular disease (a review). Vascul Pharmacol 71:40–56. doi:10.1016/j.vph.2015.03.005
16. Santilli F, D'Ardes D, Davì G (2015) Oxidative stress in chronic vascular disease: from prediction to prevention. Vascul Pharmacol 74:23–37. doi:10.1016/j.vph.2015.09.003
17. Li H, Horke S, Förstermann U (2014) Vascular oxidative stress, nitric oxide and atherosclerosis. Atherosclerosis 237:208–219. doi:10.1016/j.atherosclerosis.2014.09.001
18. Montezano AC, Dulak-Lis M, Tsiropoulou S, Harvey A, Briones AM, Touyz RM (2015) Oxidative stress and human hypertension: vascular mechanisms, biomarkers, and novel therapies. Can J Cardiol 31:631–641. doi:10.1016/j.cjca.2015.02.008
19. Fuentes E, Palomo I (2016) Role of oxidative stress on platelet hyperreactivity during aging. Life Sci 148:17–23. doi:10.1016/j.lfs.2016.02.026
20. Collado R, Ivars D, Oliver I, Tormos C, Egea M, Miguel A, Sáez GT, Carbonell F (2014) Increased oxidative damage associated with unfavorable cytogenetic subgroups in chronic lymphocytic leukemia. Biomed Res Int 2014:686392. doi:10.1155/2014/686392
21. Oh B, Figtree G, Costa D, Eade T, Hruby G, Lim S, Elfiky A, Martine N, Rosenthal D, Clarke S, Back M (2016) Oxidative stress in prostate cancer patients: a systematic review of case control studies. Prostate Int 4:71–87. doi:10.1016/j.prnil.2016.05.002
22. Zhou L, Wen J, Huang Z, Nice EC, Huang C, Zhang H, Li Q (2016) Redox proteomics screening cellular factors associated with oxidative stress in hepatocarcinogenesis. Proteomics Clin Appl. 20 Oct 2016. doi:10.1002/prca.201600089. (Epub ahead of print)

23. Marengo B, Nitti M, Furfaro AL, Colla R, Ciucis CD, Marinari UM, Pronzato MA, Traverso N, Domenicotti C (2016) Redox homeostasis and cellular antioxidant systems: crucial players in cancer growth and therapy. Oxid Med Cell Longev 2016:6235641. doi:10. 1155/2016/6235641

24. Ivanova D, Zhelev Z, Aoki I, Bakalova R, Higashi T (2016) Overproduction of reactive oxygen species—obligatory or not for induction of apoptosis by anticancer drugs. Chin J Cancer Res 28:383–396. doi:10.21147/j.issn.1000-9604.2016.04.01

25. Ivanov AV, Valuev-Elliston VT, Ivanova ON, Kochetkov SN, Starodubova ES, Bartosch B, Isagulants MG (2016) Oxidative stress during HIV infection: mechanisms and consequences. Oxid Med Cell Longev 2016:8910396 PubMed PMID:27829986

26. Elbim C, Pillet S, Prevost MH, Preira A, Girard PM, Rogine N, Hakim J, Israel N, Gougerot-Pocidalo MA (2001) The role of phagocytes in HIV-related oxidative stress. J Clin Virol 20:99–109 PubMed PMID: 11166656

27. Henchcliffe C, Beal M (2008) Mitochondrial biology and oxidative stress in Parkinson disease pathogenesis. Nature Clin Practice Neurology 4:600–609. doi:10.1038/ncpneuro0924

28. Barnham K, Masters C, Busch AJ (2004) Neurodegenerative diseases and oxidative stress. Nat Rev Drug Discov 3:205–214. doi:10.1038/nrd1330

29. Kamat PK, Kalani A, Rai S, Swarnkar S, Tota S, Nath C, Tyagi N (2016) Mechanism of oxidative stress and synapse dysfunction in the pathogenesis of Alzheimer's disease: understanding the therapeutics strategies. Mol Neurobiol 53:648–661. doi:10.1007/s12035-014-9053-6

30. Rani V, Deep G, Singh RK, Palle K, Yadav UC (2016) Oxidative stress and metabolic disorders: pathogenesis and therapeutic strategies. Life Sci 148:183–193. doi:10.1016/j.lfs. 2016.02.002

31. Forman HJ (2016) Redox signaling: an evolution from free radicals to aging. Free Radic Biol Med 97:398–407. doi:10.1016/j.freeradbiomed.2016.07.003

32. Dugas B, Debré P, Moncada S (1995) Nitric oxide, a vital poison inside the immune and inflammatory network. Res Immunol 146:664–670 PubMed PMID: 8852607

33. Erusalimsky JD, Moncada S (2007) Nitric oxide and mitochondrial signaling: from physiology to pathophysiology. Arterioscler Thromb Vasc Biol 27:2524–2531 PubMed PMID: 17885213

34. Speckmann B, Steinbrenner H, Grune T, Klotz LO (2016) Peroxynitrite: from interception to signaling. Arch Biochem Biophys 595:153–160. doi:10.1016/j.abb.2015.06.022

35. El-Benna J, Hurtado-Nedelec M, Marzaioli V, Marie JC, Gougerot-Pocidalo MA, Dang PM (2016) Priming of the neutrophil respiratory burst: role in host defense and inflammation. Immunol Rev 273:180–193. doi:10.1111/imr.12447

36. Burhans WC, Heintz NH (2009) The cell cycle is a redox cycle: linking phase-specific targets to cell fate. Free Radic Biol Med 47:1282–1293. doi:10.1016/j.freeradbiomed.2009. 05.026

37. Lionaki E, Markaki M, Tavernarakis N (2013) Autophagy and ageing: insights from invertebrate model organisms. Ageing Res Rev 12:413–428. doi:10.1016/j.arr.2012.05.001

38. Gibellini L, De Biasi S, Pinti M, Nasi M, Riccio M, Carnevale G, Cavallini GM, Sala de Oyanguren FJ, O'Connor JE, Mussini C, De Pol A, Cossarizza A (2012) The protease inhibitor atazanavir triggers autophagy and mitophagy in human preadipocytes. AIDS. 26:2017–2026. doi:10.1097/QAD.0b013e328359b8be

39. Dickinson BC, Chang CJ (2011) Chemistry and biology of reactive oxygen species in signaling or stress responses. Nat Chem Biol 7:504–511. doi:10.1038/nchembio.607

40. Rahal A, Kumar A, Singh V, Yadav B, Tiwari R, Chakraborty S, Dhama K (2014) Oxidative stress, prooxidants, and antioxidants: the interplay. Biomed Res Int. 2014:761264. doi:10. 1155/2014/761264

41. Nathan C, Shiloh MU (2000) Reactive oxygen and nitrogen intermediates in the relationship between mammalian hosts and microbial pathogens. Proc Natl Acad Sci USA 97:8841–8848 PubMed PMID: 10922044

42. Karupiah G, Hunt NH, King NJ, Chaudhri G (2000) NADPH oxidase, Nramp1 and nitric oxide synthase 2 in the host antimicrobial response. Rev Immunogenet 2:387–415 PubMed PMID: 11256747

43. Ritz D, Beckwith J (2001) Roles of thiol-redox pathways in bacteria. Annu Rev Microbiol 55:21–48. doi:10.1146/annurev.micro.55.1.21

44. Choi H, Kim S, Mukhopadhyay P, Cho S, Woo J, Storz G, Ryu SE (2001) Structural basis of redox switch in the OxyR transcription factor. Cell 105:103–113 PMID: 11301006

45. Park HS, Park D, Bae YS (2006) Molecular interaction of NADPH Oxidase 1 with betaPix and Nox Organizer 1. Biochem Biophys Res Commun 339:985–990. doi:10.1016/j.bbrc.2005.11.108

46. Burch PM, Heintz HH (2005) Redox regulation of cell-cycle re-entry: cyclin D1 as a primary target for the mitogenic effects of reactive oxygen and nitrogen species. Antioxid Redox Sign 7:741–751 PubMed PMID: 15890020

47. Havens CG, Ho A, Yoshioka N, Dowdy SF (2006) Regulation of late G1/S phase transition and APCCdh1 by reactive oxygen species. Mol Cell Biol 26:4701–4711. doi:10.1128/MCB.00303-06

48. Nakano H, Nakajima A, Sakon-Komazawa S, Piao JH, Xue X, Okumura K (2006) Reactive oxygen species mediate crosstalk between NF-kappaB and JNK. Cell Death Diff 13:730–737. doi:10.1038/sj.cdd.4401830

49. Tormos C, Javier Chaves F, Garcia MJ, Garrido F, Jover R, O'Connor JE, Iradi A, Oltra A, Oliva MR, Sáez GT (2004) Role of glutathione in the induction of apoptosis and c-fos and c-jun mRNAs by oxidative stress in tumor cells. Cancer Lett 208:103–113. doi:10.1016/j.canlet.2003.11.007

50. Jang JY, Min JH, Chae YH, Baek JY, Wang SB, Park SJ, Oh GT, Lee SH, Ho YS, Chang TS (2014) Reactive oxygen species play a critical role in collagen-induced platelet activation via SHP-2 oxidation. Antioxid Redox Sign 20:2528–2540. doi:10.1089/ars.2013.5337

51. Banchard JL, Wholey W-Y, Conlon EM, Pomposiello PJ (2007) Rapid changes in gene expression dynamics in response to superoxide reveal SoxRS-dependent and independent transcriptional networks. PLoS One. 14 Nov 2007; 2(11):e1186. Erratum in: PLoS One. 2012; 7(11). doi:10.1371/annotation/5cba04eb-5172-43a7-ad92-10efcd3858c9

52. Ghezzi P, Bonetto V (2003) Redox proteomics: identification of oxidatively modified proteins. Proteomics 3:1145–1153. doi:10.1002/pmic.200300435

53. Imlay JA (2013) The molecular mechanisms and physiological consequences of oxidative stress: lessons from a model bacterium. Nature Rev Microbiol 11:443–454

54. Bielski BHJ, Arudi RL, Sutherland MW (1983) A study of the reactivity of HO2/O2 with unsaturated fatty acids. J Biol Chem 258:4758–4761 PMID: 6833274

55. Gros L, Saparbaev MK, Laval L (2002) Enzymology of the repair of free radicals-induced DNA damage. Oncogene 21:8905–8925. doi:10.1038/sj.onc.1206005

56. Fortini P, Pascucci B, Parlanti E, D'Errico M, Simonelli V, Dogliotti E (2003) 8-Oxoguanine DNA damage: at the crossroad of alternative repair pathways. Mutat Res 531:127–139 PubMed PMID: 14637250

57. Barregard L, Møller P, Henriksen T, Mistry V, Koppen G, Rossner P Jr, Sram RJ, Weimann A, Poulsen HE, Nataf R, Andreoli R, Manini P, Marczylo T, Lam P, Evans MD, Kasai H, Kawai K, Li YS, Sakai K, Singh R, Teichert F, Farmer PB, Rozalski R, Gackowski D, Siomek A, Saez GT, Cerda C, Broberg K, Lindh C, Hossain MB, Haghdoost S, Hu CW, Chao MR, Wu KY, Orhan H, Senduran N, Smith RJ, Santella RM, Su Y, Cortez C, Yeh S, Olinski R, Loft S, Cooke MS (2013) Human and methodological sources of variability in the measurement of urinary 8-oxo-7,8-dihydro-2'-deoxyguanosine. Antioxid Redox Sign 18:2377–2391. doi:10.1089/ars.2012.4714

58. Frijhoff J, Winyard PG, Zarkovic N, Davies SS, Stocker R, Cheng D, Knight AR, Taylor EL, Oettrich J, Ruskovska T, Gasparovic AC, Cuadrado A, Weber D, Poulsen HE, Grune T, Schmidt HH, Ghezzi P (2015) Clinical relevance of biomarkers of oxidative stress. Antioxid Redox Sign 23:1144–1170. doi:10.1089/ars.2015.6317

59. Halliwell B, Gutteridge JMC (2004) Measuring reactive species and oxidative damage in vivo and in cell cultures: how should you do it and what do the results mean? Br J Pharmacol 142:231–252. doi:10.1038/sj.bjp.0705776

60. Bartosz G (2006) Use of spectroscopic probes for detection of reactive oxygen species. Clin Chim Acta 368:53–76. doi:10.1016/j.cca.2005.12.039

61. Lu C, Sung G, Lin JM (2006) Reactive oxygen species and their chemiluminescence-detection methods. Trends Anal Chem 25:985–995. doi:10.1016/j.trac.2006.07.007

62. Gomes A, Fernandes E, Lima JL (2005) Fluorescence probes used for detection of reactive oxygen species. J Biochem Biophys Methods 65:45–80. doi:10.1016/j.jbbm.2005.10.003

63. Wardman P (2007) Fluorescent and luminescent probes for measurement of oxidative and nitrosative species in cells and tissues: progress, pitfalls, and prospects. Free Radic Biol Med 43:995–1022. doi:10.1016/j.freeradbiomed.2007.06.026

64. Kalyanaraman B, Darley-Usmar V, Davies KJ, Dennery PA, Forman HJ, Grisham MB, Mann GE, Moore K, Roberts LJ 2nd, Ischiropoulos H (2012) Measuring reactive oxygen and nitrogen species with fluorescent probes: challenges and limitations. Free Radic Biol Med 52:1–6. doi:10.1016/j.freeradbiomed.2011.09.030

65. Debowska K, Debski D, Hardy M, Jakubowska M, Kalyanaraman B, Marcinek A, Michalski R, Michalowski B, Ouari O, Sikora A, Smulik R, Zielonka J (2015) Toward selective detection of reactive oxygen and nitrogen species with the use of fluorogenic probes–Limitations, progress, and perspectives. Pharmacol Rep 67:756–764. doi:10.1016/j.pharep.2015.03.016

66. Martínez-Pastor F, Mata-Campuzano M, Alvarez-Rodríguez M, Alvarez M, Anel L, de Paz P (2010) Probes and techniques for sperm evaluation by flow cytometry. Reprod Domest Anim 45(Suppl 2):67–78. doi:10.1111/j.1439-0531.2010.01622.x

67. Cottet-Rousselle C, Ronot X, Leverve X, Mayol JF (2011) Cytometric assessment of mitochondria using fluorescent probes. Cytometry A 79:405–425. doi:10.1002/cyto.a.21061

68. Liegibel UM, Abrahamse SL, Pool-Zobel BL, Rechkemmer G (2000) Application of confocal laser scanning microscopy to detect oxidative stress in human colon cells. Free Radic Res 32:535–547 PubMed PMID: 10798719

69. Manshian BB, Abdelmonem AM, Kantner K, Pelaz B, Klapper M, Nardi Tironi C, Parak WJ, Himmelreich U, Soenen SJ (2016) Evaluation of quantum dot cytotoxicity: interpretation of nanoparticle concentrations versus intracellular nanoparticle numbers. Nanotoxicology 10:1318–1328. doi:10.1080/17435390.2016.1210691

70. Ploppa A, George TC, Unertl KE, Nohe B, Durieux ME (2011) ImageStream cytometry extends the analysis of phagocytosis and oxidative burst. Scand J Clin Lab Invest 71:362–369. doi:10.3109/00365513.2011.572182

71. Moktar A, Singh R, Vadhanam MV, Ravoori S, Lillard JW, Gairola CG, Gupta RC (2011) Cigarette smoke condensate-induced oxidative DNA damage and its removal in human cervical cancer cells. Int J Oncol 39:941–947. doi:10.3892/ijo.2011.1106

72. Keston AS, Brandt R (1965) The fluorometric analysis of ultramicro quantities of hydrogen peroxide. Anal Biochem 1:1–5 PMID: 14328641

73. van Eeden SF, Klut ME, Walker BA, Hogg JC (1999) The use of flow cytometry to measure neutrophil function. J Immunol Methods 232:23–43 PMID: 10618507

74. Caldefie-Chézet F, Walrand S, Moinard C, Tridon A, Chassagne J, Vasson MP (2002) Is the neutrophil reactive oxygen species production measured by luminol and lucigenin chemiluminescence intra or extracellular? comparison with DCFH-DA flow cytometry and cytochrome c reduction. Clin Chim Acta 319:9–17 PMID: 11922918

75. Bourré L, Thibaut S, Briffaud A, Rousset N, Eléouet S, Lajat Y, Patrice T (2002) Indirect detection of photosensitizer ex vivo. J Photochem Photobiol, B Biol 67:23–31 PMID: 12007464

76. Silveira LR, Pereira-da-Silva L, Juel C, Hellstein Y (2003) Formation of hydrogen peroxide and nitric oxide in rat skeletal muscle cells during contractions. Free Radic Biol Med 35:455–464 PMID: 12927595

77. Tampo Y, Kotamraju S, Chitambar CR, Kalivendi SV, Keszler A, Joseph J, Kalyanaraman B (2003) Oxidative stress-induced iron signaling is responsible for peroxide-dependent oxidation of dichlorodihydrofluorescein in endothelial cells: role of transferrin receptor-dependent iron uptake in apoptosis. Circ Res 92:56–63 [PubMed: 12522121]

78. Kotamraju S, Tampo Y, Keszler A, Chitambar CR, Joseph J, Haas AL, Kalyanaraman B (2003) Nitric oxide inhibits H_2O_2-induced transferrin receptor-dependent apoptosis in endothelial cells: role of ubiquitin–proteasome pathway. Proc Natl Acad Sci U S A. 100:10653–10658 PMID: 12522121

79. Kotamraju S, Kalivendi SV, Konorev E, Chitambar CR, Joseph J (2004) Kalyanaraman B (2004) Oxidant induced iron signaling in doxorubicin-mediated apoptosis. Methods Enzymol 378:362–382. doi:10.1016/S0076-6879(04)78026-X

80. https://www.thermofisher.com/it/en/home/references/molecular-probes-the-handbook/probes-for-reactive-oxygen-species-including-nitric-oxide.html

81. Crow JP (1997) Dichlorodihydrofluorescein and dihydrorhodamine 123 are sensitive indicators of peroxynitrite in vitro: implications for intracellular measurement of reactive nitrogen and oxygen species. Nitric Oxide 1:145–157. doi:10.1006/niox.1996.0113

82. Miller EW, Abers AE, Pralle A, Isacoff EY, Chang CJ (2005) Boronate-based fluorescent probes for imaging cellular hydrogen peroxide. J Am Chem Soc 127:16652–16659. doi:10.1021/ja054474f

83. Dickenson BC, Huynh C, Chang CJ (2010) A palette of fluorescent probes with varying emission colors for imaging hydrogen peroxide signaling in living cells. J Am Chem Soc 132:5906–5915. doi:10.1021/ja1014103

84. Guo H, Aleyasin H, Dickinson BC, Haskew-Layton RE, Ratan RR (2014) Recent advances in hydrogen peroxide imaging for biological applications. Cell Biosci 4:64. doi:10.1186/2045-3701-4-64

85. Guo HC, Aleyasin H, Howard SS, Dickinson BC, Lin VS, Haskew-Layton RE, Xu C, Chen Y, Ratan RR (2013) Two-photon fluorescence imaging of intracellular hydrogen peroxide with chemoselective fluorescent probes. J Biomed Opt 18:106002. doi:10.1117/1.JBO.18.10.106002

86. Albers AE, Okreglak VS, Chang CJ (2006) A FRET-based approach to ratiometric fluorescence detection of hydrogen peroxide. J Am Chem Soc 128:9640–9641. doi:10.1021/ja063308k

87. Han Z, Liang X, Ren X, Shang L, Yin Z (2016) A 3,7-dihydroxyphenoxazine-based fluorescent probe for selective detection of intracellular hydrogen peroxide. Chem Asian J 11:818–822. doi:10.1002/asia.201501304

88. Dickinson BC, Chang CJ (2008) A targetable fluorescent probe for imaging hydrogen peroxide in the mitochondria of living cells. J Am Chem Soc 130:9638–9639. doi:10.1021/ja802355u

89. Xu J, Zhang Y, Yu H, Gao X, Shao S (2016) Mitochondria-targeted fluorescent probe for imaging hydrogen peroxide in living cells. Anal Chem 88:1455–1461. doi:10.1021/acs.analchem.5b04424

90. Dickinson BC, Tang Y, Chang ZY, Chang CJ (2011) A nuclear-localized fluorescent hydrogen peroxide probe for monitoring sirtuin-mediated oxidative stress responses in vivo. Chem Biol 18:943–948. doi:10.1016/j.chembiol.2011.07.005

91. Benov L, Sztejnberg L, Fridovich I (1998) Critical evaluation of the use of hydroethidine as a measure of superoxide anion radical. Free Radic Biol Med 25:826–831 PMID: 9823548

92. Rothe G, Valet G (1990) Flow cytometric analysis of respiratory burst activity in phagocytes with hydroethidine and 2,7-dichlorofluorescin. J Leukoc Biol 47:440–448 PMID: 2159514

93. Walrand S, Valeix S, Rodriguez C, Ligot P, Chassagne J, Vasson MP (2003) Flow cytometry study of polymorphonuclear neutrophil oxidative burst: a comparison of three fluorescent probes. Clin Chim Acta 331:103–110 PMID: 12691870

94. Carter WO, Narayanan PK, Robinson JP (1994) Intracellular hydrogen peroxide and superoxide anion detection in endothelial cells. J Leukoc Biol 55:253–258 PMID: 8301222

95. Barbacanne MA, Souchard JP, Darblade B, Iliou JP, Nepveu F, Pipy B, Bayard F, Arnal JF (2000) Detection of superoxide anion released extracellularly by endothelial cells using cytochrome c reduction, ESR, fluorescence and lucigenin-enhanced chemiluminescence techniques. Free Radic Biol Med 29:388–396 PMID: 11020659

96. Munzel T, Afanas'ev IB, Kleschyov AL, Harrison DG (2002) Detection of superoxide in vascular tissue. Arterioscler Thromb Vasc Biol 22:1761–1768 PMID: 12426202

97. Tarpey MM, Wink DA, Grisham MB (2004) Methods for detection of reactive metabolites of oxygen and nitrogen: in vitro and in vivo considerations. Am J Physiol Regul Integr Comp Physiol 286:R431–444. doi:10.1152/ajpregu.00361.2003

98. Guo TL, Miller MA, Shapiro IM, Shenker BJ (1998) Mercuric chloride induces apoptosis in human T lymphocytes: evidence of mitochondrial dysfunction. Toxicol Appl Pharmacol 153:250–257. doi:10.1006/taap.1998.8549

99. Le SB, Hailer MK, Buhrow S, Wang Q, Flatten K, Pediaditakis P, Bible KC, Lewis LD, Sausville EA, Pang YP, Ames MM, Lemasters JJ, Holmuhamedov EL, Kaufmann SH (2007) Inhibition of mitochondrial respiration as a source of adaphostin-induced reactive oxygen species and cytotoxicity. J Biol Chem 282:8860–8872. doi:10.1074/jbc. M611777200

100. De Biasi S, Gibellini L, Bianchini E, Nasi M, Pinti M, Salvioli S, Cossarizza A (2016) Quantification of mitochondrial reactive oxygen species in living cells by using multi-laser polychromatic flow cytometry. Cytometry A 89:1106–1110. doi:10.1002/cyto.a.22936

101. Robinson KM, Janes MS, Beckman JS (2008) The selective detection of mitochondrial superoxide by live cell imaging. Nat Protoc 3:941–947. doi:10.1038/nprot.2008.56

102. Zielonka J, Kalyanaraman B (2010) Hydroethidine- and MitoSOX-derived red fluorescence is not a reliable indicator of intracellular superoxide formation: another inconvenient truth. Free Radic Biol Med 48:983–1001. doi:10.1016/j.freeradbiomed.2010.01.028

103. Ahn HY, Fairfull-Smith KE, Morrow BJ, Lussini V, Kim B, Bondar MV, Bottle SE, Belfield KD (2012) Two-photon fluorescence microscopy imaging of cellular oxidative stress using profluorescent nitroxides. J Am Chem Soc 134:4721–4730. doi:10.1021/ ja210315x

104. DeLoughery Z, Luczak MW, Zhitkovich A (2014) Monitoring Cr intermediates and reactive oxygen species with fluorescent probes during chromate reduction. Chem Res Toxicol 27:843–851. doi:10.1021/tx500028x

105. Plaza Davila M, Martin Muñoz P, Tapia JA, Ortega Ferrusola C, Balao da Silva CC, Peña FJ (2015) Inhibition of mitochondrial complex i leads to decreased motility and membrane integrity related to increased hydrogen peroxide and reduced ATP production, while the inhibition of glycolysis has less impact on sperm motility. PLoS One 10(9):e0138777. doi:10.1371/journal.pone.0138777

106. Kojima H, Sakurai K, Kikuchi K, Kawahara S, Kirino Y, Nagoshi H, Hirata Y, Nagano T (1998) Development of a fluorescent indicator for nitric oxide based on the fluorescein chromophore. Chem Pharm Bull (Tokyo) 46:373–375 PMID: 9501473

107. Kojima H, Nakatsubo N, Kikuchi K, Kawahara S, Kirino Y, Nagoshi H, Hirata Y, Nagano T (1998) Detection and imaging of nitric oxide with novel fluorescent indicators: diaminofluoresceins. Anal Chem 70:2446–2453 PMID: 9666719

108. Leikert JF, Räthel TR, Müller C, Vollmar AM, Dirsch VM (2001) Reliable in vitro measurement of nitric oxide released from endothelial cells using low concentrations of the fluorescent probe 4,5-diaminofluorescein. FEBS Lett 506:131–134 PMID: 11591386

109. Xian JA, Guo H, Li B, Miao YT, Ye JM, Zhang SP, Pan XB, Ye CX, Wang AL, Hao XM (2013) Measurement of intracellular nitric oxide (NO) production in shrimp haemocytes by flow cytometry. Fish Shellfish Immunol 35:2032–2039. doi:10.1016/j.fsi.2013.10.014

110. Kolpen M, Bjarnsholt T, Moser C, Hansen CR, Rickelt LF, Kühl M, Hempel C, Pressler T, Høiby N, Jensen PØ (2014) Nitric oxide production by polymorphonuclear leucocytes in infected cystic fibrosis sputum consumes oxygen. Clin Exp Immunol 177:310–319. doi:10. 1111/cei.12318

111. Balaguer S, Diaz L, Gomes A, Herrera G, O'Connor JE, Urios A, Felipo V, Montoliu C (2015) Real-time cytometric assay of nitric oxide and superoxide interaction in peripheral blood monocytes: a no-wash, no-lyse kinetic method. Cytometry B Clin Cytom. doi:10.1002/cyto.b.21237. (Epub ahead of print)

112. Wardman P (2008) Methods to measure the reactivity of peroxynitrite-derived oxidants toward reduced fluoresceins and rhodamines. Methods Enzymol 441:261–282. doi:10.1016/S0076-6879(08)01214-7

113. Kuypers FA, van den Berg JJ, Schalkwijk C, Roelofsen B, Op den Kamp JA (1987) Parinaric acid as a sensitive fluorescent probe for the determination of lipid peroxidation. Biochim Biophys Acta 921:266–274 PMID: 3651488

114. Hedley D, Chow S (1992) Flow cytometric measurement of lipid peroxidation in vital cells using parinaric acid. Cytometry A 13:686–692. doi:10.1002/cyto.990130704

115. Drummen GP, Makkinje M, Verkleij AJ, Op den Kamp JA, Post JA (2004) Attenuation of lipid peroxidation by antioxidants in rat-1 fibroblasts: comparison of the lipid peroxidation reporter molecules cis-parinaric acid and C11-BODIPY(581/591) in a biological setting. Biochim Biophys Acta 1636:136–150 PMID: 15164761

116. Yoshida Y, Shimakawa S, Itoh N, Niki E (2003) Action of DCFH and BODIPY as a probe for radical oxidation in hydrophilic and lipophilic domain. Free Radic Res 37:861–872 PMID: 14567446

117. Brouwers JF, Gadella BM (2003) In situ detection and localization of lipid peroxidation in individual bovine sperm cells. Free Radic Biol Med 35:1382–1391 PMID: 14642386

118. Cheloni G, Slaveykova VI (2013) Optimization of the C11-BODIPY(581/591) dye for the determination of lipid oxidation in Chlamydomonas reinhardtii by flow cytometry. Cytometry A 83:952–961. doi:10.1002/cyto.a.22338

119. Peluso I, Adorno G, Raguzzini A, Urban L, Ghiselli A, Serafini M (2013) A new flow cytometry method to measure oxidative status: the Peroxidation of Leukocytes Index Ratio (PLIR). J Immunol Methods 390:113–120. doi:10.1016/j.jim.2013.02.005

120. Donato MT, Martínez-Romero A, Jiménez N, Negro A, Herrera G, Castell JV, O'Connor JE, Gómez-Lechón MJ (2009) Cytometric analysis for drug-induced steatosis in HepG2 cells. Chem Biol Interact 181:417–423. doi:10.1016/j.cbi.2009.07.019

121. Makrigiorgos GM, Kassis AI, Mahmood A, Bump EA, Savvides P (1997) Novel fluorescein-based flow–cytometric method for detection of lipid peroxidation. Free Radic Biol Med 22:93–100 PMID: 8958133

122. Maulik G, Kassis AI, Savvides P, Makrigiorgos GM (1998) Fluoresceinated phospho-ethanolamine for flow–cytometric measurement of lipid peroxidation. Free Radic Biol Med 26:645–653 PMID: 9801063

123. Lee SH, Blair IA (2000) Characterization of 4-oxo-2-nonenal as a novel product of lipid peroxidation. Chem Res Toxicol 13:698–702 PMID: 10956056

124. Csala M, Kardon T, Legeza B, Lizák B, Mandl J, Margittai É, Puskás F, Száraz P, Szelényi P, Bánhegyi G (2015) On the role of 4-hydroxynonenal in health and disease. Biochim Biophys Acta 1852:826–838. doi:10.1016/j.bbadis.2015.01.015

125. Toyokuni S, Miyake N, Hiai H, Hagiwara M, Kawakishi S, Osawa T, Uchida K (1995) The monoclonal antibody specific for the 4-hydroxy-2-nonenal histidine adduct. FEBS Lett 359:189–191 PMID: 7867796

126. Martin Muñoz P, Ortega Ferrusola C, Vizuete G, Plaza Dávila M, Rodriguez Martinez H, Peña FJ (2015) Depletion of intracellular thiols and increased production of 4-hydroxynonenal that occur during cryopreservation of stallion spermatozoa lead to caspase activation, loss of motility, and cell death. Biol Reprod 93:143. doi:10.1095/biolreprod.115.132878

127. https://www.emdmillipore.com/US/en/product/OxyDNA-Assay-Kit,-Fluorometric,   EMD_BIO-500095

128. Nagy S, Kakasi B, Bercsényi M (2016) Flow cytometric detection of oxidative DNA damage in fish spermatozoa exposed to cadmium—short communication. Acta Vet Hung 64:120–124. doi:10.1556/004.2016.013

129. Esperanza M, Cid Á, Herrero C, Rioboo C (2015) Acute effects of a prooxidant herbicide on the microalga Chlamydomonas reinhardtii: screening cytotoxicity and genotoxicity endpoints. Aquat Toxicol 165:210–221. doi:10.1016/j.aquatox.2015.06.004

130. Zribi N, Feki Chakroun N, El Euch H, Gargouri J, Bahloul A, Ammar Keskes L (2010) Effects of cryopreservation on human sperm deoxyribonucleic acid integrity. Fertil Steril 93:159–166. doi:10.1016/j.fertnstert.2008.09.038

131. Cambi M, Tamburrino L, Marchiani S, Olivito B, Azzari C, Forti G, Baldi E, Muratori M (2013) Development of a specific method to evaluate 8-hydroxy, 2-deoxyguanosine in sperm nuclei: relationship with semen quality in a cohort of 94 subjects. Reproduction 145:227–235. doi:10.1530/REP-12-0404

132. Aguilar C, Meseguer M, García-Herrero S, Gil-Salom M, O'Connor JE, Garrido N (2010) Relevance of testicular sperm DNA oxidation for the outcome of ovum donation cycles. Fertil Steril 94:979–988. doi:10.1016/j.fertnstert.2009.05.015

133. Meseguer M, Martínez-Conejero JA, O'Connor JE, Pellicer A, Remohí J, Garrido N (2008) The significance of sperm DNA oxidation in embryo development and reproductive outcome in an oocyte donation program: a new model to study a male infertility prognostic factor. Fertil Steril 89:1191–1199. doi:10.1016/j.fertnstert.2007.05.005

134. Balao da Silva CM, Ortega-Ferrusola C, Morrell JM, Rodriguez Martínez H, Peña FJ (2016) Flow cytometric chromosomal sex sorting of stallion spermatozoa induces oxidative stress on mitochondria and genomic DNA. Reprod Domest Anim 51:18–25. doi:10.1111/rda.12640

135. Durand RE, Olive PL (1983) Flow cytometry techniques for studying cellular thiols. Radiat Res 95:456–470 PubMed PMID: 6193555

136. Treumer J, Valet G (1986) Flow-cytometric determination of glutathione alterations in vital cells by o-phthaldialdehyde (OPT) staining. Exp Cell Res 163:518–524 PMID: 2420623

137. O'Connor JE, Kimler BF, Morgan MC, Tempas KJ (1988) A flow cytometric assay for intracellular nonprotein thiols using mercury orange. Cytometry A 9:529–532 PMID: 3208619

138. http://www.thermofisher.com/it/en/home/references/molecular-probes-the-handbook/assays-for-cell-viability-proliferation-and-function/probes-for-cell-adhesion-chemotaxis-multidrug-resistance-and-glutathione.html#head5

139. Nair S, Singh SV, Krishan A (1991) Flow cytometric monitoring of glutathione content and anthracycline retention in tumor cells. Cytometry A 12:336–342. doi:10.1002/cyto.990120408

140. Hedley DW, Chow S (1994) Evaluation of methods for measuring cellular glutathione content using flow cytometry. Cytometry A 15:349–358. doi:10.1002/cyto.990150411

141. Chow S, Hedley D (1995) Flow cytometric determination of glutathione in clinical samples. Cytometry 21:68–71. doi:10.1002/cyto.990210113

142. Skindersoe ME, Kjaerulff S (2014) Comparison of three thiol probes for determination of apoptosis-related changes in cellular redox status. Cytometry A 85:179–187. doi:10.1002/cyto.a.22410

143. O'Connor JE, Herrera G, Corrochano V (1998) Flow versus flux: functional assays by flow cytometry. In: Slavík J (ed) Fluorescence and Fluorescent Probes II. Plenum Press, New York, pp 47–54

144. O'Connor JE, Callaghan RC, Escudero M, Herrera G, Martínez A, Monteiro MD, Montolíu H (2001) The relevance of flow cytometry for biochemical analysis. IUBMB Life 51:231–239. doi:10.1080/152165401753311771

145. Grzelak A, Rychlik B, Bartosz G (2000) Reactive oxygen species are formed in cell culture media. Acta Biochim Pol 47:1197–1198 PMID: 11996110

146. Petasne RG, Zika RG (1987) Fate of superoxide in coastal sea water. Nature 325:516–518

147. Van Baalen C, Marler JE (1966) Occurrence of hydrogen peroxide in sea water. Nature 211:951

148. Subramaniam R, Fan XJ, Scivittaro V, Yang J, Ha CE, Petersen CE, Surewicz WK, Bhagavan NV, Weiss MF, Monnier VM (2002) Cellular oxidant stress and advanced glycation endproducts of albumin: caveats of the dichlorofluorescein assay. Arch Biochem Biophys 400:15–25. doi:10.1006/abbi.2002.2776

149. Keller A, Mohamed A, Drose S, Brandt U, Fleming I, Brandes RP (2004) Analysis of dichlorodihydrofluorescein and dihydrocalcein as probes for the detection of intracellular reactive oxygen species. Free Radic Res 38:1257–1267. doi:10.1080/10715760400022145

150. Swift LM, Sarvazyan N (2000) Localization of dichlorofluorescin in cardiac myocytes: implications for assessment of oxidative stress. Am J Physiol Heart Circ Physiol 278:H982–990 PMID: 10710368

151. Jakubowski W, Bartosz G (1997) Estimation of oxidative stress in Saccharomyces cerevisae with fluorescent probes. Int J Biochem Cell Biol 29:1297–1301 PMID: 9451827

152. Saengkhae C, Loetchutinat C, Garnier-Suillerot A (2003) Kinetic analysis of fluorescein and dihydrofluorescein effluxes in tumour cells expressing the multidrug resistance protein, MRP1. Biochem Pharmacol 65:969–977 PMID: 12623128

153. Kooy NW, Royall JA, Ischiropoulos H, Beckman JS (1994) Peroxynitrite mediated oxidation of dihydrorhodamine 123. Free Radic Biol Med 16:149–156 PMID: 8005510

154. Kooy NW, Royall JA, Ischiropoulos H (1997) Oxidation of 2,7-dichlorofluorescin by peroxynitrite. Free Radic Res 27:245–254 PMID: 9350429

155. Chignell CF, Sik RH (2003) A photochemical study of cells loaded with 2,7-dichlorofluorescin: implications for the detection of reactive oxygen species generated during UVA irradiation. Free Radic Biol Med 34:1029–1034 PMID: 12684087

156. Sikora A, Zielonka J, Lopez M, Joseph J, Kalyanaraman B (2009) Direct oxidation of boronates by peroxynitrite: mechanism and implications in fluorescence imaging of peroxynitrite. Free Radic Biol Med 47:1401–1407. doi:10.1016/j.freeradbiomed.2009.08.006

157. Zhao H, Kalivendi S, Zhang H, Joseph J, Nithipatikom K, Vásquez-Vivar J, Kalyanaraman B (2003) Superoxide reacts with hydroethidine but forms a fluorescent product that is distinctly different from ethidium: potential implications in intracellular fluorescence detection of superoxide. Free Radic Biol Med 34:1359–1368 PMID: 12757846

158. Zhao H, Joseph J, Fales HM, Sokoloski EA, Levine RL, Vasquez-Vivar J, Kalyanaraman B (2005) Detection and characterization of the product of hydroethidine and intracellular superoxide by HPLC and limitations of fluorescence. Proc Natl Acad Sci USA 102:5727–5732. doi:10.1073/pnas.0501719102

159. Papapostolou I, Patsoukis N, Georgiou CD (2004) The fluorescence detection of superoxide radical using hydroethidine could be complicated by the presence of heme proteins. Anal Biochem 332:290–298. doi:10.1016/j.ab.2004.06.022

160. Roychowdhury S, Luthe A, Keilhoff G, Wolf G, Horn TF (2002) Oxidative stress in glial cultures: detection by DAF-2 fluorescence used as a tool to measure peroxynitrite rather than nitric oxide. Glia 38:103–114 PMID: 11948804

161. Jourd'heuil D (2002) Increased nitric oxide-dependent nitrosylation of 4,5-diaminofluorescein by oxidants: implications for the measurement of intracellular nitric oxide. Free Radic Biol Med 33:676–684 PMID: 12208354

162. Nagata N, Momose K, Ishida Y (1999) Inhibitory effects of catecholamines and anti-oxidants on the fluorescence reaction of 4,5-diaminofluorescein, DAF-2, a novel indicator of nitric oxide. J Biochem 125:658–661 PMID: 10101276

163. Zhang X, Kim WS, Hatcher N, Potgieter K, Moroz LL, Gillette R, Sweedler JV (2002) Interfering with nitric oxide measurements. 4,5-diaminofluorescein reacts with dehy-droascorbic acid and ascorbic acid. J Biol Chem 277:48472–48478. doi:10.1074/jbc.M209130200

164. Qian SY, Buettner GR (1999) Iron and dioxygen chemistry is an important route to initiation of biological and free radical oxidations: an electron paramagnetic resonance spin trapping study. Free Radic Biol Med 26:1447–1456 PMID: 10401608

165. Drummen GP, van Liebergen LC, Op den Kamp JA, Post JA (2002) C11-BODIPY (581/591), an oxidation-sensitive fluorescent lipid peroxidation probe: (micro)spectroscopic characterization and validation of methodology. Free Radic Biol Med 33:473–490 PMID: 12160930

166. van der Ven AJ, Mier P, Peters WH, Dolstra H, van Erp PE, Koopmans PP, van der Meer JW (1994) Monochlorobimane does not selectively label glutathione in peripheral blood mononuclear cells. Anal Biochem 217:41–47 PMID:7515598

167. Nimse SB, Palb D (2015) Free radicals, natural antioxidants, and their reaction mechanisms RSC Adv 5: 27986–28006

168. Aitken RJ, Buckingham D, Harkiss D (1993) Use of a xanthine oxidase free radical generating system to investigate the cytotoxic effects of reactive oxygen species on human spermatozoa. J Reprod Fertil 97:441–450 PMID: 8388958

169. Alexandre J, Nicco C, Chéreau C, Laurent A, Weill B, Goldwasser F, Batteux F (2006) Improvement of the therapeutic index of anticancer drugs by the superoxide dismutase mimic mangafodipir. J Natl Cancer Inst 98:236–244. doi:10.1093/jnci/djj049

170. Hall S, McDermott C, Anoopkumar-Dukie S, McFarland AJ, Forbes A, Perkins AV, Davey AK, Chess-Williams R, Kiefel MJ, Arora D, Grant GD (2016) Cellular effects of pyocyanin, a secreted virulence factor of Pseudomonas aeruginosa. Toxins 8:E236. doi:10. 3390/toxins8080236

171. Harris C, Hansen JM (2012) Oxidative stress, thiols, and redox profiles. Methods Mol Biol 889:325–346. doi:10.1007/978-1-61779-867-2_21

172. Siwik DA, Tzortzis JD, Pimental DR, Chang DL, Pagano PJ, Singh K, Sawyer DB, Colucci WS (1999) Inhibition of copper-zinc superoxide dismutase induces cell growth, hypertrophic phenotype, and apoptosis in neonatal rat cardiac myocytes in vitro. Circ Res 85:147–153 PMID: 10417396

173. Titov VY, Osipov AN (2016) Nitrite and nitroso compounds can serve as specific catalase inhibitors. Redox Rep Apr 14:1–7. (Epub ahead of print) PMID: 27075937

174. Serafim RA, Primi MC, Trossini GH, Ferreira EI (2012) Nitric oxide: state of the art in drug design. Curr Med Chem 19:386–405 PMID: 22335514

175. Yuan S, Patel RP, Kevil CG (2015) Working with nitric oxide and hydrogen sulfide in biological systems. Am J Physiol Lung Cell Mol Physiol 308:L403–415. doi:10.1152/ajplung.00327.2014

176. Khodade VS, Kulkarni A, Sen Gupta A, Sengupta K, Chakrapani H (2016) A small molecule for controlled generation of peroxynitrite. Org Lett 18:1274–1277. doi:10.1021/acs.orglett.6b00186

177. Kim EJ, Lee HJ, Lee J, Youm HW, Lee JR, Suh CS, Kim SH (2015) The beneficial effects of polyethylene glycol-superoxide dismutase on ovarian tissue culture and transplantation. J Assist Reprod Genet 32:1561–1569. doi:10.1007/s10815-015-0537-8

178. Hrušková K, Potůčková E, Hergeselová T, Liptáková L, Hašková P, Mingas P, Kovaříková P, Šimůnek T, Vávrová K (2016) Aroylhydrazone iron chelators: tuning antioxidant and antiproliferative properties by hydrazide modifications. Eur J Med Chem 120:97–110. doi:10.1016/j.ejmech.2016.05.015

179. Herrera G, Martínez A, O'Connor JE, Blanco M (2003) UNIT 11.16 Functional assays of oxidative stress using genetically engineered Escherichia coli strains. Current Protocols Cytometry Published Online: 1 May 2003. doi:10.1002/0471142956.cy1116s24

180. Herrera G, Martinez A, Blanco M, O'Connor JE (2002) Assessment of Escherichia coli B with enhanced permeability to fluorochromes for flow cytometric assays of bacterial cell function. Cytometry A 49:62–69. doi:10.1002/cyto.10148

181. Alvarez-Barrientos A, O'Connor JE, Nieto-Castillo R, Moreno-Moreno AB, Prieto P (2001) Use of flow cytometry and confocal microscopy techniques to investigate early $CdCl_2$-induced nephrotoxicity in vitro. Toxicol In Vitro 15:407–412. doi:10.1016/S0887-2333(01) 00044-3

182. Pinti M, Gibellini L, De Biasi S, Nasi M, Roat E, O'Connor JE, Cossarizza A (2011) Functional characterization of the promoter of the human Lon protease gene. Mitochondrion 11:200–206. doi:10.1016/j.mito.2010.09.010

Flow Cytometry in Multi-center and Longitudinal Studies

Anis Larbi

Abstract In the era of consortium-based studies, "omics," and data sharing, flow cytometry needs to match other technological platforms in terms of standard operating procedures, reduced variability, and reproducibility. While tools such as gene-expression platforms have proven robustness and reproducibility, flow cytometry still relies heavily on scientists for the development of antibody panels as well as detailed experimental procedures. This is expected to remain for several decades, as the limit of markers to be assessed in a single staining does not allow attainment of the "omic" level of other technological platforms. This chapter presents a non-exhaustive series of multi-centric and longitudinal studies and their integration of flow cytometry measures. We also discuss recommendations made by key consortia to minimize variability in flow cytometry experiments. This chapter also aims to raise awareness of factors that may influence flow cytometry data obtained in large studies.

Keywords Multi-center studies · Standardization · Variability · Recommendations · Data analysis

A. Larbi (✉)
Biology of Aging Laboratory, Singapore Immunology Network, Agency for Science Technology and Research (A*STAR), Singapore, Singapore
e-mail: Anis_Larbi@immunol.a-star.edu.sg

A. Larbi
A*STAR Flow Cytometry, SIgN Immunomonitoring Platform Facility, A*STAR, Singapore, Singapore

A. Larbi
Faculty of Medicine, University of Sherbrooke, Sherbrooke, QC, Canada

A. Larbi
Department of Biology, Faculty of Sciences, ElManar University, Tunis, Tunisia

© Springer Nature Singapore Pte Ltd. 2017
J.P. Robinson and A. Cossarizza (eds.), *Single Cell Analysis*,
Series in BioEngineering, DOI 10.1007/978-981-10-4499-1_5

1 Introduction

Following infection by human immunodeficiency virus (HIV) the immune system displays progressive failure, culminating in the acquired immunodeficiency syndrome (AIDS), which increases susceptibility to opportunistic infections and cancers. Asymptomatic infections such as cytomegalovirus (CMV) become problematic in HIV-seropositive individuals. Without treatment, survival may not exceed a decade, while tailored therapy enables several decades of life without AIDS. HIV preferably infects $CD4^+$ T cells and via several mechanisms leads to CD4 depletion [1]. The follow-up of CD4 counts has been a simple but powerful tool to assess HIV infection status. As of today, it is the best example for successful implementation of flow cytometry in clinical practice with a longitudinal approach. HIV patients are to be followed for life to ensure that treatments are adapted to control infection. Following primary infection, the number of $CD4^+$ T cells is reduced by twofold in the first weeks (from an average of 1100 cells/mm^3 blood). Untreated individuals will then experience a steady decrease in CD4 count to reach as low as 200 cells/mm^3 blood, while HIV RNA copies usually stabilize at circa 500 copies/mL plasma. When $CD4^+$ T-cell numbers reach the lowest values (<200 cells/mm^3 blood) and HIV RNA copies are high (>500 copies/mL plasma), patients will very likely experience symptoms of AIDS. Since the emergence of HIV, many protocols have been developed to enable low-income high–HIV prevalence countries to detect and follow HIV infection. Seropositivity can be revealed by an enzyme-linked immunosorbent assay (ELISA) to detect anti-HIV-1 antibodies. While assessing RNA copies is a good way to follow up on viral spread and immune control, it is more difficult to implement such molecular techniques in field biology. Since the early days of HIV recognition, flow cytometry has been a key technology to follow the evolution of the infection by offering a cheaper, accurate, and easy-to-implement technology. Based on the tremendous amount of research and discoveries in the field of HIV, it is clear that flow cytometry has contributed greatly to those advances and still continues with the expansion of immunophenotyping capacities. As illustrated by the example of HIV, this chapter will focus on the implementation of flow cytometry for large-scale, multi-center, and longitudinal studies. The contribution of flow cytometry in successful studies will be highlighted along with some of the limitations faced by scientists and clinicians to reach successful integration of flow cytometry in such studies.

2 Flow Cytometry in Multi-center and Longitudinal Studies

We are showcasing in this section some of the first multi-centric and longitudinal clinical studies. This will also cover some efforts done to standardize protocols, panels, and data analysis to increase reliability of flow cytometry data. We arbitrarily selected studies of different types to show the reader the diversity and limitations of current practices. The use of flow cytometry in larger numbers of patients in a longitudinal setting has rarely been reported in the 20th century.

Apart from the HIV-related longitudinal studies, one of the first studies in the 90s implementing flow cytometry in clinics consisted of following hemostatic parameters in women (n = 19) during the first, second, and third trimester of pregnancy [2]. Among these, platelet function was examined using an anti-GPM-140 antibody, and the percentage of activated platelets was found to be within the normal reference range, even in late pregnancy. The CVs were not tested at that time, so it could be that a high CV did not enable the identification of deregulation during pregnancy. This shows the failure of some of the early studies to reach the level of quality required. The hurdles for the implementation of procedures are multifactorial: (i) expertise in flow cytometry, (ii) logistics/time, and (iii) funding. Groups have started to integrate procedures in order to ensure the reliability of data for longitudinal or multi-center studies. For example, groups performed multi-centric flow-cytometry screening using different cytometers [3], or using different antibody cocktails [4], or with minimal quality controls [5], in order to identify pitfalls in clinical flow cytometry studies. Those studies revealed that implementation of only one corrective action for multiple sources of variability is not sufficient to reach the level of quality required for clinical studies. At the end of the 90s Wikby et al. reported that an inverted CD4:CD8 ratio was associated with higher mortality in elderly Swedes representative of the normal population of the same age. In a follow-up study, the same group identified individuals who converted to this risk profile and were able to validate the clinical outcome (mortality) [6]. In addition to the CD4:CD8 ratio, poor T-cell proliferation from cryopreserved PBMC samples was part of the composite score (immune risk profile) associated with mortality. This follow-up study was utilizing the same staining, flow cytometer, and analysis pipeline as the original one. This shows that a combination of corrective actions could be very valuable in a clinical setting. Eighteen years after this initial study in Sweden, research has progressed; a recent report confirmed the importance of the CD4:CD8 ratio in older individuals for its ability to predict mortality [7]. In an independent cohort of 235 individuals aged 81.5 years or older from the BELFRAIL cohort (Belgium), the hazard for all-cause mortality adjusted for age, comorbidity, and CMV serostatus increased 1.53-fold with every increment in the CD4:CD8 ratio from $r < 1$, to $1 < r < 5$ and $r > 5$. This reinforces the idea that well-controlled flow cytometry experiments can be of significant clinical importance and validated in other clinical settings.

At the beginning of the 21st century, a series of consortia emerged proposing panels for the identification of immune-cell populations or aiming to harmonize panels, as in the case of minimal residual disease (MRD). Minimal residual disease profiling provides prognostic information. Moreover, MRD assessment would enable clinicians to tailor therapy and thus reduce treatment dose for good responders based on MRD status. By measuring aberrant immune-cell phenotypes The United Kingdom Flow MRD group has reported a standardized protocol for B-ALL [8] used by various laboratories across the UK. Their 4-color protocol was of high sensitivity, provided similar diagnosis compared to molecular data, and was validated by the UK MRD group. The standardized protocol was highly reproducible between laboratories based on the way the group defined acceptable variability.

Performing phenotyping on PBMCs is probably the standard and simplest model for multi-center studies. However, the recent advances in immunology have shown that the immune system is largely present in the tissues, while a minority of immune cells are in the periphery. In the field of immunotherapy against cancers, researchers have shown that investigating intra-tumoral immunity is important. The same applies to other research fields, but very few studies were able to conduct standardized immunophenotyping in tissues across different laboratories. In 2015, a group explored the feasibility of flow cytometry analysis of rectal biopsies for the investigation of gut-associated immunity in different clinical sites [9]. The conclusion of this study: *Standardized protocols to collect, stain, and analyze MMC and PBMC, including centralized analysis, can reduce but not exclude variability in reporting flow data within multi-site studies. Based on these data, centralized processing, flow cytometry, and analysis of samples may provide more robust data across multi-site studies. Centralized processing requires either shipping of fresh samples or cryopreservation and the decision to perform centralized versus site processing needs to take into account the drawbacks and restrictions associated with each method.* The authors reached this conclusion after critically analyzing the data generated. The design of the study was quite elegant, as it included:

- a batch of cryopreserved PBMCs from the same donor to be tested across study sites as internal controls
- PBMCs from the patients undergoing rectal biopsy
- similar protocol for rectal biopsy acquisition
- similar protocol for enzymatic digestion and leukocyte isolation
- usage of same reagents (concentration, lot #)
- determination of target MFI and a linear range for each parameter
- photomultiplier tube (PMT) target MFI values across the two flow cytometers at the two sites over the course of the study
- inclusion of a fluorescence-minus-one (FMO) control
- analysis of the data by an independent laboratory
- test of variance across sites using control samples
- appropriate statistical analysis.

The recommendation by the authors to perform flow cytometry analysis in a central laboratory is explained by several factors. One of the study sites showed very high variance in the control samples during the course of the study, which may be one way to select the right laboratory in which to conduct the assays. While the overall immune profile was very similar in both sites, alteration in the frequency of the subpopulations was an issue for this clinical study. The study showed FMO controls not to be necessary in each experiment. Their recommendation to use a centralized flow cytometry laboratory should be taken with care, as this study was comparing different individuals and the control PBMC samples showed clearly that the two sites reach similar overall phenotypes but with variability during the course of the study. Additionally, it will be necessary to determine whether the expected change in marker expression in an intervention trial is larger than the inter-site variability. This can be tested prior to making the decision.

The SPIROMICS group (subpopulations and intermediate outcome measures in COPD study) chose the centralized approach for flow cytometry acquisition of the samples collected from six clinical sites [10]. This study combined a multi-center, longitudinal, multiple body-site sampling (sputum, bronchoalveolar lavage (BAL), and peripheral blood). Samples were stained at all study sites, fixed, and then shipped to the centralized facility to be acquired on a single instrument. The 12-color panel was tested in a total of 269 samples to identify $CD4^+$ and $CD8^+$ T cells, B cells, monocytes, macrophages, neutrophils, and eosinophils. One key parameter to enable success of such a setup is the communication between the several study sites and the centralized facility. Sources of variability seem to have been reduced using this setting; however, it is difficult to evaluate as samples from the various study sites differed and did not allow direct comparison. In this study, batches with >85% usable data were considered acceptable. The study design allowed a minimal number of variables identified by the multicenter AIDS cohort study to be critical: (i) flow cytometer model, (ii) antibody/fluorochrome selection, (iii) sample preparation, and (iv) data analysis [11]. Of note, the antibody cocktails were prepared at the central facility and shipped to the different clinical sites where samples were stained and fixed before shipment back to the central facility. One limitation in such a process is the potential alteration of antibody staining by the fixative agent (paraformaldehyde) with time [12]. Studies have shown that staining remains stable for a few days after fixation. The fixation still affects surface markers differently and must be tested before making a decision on utilizing fixative agents before acquisition. A major issue in the SPIROMICS study is the lack of a live/dead exclusion marker.

In our laboratory we have assessed the profile of immune cells in whole blood in a longitudinal manner. We used 11 markers (CD66b, CD16, CD38, CD14, CD56, CD8, CD123, HLA-DR, CD4, CD19, and CD3) and were able to identify 19 clusters using t-distributed stochastic neighbor embedding (t-SNE), a technique for dimensionality reduction, to avoid the bias of any gating strategy [13]. The same individuals donated blood at baseline and 2, 7, and 28 days later. Analysis showed the presence of 19 clusters when the four time points were merged (Fig. 1a). The phenograph enables visualization of the expression of the individual markers in a heatmap-like display. The markers are grouped based on the clustering. When the heatmaps are analyzed individually we observe variations in the dendrogram of the different time points. While the overall cell populations are distinguished, there are differences in the intensity of expression of the markers as well as in the frequency of cells detected (Fig. 1b). We highlighted the neutrophil population (black line), which consisted of six mini-clusters. The neutrophil clusters are maintained across the longitudinal measures despite slight differences in the intensity of the markers, suggesting that automated analysis is an attractive method for unbiased analysis of

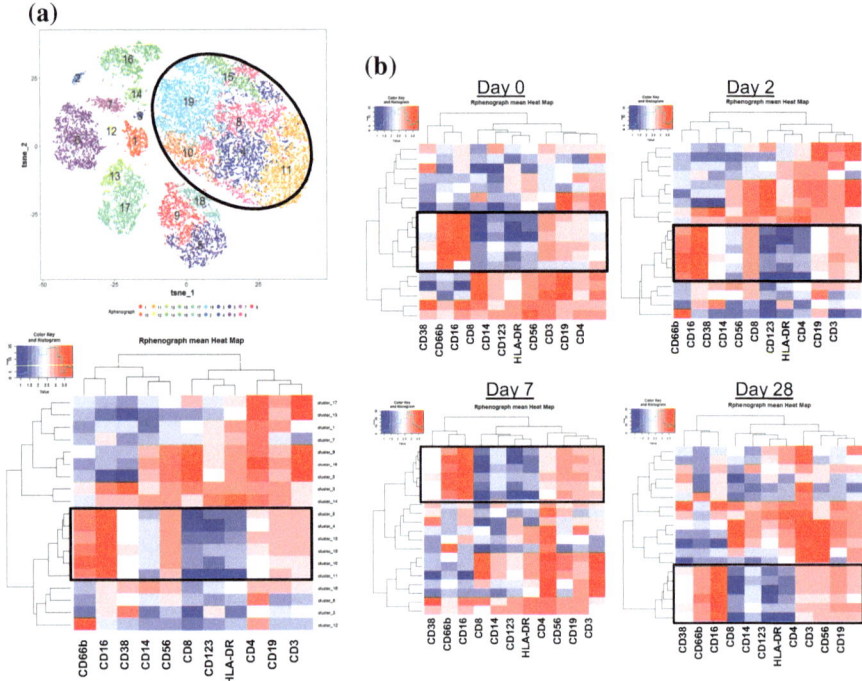

Fig. 1 a t-SNE representation of whole-blood staining from a healthy volunteer. Whole blood was stained with CD66b, CD16, CD38, CD14, CD56, CD8, CD123, HLA-DR, CD4, CD19, and CD3 antibodies. Blood samples at 0, 2, 7, and 28 days were merged for analysis. The resulting 19 clusters are shown by color in the phenograph with the respective markers and cluster identified by t-SNE. In black are the neutrophils clusters (cluster 4, 8, 10, 11, 15, and 19). **b** The individual phenographs are shown for each time point. The neutrophil clusters are included in the *black box*. The clustering and phenograph differ slightly between time points; the order of the markers has been adjusted

flow cytometry data. The data also suggest a physiological variation in the frequency of the different populations, and this observation for immune cells such as neutrophils may be exacerbated in rare populations.

3   Testing Flow Cytometry(-ists)

One of the main issues in research is the ability to replicate results and translate discoveries to other populations or to similar populations in other settings. The study of biological processes, disease progression, and response to therapy can be limited by the techniques. For instance, despite the plethora of transcriptomic data generated in clinical studies, the fact that each study is utilizing a different genomic platform, different protocols for RNA extraction, and other such technical differences hinders and complicates discovery replication. While the core message may still be replicated, there are clear limitations. The same applies to other technological platforms; we will focus here on flow cytometry.

The development of flow cytometry during the past few decades and utilization of antibodies to detect the presence of molecules expressed by cells pushed flow cytometry at the forefront of technologies and supplemented other technology platforms for testing hypotheses. The advantage of flow cytometry is the single-cell information obtained, compared to the information on bulk populations provided by Western blotting or other techniques. With as little as one drop of blood one could assess the presence of the various components of the immune system. This is another and important advantage of flow cytometry. For two decades the level of information obtained has been constantly increasing. As of now, cytometers enable one to obtain information using up to 30 channels, providing information on the immune system like never before. The increasing availability of reagents to reach this complexity is also an important factor contributing to the expansion of flow cytometry in clinical settings. To match the complexity of fcs files, new tools have been developed in recent years to maximize the analysis of datasets.

Some attempts have been made to test the variability in flow cytometry data generated from different laboratories. For instance, our laboratory participated in the MHC Multimer Proficiency Panel that aimed to test the variability in detection of antigen-specific CD8$^+$ T cells. Such studies have been conducted for many years by the Cancer Immunotherapy Consortium of the Cancer Research Institute (USA) and the Association for Cancer Immunotherapy (CIMT, Europe). Up to 30 laboratories participate every year in the proficiency panel testing, consisting of PBMCs shipped in liquid nitrogen with a temperature logger, MHC multimers (EBV HLA-A*0201/GLCTLVAML, EBV HLA-A*0201/CLGGLLTMV, EBV HLA-B*0702/RPPIFIRRL, and CMV HLA-B*0702/TPRVTGGGAM, and a negative control multimer) to be used by all participating laboratories. No standard operating procedure (SOP) was imposed regarding equipment use or PMT voltage selection, but a protocol was provided for tetramer staining. We discuss here only one dataset, the frequency of CMV-specific CD8$^+$ T cells, which was expected to be 0.8% [14]. Of the 30 laboratories worldwide, five accurately reached the value 0.8% for CMV-specific CD8$^+$ T cells from the same donor (100% accuracy) while 13 laboratories report frequencies with at least 33% difference from the expected value (Fig. 2a). The full dataset is available [14]. This report clearly shows the variability induced by the different operating procedures in place. This also suggests an urgent need for standardizing flow cytometry protocols when multi-center studies are to be conducted.

The EuroFlow consortium [15] aimed at establishing standard operating procedures for instrument setup, design of antibody panels, sample preparation, compensation of fluorescence overlap, and development of tools for evaluation of the reagents used. The consortium developed an 8-color immunophenotying panel tested across various laboratories [16]. The process included the following steps:

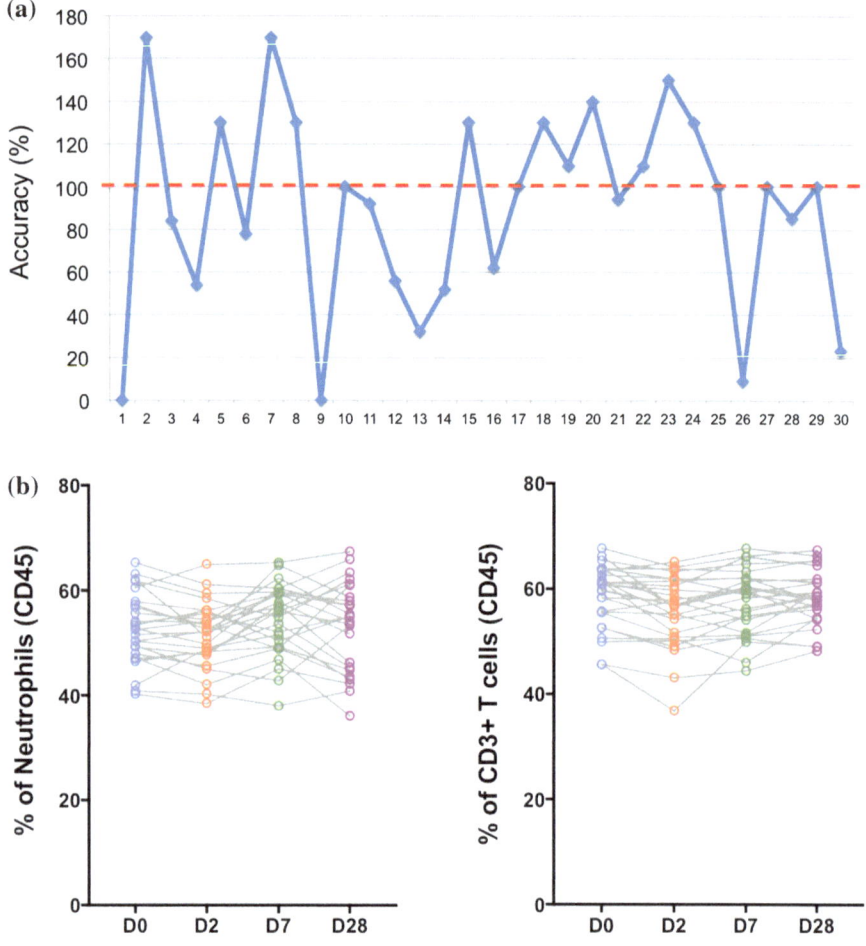

Fig. 2 **a** Accuracy for the detection of CMV-specific CD8$^+$ T cells in the 2015 Proficiency study. PBMCs were shipped to participating laboratories together with CMV dextramers. Thirty laboratories participated in the study (x axis) and reported the frequency of CMV-specific T cells obtained. The accuracy of the expected value (0.8%, *red dotted line*) was plotted for each laboratory. **b** Whole-blood samples from healthy donors were stained with CD45, CD66b, and CD3. The frequency of neutrophils (CD66b$^+$) and T cells (CD3$^+$) was calculated, with CD45$^+$ as parent population. Samples collected at baseline were compared with the same donor's sample at 2, 7, and 28 days

- selection of flow cytometry instruments and their optical configurations
- selection of fluorochromes
- placement of PMT voltages for fluorescence measurements
- monitoring of instrument performance
- automated baseline settings and instrument monitoring
- fluorescence compensation standards and controls

- red-cell lysis and cell staining
- sample acquisition in the flow cytometer
- data analysis.

Those steps took into account the specificities of each study site (n = 8), including flow cytometer configurations for the efficient testing and classification of hematological malignancies (leukemias and lymphomas). The 8-color panel (CD20, CD45, CD8, CD27, CD4, CD19, CD14, CD3) aimed at identifying the following populations: B-cells, $CD4^+$ $CD27^+$ T cells, $CD4^+$ $CD27^-$ T cells, $CD8^+$ $CD27^+$ T cells, $CD8^+$ $CD27^-$ T cells, $CD3^+$ $CD4^-$ $CD8^-$ T cells, NK cells, and monocytes. The same samples stabilized (TransFix: Cytomark, UK) as well as peripheral blood mononuclear cells (PBMCs) from healthy volunteers (n = 30) were shipped to the study sites for flow cytometry analysis. The result of the experiments revealed mean fluorescence intensity (MFI) CVs ranging from 11.1 to 43.8% for the stabilized sample. Despite those differences, principal component analysis showed that the populations clustered similarly when files from the different study sites were merged. The data from the EuroFlow consortium clearly suggested that standardization of the procedures enables researchers to perform multi-center studies and obtain high-quality, reliable data. A similar approach has been taken by the ONE study [17], with six leukocyte panels each comprising 7–9 markers. The staining of whole blood in this case should be performed within 4 h of sample collection. The variability for the various immune-cell subsets ranged from 0.05 to 30% between the different sites. The intra-assay variability ranged from 0.05 to 20% for most of the cells. To the variability observed one should also add the expected physiological variation in the frequency of immune cells. Homeostasis is a very tightly regulated process that is, however, influenced by many environmental factors. We present data on the same individuals' blood samples tested three times in the same week (Fig. 2b). The above-mentioned studies are representative of the efforts in the European Union to harmonize multi-center studies. This type of effort is needed owing to the fact that a number of pan-EU projects are multi-centric.

Recently, this has been pushed to the next level by the Human ImmunoPhenotyping Consortium in the USA [18]. Five eight-color panels were developed to identify T cells, Tregs, B cells, DC/Mono/NK, and Th1/2/17 cells. The antibody panels were utilized as lyoplates (BD Biosciences, USA) across different laboratories to further reduce variability introduced by pipetting, and automated analysis was implemented (Table 1). The results of this investigation were:

- laser power and filter configuration can contribute to site-to-site variability
- lyophilized reagents did not significantly ameliorate resolution of "dim" subsets
- gating strategy should be taught to the participating centers, although centralization of analysis is preferred
- automated analysis introduces less bias
- discrepancies between manual and automated analysis often occurred with rare-cell populations
- discrepancies between manual and automated analysis can indicate poor adherence to SOPs.

Table 1 Panels developed and utilized by the human immunophenotyping consortium

	T cell	Treg	B cell	DC/mono/NK	Th1/2/17
FITC	Dead	Dead	Dead	Dead	Dead
PE	CCR7 (150503)	CD25 (2A3)	CD24 (ML5)	CD56 (B159)	CXCR3 (1C6/CXCR3)
PerCP-Cy5.5	CD4 (SK3)	CD4 (SK3)	CD19 (SJ25C1)	CD123 (7G3)	CD4 (SK3)
PE-Cy7	CD45RA (L48)	CCR4 (1G1)	CD27 (M-T271)	CD11c (B-LY6)	CCR6 (11A9)
APC	CD38 (HIT2)	CD127 (HIL-7R-M21)	CD38 (HIT2)	CD16 (B73.1)	CD38 (HIT2)
APC-H7	CD8 (SK1)	CD45RO (UCHL1)	CD20 (2H7)	CD3$^+$ 19 + 20 (SK7, SJ25C1, 2H7)	CD8 (SK1)
V450	CD3 (UCHT1)	CD3 (UCHT1)	CD3 (UCHT1)	CD14 (MPHIP9)	CD3 (UCHT1)
V500	HLA-DR (G46-6)	HLA-DR (G46-6)	IgD (IA6-2)	HLA-DR (G46-6)	HLA-DR (G46-6)

Antibodies were used in their lyophilized form [18]

The National Institute of Allergy and Infectious Diseases Division of AIDS (NIAID DAIDS) Immunology Quality Assessment (IQA) Program has shown how multiple laboratories demonstrate good skills in lymphocyte subset phenotyping [19]. The IQA program included on average more than 75 laboratories, and a review of 10 years of proficiency testing provides very valuable information. Analysis of the IQA program data using various statistical models was used to assess the accuracy, precision, and evolution of these for the measure of CD4$^+$ and CD8$^+$ percentages and absolute counts. The data showed that participating laboratories improved over time to reach the expected frequencies. The number of outlier laboratories was reduced during the 10-year follow-up. The IQA review demonstrated a 42% reduction of variability. The improved laboratory performance in reducing overall measurement variability over time shows that implementation of standard operating procedures, education, and proper training are efficient.

4 Can You Analyze the 1000+ .fcs Files?

A recurring item in the above-described clinical studies and their conclusions is data analysis. There are several—justified—concerns regarding the data analysis strategy, the management of large datasets to make meaningful interpretations. For maximizing reproducibility of flow cytometry data, the use of automated analysis has developed significantly over the past years. An example is the setup of the ReFlow framework for immune monitoring in cancer immunotherapy [20]. The combination of statistics, machine learning, and data visualization minimizes subjective analysis. This data management system avoids errors in processing and interpretation of the data. However, one should ensure that the data management platform can integrate other datasets, generated from other technology platforms such as transcriptomics and epigenetics [21].

Once data management is tackled, several actions are to be performed in order to ensure that the data can be utilized. First, a significant number of fcs files contain abnormalities in the signals. These can be due to equipment instability during the course of acquisition, sample clogging, or sample cellular concentration. Several tools have been developed to clear fcs files of unwanted information that may affect the output. We have developed flowAI [22] to reduce false discoveries due to inter-sample differences in quality and properties. It provides an automatic method that adopts algorithms for the detection of anomalies as well as an interactive method with a graphical user interface implemented into an R Shiny application. The two methods check and remove suspected anomalies from abrupt changes in the flow rate, instability of signal acquisition, and outliers in the lower limit and margin events in the upper limit of the dynamic range. The software presented is an intuitive solution seeking to improve the results not only of manual but also and in particular of automatic analysis of FCM data. Other tools have been developed, such as flowClean [23], QUALiFiER [24], and FlowQ [25].

Several studies have shown the power of automated analysis and the necessity of moving towards this approach [18]. The Flow Cytometry: Critical Assessment of Population Identification Methods (FlowCAP) project aims at advancing the development of computational methods for the identification of cell populations in flow cytometry data [26]. The flow cytometry community is enriched every day by bioinformaticians, mathematicians, statisticians, and other computer science experts to push the limits of data analysis, integration, and interpretation. The current challenge is the consistent labeling of cell populations across multiple samples. The strategies mentioned in the previous sections to generate reliable flow cytometry data in multi-center and longitudinal settings should also enable one to maximize cell population labeling. Several tools have been developed to perform automated analysis of flow cytometry data; these are described in [27]. Many comparative studies, including those conducted by the FlowCAP project, demonstrate that many of these tools are mature and can be used reliably. As of now, more than 30 different packages have been developed for automated gating, and used in combination could be useful to validate and publish resource information [28].

5   A Future for Multi-center and Longitudinal Studies?

Many attempts have been made to integrate recommendations in order to minimize challenges to the reliability of flow cytometry data. It is tempting to propose here a full set of procedures to avoid issues in the generation and interpretation of flow cytometry data. However, each study has its specificities, and as shown in the different examples in this chapter, some procedures may prove difficult to implement in clinical settings. Probably the best recommendation is to adapt the clinical study to the experimental need, when feasible. Many solutions are described in the chapter and others not cited here could have been mentioned. For instance, in longitudinal studies, it is preferable to operate on the same equipment, especially if "dim" markers are part of the antibody panel or if the changes to be observed are relatively small. Another issue in most of the clinical studies presented is the subjectivity of "acceptable variability." Some groups are more stringent than others; variability <10% was acceptable in some studies, a higher level in others. Variability is an issue, especially for small-scale studies. This is less of a problem for large-scale studies where the sample size and statistical power will overcome, at least in part, the variability issue. Ideally, the sample size calculation should take into account the expected variability in the flow cytometry experiments A recurring issue in most of the clinical studies showcased in this chapter is the lack of validation. Only a few studies are validated in other independent cohorts. Validation will bring flow cytometry in the forefront of technologies to be implemented in large-scale studies. Concerted efforts are showing promising results [29], but more emphasis should be put towards bringing flow cytometry to the same level as other technologies such as gene expression data and proteomics. In the past few years the development of mass cytometry has tremendously helped understanding the immune system. A recent report evaluated standardization and quality-control methods applied to mass-cytometry experiments [30]. The spiking of a known sample in PBMC and bronchoalveolar lavage samples from HIV patients was used as (1) quality control for antibodies in the panel, (2) identification of batch effects, and (3) implementation of a defined gating strategy. Although mass cytometry is much more recent than fluorescence flow cytometry, it will greatly benefit from the several decades of attempts to harmonize and standardize measures. Longitudinal studies are a unique opportunity to understand immune homeostasis across lifespans as well as in health and diseases, while multi-center studies are important to compare groups of individuals that can differ by ethnicity, health condition, and environment. Flow cytometry has an important role in achieving this enormous task.

Acknowledgments Anis Larbi is supported by the Singapore Immunology Network, the Agency for Science Technology and Research (Clinical Immunomonitoring Platform grant #H16/99/b0/0111) and the Joint Council Office Development Program (grant #1434m0011). Anis Larbi is an emeritus Marylou Ingram ISAC Scholar.

References

1. Cossarizza A, Ortolani C, Mussini C, Borghi V, Guaraldi G, Mongiardo N, Bellesia E, Franceschini MG, De Rienzo B, Franceschi C (1995) Massive activation of immune cells with an intact T cell repertoire in acute human immunodeficiency virus syndrome. J Infect Dis 172:105–112
2. Gatti L, Tenconi PM, Guarneri D, Bertulessi C, Ossola MW, Bosco P, Gianotti GA (1994) Hemostatic parameters and platelet activation by flow-cytometry in normal pregnancy: a longitudinal study. Int J Clin Lab Res 24:217–219
3. Bradstock K, Matthews J, Benson E, Page F, Bishop J (1994) Prognostic value of immunophenotyping in acute myeloid leukemia. Australian Leukaemia Study Group. Blood 84:1220–1225

4. Benevolo G, Stacchini A, Spina M, Ferreri AJ, Arras M, Bellio L, Botto B, Bulian P, Cantonetti M, Depaoli L, Di Renzo N, Di Rocco A, Evangelista A, Franceschetti S, Godio L, Mannelli F, Pavone V, Pioltelli P, Vitolo U, Pogliani EM (2012) Final results of a multicenter trial addressing role of CSF flow cytometric analysis in NHL patients at high risk for CNS dissemination. Blood 120:3222–3228
5. Della Porta MG1, Picone C, Pascutto C, Malcovati L, Tamura H, Handa H, Czader M, Freeman S, Vyas P, Porwit A, Saft L, Westers TM, Alhan C, Cali C, van de Loosdrecht AA, Ogata K (2012) Multicenter validation of a reproducible flow cytometric score for the diagnosis of low-grade myelodysplastic syndromes: results of a European LeukemiaNET study. Haematologica 97:1209–1217
6. Wikby A, Maxson P, Olsson J, Johansson B, Ferguson FG (1998) Changes in CD8 and CD4 lymphocyte subsets, T cell proliferation responses and non-survival in the very old: the Swedish longitudinal OCTO-immune study. Mech Ageing Dev 102:187–198
7. Adriaensen W, Pawelec G, Vaes B, Hamprecht K, Derhovanessian E, van Pottelbergh G, Degryse JM, Matheï C (2016) CD4:8 ratio above 5 is associated with all-cause mortality in cmv-seronegative very old women: results from the BELFRAIL study. J Gerontol A Biol Sci Med Sci (in press)
8. Irving J, Jesson J, Virgo P, Case M, Minto L, Eyre L, Noel N, Johansson U, Macey M, Knotts L, Helliwell M, Davies P, Whitby L, Barnett D, Hancock J, Goulden N, Lawson S; UKALL Flow MRD Group; UK MRD steering Group (2009) Establishment and validation of a standard protocol for the detection of minimal residual disease in B lineage childhood acute lymphoblastic leukemia by flow cytometry in a multi-center setting. Haematologica 94:870–874
9. McGowan I, Anton PA, Elliott J, Cranston RD, Duffill K, Althouse AD, Hawkins KL, De Rosa SC (2015) Exploring the feasibility of multi-site flow cytometric processing of gut associated lymphoid tissue with centralized data analysis for multi-site clinical trials. PLoS ONE 10:e0126454
10. Freeman CM, Crudgington S, Stolberg VR, Brown JP, Sonstein J, Alexis NE, Doerschuk CM, Basta PV, Carretta EE, Couper DJ, Hastie AT, Kaner RJ, O'Neal WK, Paine R 3rd, Rennard SI, Shimbo D, Woodruff PG, Zeidler M, Curtis JL (2015) Design of a multi-center immunophenotyping analysis of peripheral blood, sputum and bronchoalveolar lavage fluid in the Subpopulations and Intermediate Outcome Measures in COPD Study (SPIROMICS). J Transl Med 13:19
11. Giorgi JV, Cheng HL, Margolick JB, Bauer KD, Ferbas J, Waxdal M, Schmid I, Hultin LE, Jackson AL, Park L (1990) Quality control in the flow cytometric measurement of T-lymphocyte subsets: the multicenter AIDS cohort study experience. The Multicenter AIDS Cohort Study Group. Clin Immunol Immunopathol 55:173–186
12. Stewart JC1, Villasmil ML, Frampton MW (2007) Changes in fluorescence intensity of selected leukocyte surface markers following fixation. Cytometry A 71:379–385
13. Mair F, Hartmann FJ, Mrdjen D, Tosevski V, Krieg C, Becher B. The end of gating? An introduction to automated analysis of high dimensional cytometry data. Eur J Immunol. 46:34–43
14. http://www.immudex.com/media/46328/mhc_multimer_proficiency_panel_2015.pdf
15. https://www.euroflow.org/usr/pub/pub.php
16. Kalina T, Flores-Montero J, van der Velden VH, Martin-Ayuso M, Böttcher S, Ritgen M, Almeida J, Lhermitte L, Asnafi V, Mendonça A, de Tute R, Cullen M, Sedek L, Vidriales MB, Pérez JJ, te Marvelde JG, Mejstrikova E, Hrusak O, Szczepański T, van Dongen JJ, Orfao A (2012) EuroFlow standardization of flow cytometer instrument settings and immunophenotyping protocols. Leukemia 26:1986–2010
17. Streitz M, Miloud T, Kapinsky M, Reed MR, Magari R, Geissler EK, Hutchinson JA, Vogt K, Schlickeiser S, Kverneland AH, Meisel C, Volk HD, Sawitzki B (2013) Standardization of whole blood immune phenotype monitoring for clinical trials: panels and methods from the ONE study. Transplant Res 2:17

18. Finak G, Langweiler M, Jaimes M, Malek M, Taghiyar J, Korin Y, Raddassi K, Devine L, Obermoser G, Pekalski ML, Pontikos N, Diaz A, Heck S, Villanova F, Terrazzini N, Kern F, Qian Y, Stanton R, Wang K, Brandes A, Ramey J, Aghaeepour N, Mosmann T, Scheuermann RH, Reed E, Palucka K, Pascual V, Blomberg BB, Nestle F, Nussenblatt RB, Brinkman RR, Gottardo R, Maecker H, McCoy JP (2016) Standardizing flow cytometry immunophenotyping analysis from the human immunophenotyping consortium. Sci Rep 6:20686

19. Bainbridge J, Wilkening CL, Rountree W, Louzao R, Wong J, Perza N, Garcia A, Denny TN (2014) The immunology quality assessment proficiency testing program for CD3$^+$ 4$^+$ and CD3$^+$ 8$^+$ lymphocyte subsets: a ten year review via longitudinal mixed effects modeling. J Immunol Methods 409:82–90

20. White S, Laske K, Welters MJ, Bidmon N, van der Burg SH, Britten CM, Enzor J, Staats J, Weinhold KJ, Gouttefangeas C, Chan C (2015) managing multi-center flow cytometry data for immune monitoring. Cancer Inf 13:111–122

21. Schadt EE, Linderman MD, Sorenson J, Lee L, Nolan GP (2016) Computational solutions to large-scale data management and analysis. Nat Rev Genet 11:647–657

22. Monaco G, Chen H, Poidinger M, Chen J, de Magalhães JP, Larbi A (2016) flowAI: automatic and interactive anomaly discerning tools for flow cytometry data. Bioinformatics 32:2473–2480

23. Fletez-Brant K, Špidlen J, Brinkman RR, Roederer M, Chattopadhyay PK (2016) flowClean: Automated identification and removal of fluorescence anomalies in flow cytometry data. Cytometry A 89:461–471

24. Finak G, Jiang W, Pardo J, Asare A, Gottardo R (2012) QUAliFiER: an automated pipeline for quality assessment of gated flow cytometry data. BMC Bioinform 13:252

25. https://www.bioconductor.org/packages/release/bioc/html/flowQ.html

26. Brinkman RR, Aghaeepour N, Finak G, Gottardo R, Mosmann T, Scheuermann RH. Automated analysis of flow cytometry data comes of age. Cytometry A 89:13–15

27. Kvistborg Pia, Gouttefangeas Cécile, Aghaeepour Nima, Cazaly Angelica, Chattopadhyay Pratip K, Chan Cliburn, Eckl Judith, Finak Greg, Hadrup Sine Reker, Maecker Holden T, Maurer Dominik, Mosmann Tim, Qiu Peng, Scheuermann Richard H, Welters Marij JP, Ferrari Guido, Brinkman Ryan R, Britten Cedrik M (2016) Thinking outside the gate: single-cell assessments in multiple dimensions. Immunity 42:591–592

28. Guilliams M, Dutertre CA, Scott CL, McGovern N, Sichien D, Chakarov S, Van Gassen S, Chen J, Poidinger M, De Prijck S, Tavernier SJ, Low I, Irac SE, Mattar CN, Sumatoh HR, Low GH, Chung TJ, Chan DK, Tan KK, Hon TL, Fossum E, Bogen B, Choolani M, Chan JK, Larbi A, Luche H, Henri S, Saeys Y, Newell EW, Lambrecht BN, Malissen B, Ginhoux F (2016) Unsupervised high-dimensional analysis aligns dendritic cells across tissues and species. Immunity 45:669–684

29. Aghaeepour N, Chattopadhyay P, Chikina M, Dhaene T, Van Gassen S, Kursa M, Lambrecht BN, Malek M, McLachlan GJ, Qian Y, Qiu P, Saeys Y, Stanton R, Tong D, Vens C, Walkowiak S, Wang K, Finak G, Gottardo R, Mosmann T, Nolan GP, Scheuermann RH, Brinkman RR (2016) A benchmark for evaluation of algorithms for identification of cellular correlates of clinical outcomes. Cytometry A 89:16–21

30. Kleinsteuber K, Corleis B, Rashidi N, Nchinda N, Lisanti A, Cho JL, Medoff BD, Kwon D, Walker BD (2016) Standardization and quality control for high-dimensional mass cytometry studies of human samples. Cytometry A 89:903–913

Validation—The Key to Translatable Cytometry in the 21st Century

Virginia Litwin, Cherie Green and Alessandra Vitaliti

Abstract With the primary goal of translating scientific advances and innovative technologies to applications that directly benefit patient care and treatment, Translational Science has justifiably generated considerable enthusiasm. Unfortunately, this enthusiasm has been somewhat tampered by the reality that the outcomes have not fully met the expectations of accelerating the transition of new discoveries from the bench to the bedside. This failure to deliver can be attributed, in part, to the lack of reproducible standards and procedures. Thus it stands to reason that suggestions for increasing the success rate of the Translation Science initiative include the implementation of more robust methods, standardization and validation. This chapter will explore opportunities in the translational science space and key concepts of analytical method validation as applied to cytometry.

Keywords Validation · Standards · Translational science · Biomarkers · Optimization · Specificity · Precision · Sensitivity · Stability assessment

V. Litwin (✉)
Hematology/Flow Cytometry, Covance Central Laboratory Services,
8211 SciCor Drive, Indianapolis, IN 46214, USA
e-mail: virginia.litwin@covance.com

C. Green
Flow Cytometry Biomarkers Development Sciences, Genentech, Inc., a Member of the Roche Group, 1 DNA Way—MS-46-1A, South San Francisco, CA 94080, USA
e-mail: green.cherie@gene.com

A. Vitaliti
BioMarker Development, Novartis Pharma AG, 4002 Basel, Switzerland
e-mail: alessandra.vitaliti@novartis.com

© Springer Nature Singapore Pte Ltd. 2017
J.P. Robinson and A. Cossarizza (eds.), *Single Cell Analysis*,
Series in BioEngineering, DOI 10.1007/978-981-10-4499-1_6

1 Introduction

Analytical method validation may sound like the dullest of topics whereas translational science and drug development seem to be exciting topics; paradoxically, in reality the two are closely intertwined.

The goal of translational and clinical science is to accelerate the transition of scientific advances from "bench to bedside." Translational science also provides scientists with a path to ensure that their work will have impact. That said, the reality is that only scientific findings based on robust data have the potential to lead to meaningful scientific advances and thereby successfully transition from bench to bedside. Analytical method validation provides a means to ensure that data are credible and reproducible, i.e., translatable.

2 The Translational Space—Opportunities and Challenges

Translational science is of considerable importance to a range of stakeholders; among them are academic centers, private foundations, the industry sector (pharmaceutical, diagnostic, technological), disease-focused and patient-advocacy organizations, independent hospitals and health systems, and government agencies such as the Food and Drug Administration (FDA) and the National Institutes of Health (NIH). While a variety of definitions can be found for the term translational science, most all include the concept of translating basic research and biomedical discoveries into clinical practice, thus accelerating the progression of scientific advances from the bench to the bedside.

Opportunities in translational science have been increasing since its inception around 2000, and even more so since Elias A. Zerhouni, then NIH director, published his seminal paper on translational and clinical science in 2005 and later established the National Center for Advancing Translational Sciences (NCATS) and the Clinical and Translational Science Award (CTSA) Consortium (Table 1) [1–3]. The NCATS network includes more than 60 academic institutions and non-profit organizations, which serve as catalysts for advancing translational and clinical science by supporting basic and clinical research activities, as well as the development of innovative technologies. Other important initiatives include the Academic Drug Discovery Consortium (ADDC), whose mission includes providing educational, technological, and advisory support to investigators with the goal of promoting private-public partnerships [4].

For the cytometry community in particular, there are abundant opportunities in the translational science space. These opportunities range from general technological advances, to technological advances specifically geared toward accelerating drug-screening activities, to the identification of new diseases (acquired or genetic), to the identification of new therapeutic targets and compounds, to the identification of new biological processes that may translate to new drug targets or novel biomarkers [5–7]. As the pharmaceutical industry focuses more resources on cancer immunotherapy, autoimmunity, vaccine development, and chronic viral diseases, additional translational opportunities between the cytometry community and the biopharmaceutical sector are emerging [8].

Table 1 Translational Science Centers. This (non-exhaustive) list provides some examples of institutes providing resources to support research investigators in transitioning discoveries from bench to bedside

Translational Science Centers	Institution	Website	Mission
National Center for Advancing Translational Sciences	National Institutes of Health	www.ncats.nih.gov	The National Center for Advancing Translational Sciences Strategic Plan highlights four major themes: translational science, collaboration and partnerships, education and training, and stewardship.
Center for Clinical and Translational Science	The Rockefeller University	www.rockefeller.edu/ccts	The Rockefeller University Center for Clinical and Translational Science is devoted to maximizing the bidirectional opportunities for clinical and translational research. The Center is funded in part by an NIH Center for Clinical and Translational Science Award.
Harvard Catalyst	Harvard University	www.catalyst.harvard.edu	Harvard Catalyst is organized into 11 programs, which serve the diverse needs of the Harvard clinical and translational research community. Harvard Catalyst is funded in part by an NIH Center for Clinical and Translational Science Award.
Center for Translational Neuromedicine	University of Rochester	www.urmc.rochester.edu/ctn.aspx	The Center for Translational Neuromedicine focuses on the development of new approaches for treating neurological diseases, primarily using cell and gene therapy.
Tufts Clinical and Translational Science Institute	Tufts University	www.tuftsctsi.org	Tufts CTSI was established with a Clinical and Translational Science Award from the National Institutes of Health. From bench to bedside, to clinical practice, to care delivery and public health, to public policy and beyond, Tufts CTSI is committed to fostering collaboration and innovation across the translational spectrum.

(continued)

Table 1 (continued)

Translational Science Centers	Institution	Website	Mission
Diabetes Translational Research Center	Indiana University	www.medicine.iupui.edu/DTRC	The mission of the Center is to organize research that improves both the prevention of diabetes and the delivery of diabetes care.
Center for Clinical and Translational Research	Seattle Children's Hospital	www.seattlechildrens.org/research/clinical-and-translational-research	The Center for Clinical and Translational Research is home to more than 400 research faculty and staff members from over 30 subdivisions and is the hub for clinical investigation and therapeutic development at Seattle Children's Research Institute.
The European Society for Translational Medicine	Global Translational Medicine Consortium	www.eutranslationalmedicine.org	The European Society for Translational Medicine is a global non-profit and neutral healthcare organization whose principal objective is to enhance world-wide healthcare by using translational medicine approaches, resources, and expertise.
Centre for Translational Research and Diagnostics	Cancer Science Institute of Singapore	www.csi.nus.edu.sg/ws/research/core-facilities-support-technologies/centre-for-translational-research-and-diagnostics-ctrad	The Centre for Translational Research and Diagnostics has three major facilities: (1) the NUHS Tissue Repository, (2) the Translational Interface molecular pathology facility, and (3) the Diagnostic Molecular Oncology Centre with expertise in clinical sample and data management, translational research, and clinical trial support.

Although the intersection of basic research, technology, drug discovery, and drug development is exciting and promises ultimately to facilitate the generation of better therapies, it is also a frustrating space for all involved. In a 2010 review in Drug Discovery Today, Martin Wehling delivered a sobering status report on the outcomes in translational science using drug development successes as the benchmark [9]. He cites the lack of reproducible standards and procedures not only as one of the major root causes of the disappointing results but also as a potential corrective action. Others have also suggested that a way to improve the robustness (translatability) of preclinical observations is to bring best practices from industry regarding standardization and process validation to research and non-regulated environments [10]. Francis Collins, the current NIH director, has specifically called out the need for better methodology and validation [11]. The application of robust analytical method validation will, without question, lead to more success in the translational space.

3 The Translational Approach in Drug Development

More than a decade ago, the translational approach was embraced by the pharmaceutical industry in order to increase the success rate, and decrease the cost, of bringing drugs to market. A major aspect of this approach was the implementation of new technologies. Flow cytometry, mass cytometry, and imaging technologies are becoming critical in the development and characterization of biologics and immunotherapies. The versatility of these technologies enables a wide variety of applications along the entire cycle of the drug-development process from target identification to pharmacodynamic (PD) biomarker monitoring associated with a therapeutic intervention.

Cytometry biomarker applications include intracellular pathway analysis (PhosFlow), intracellular cytokine staining, receptor occupancy, cell depletion, repletion, or modulation, and surface-marker analysis [12–15]. The most common application of flow cytometry in clinical evaluation is the evaluation of the immune-cell subsets in the peripheral blood and bone marrow. Data from the evaluation of the changes in the balance of immune-cell subsets contributes to the understanding of compound safety and efficacy.

Throughout the drug-development process, cytometry data are used to make critical decisions such as compound selection, whether to move a compound forward, and dose selection. Incorrect decisions will be made if the analytical methods used to generate the data are not robust and well characterized, i.e., validated as appropriate to the intended use of the data and the regulatory requirements associated with that use (fit for purpose).

4 Fit-for-Purpose Method Validation

No analytical method will ever be perfect, meaning that the measurement will never be 100% accurate, 100% of the time. Analytical method validation provides an understanding of how close to perfect the method is, and thus a context for how to interpret the data.

Table 2 Method Validation Parameters and Related Terminology

Validation Parameter/Term	Definition	Reference	Achievable with Cell-Based Assays?
Accuracy	The closeness of the agreement between the result of a measurement and a true value of the measurand.	International Standard ISO 15189:2007 (E). Medical laboratories – Particular requirements for quality and competence. Second Edition. 2007-4-15	No
Assay	Measurement procedure.	Clinical Laboratory Standards Institute EPO-05-A3, Vol. 34, No. 13. Evaluation of Precision of Quantitative Method Procedures; Approved Guideline–Third Edition. October 2014	Yes
Harmonization	Process of recognizing, understanding, and explaining differences while taking steps to achieve worldwide uniformity.	Clinical Laboratory Standards Institute EP6-A, Vol. 23 No. 16. Evaluation of the Linearity of Quantitative Measurement Procedures: A Statistical Approach; Approved Guideline, April 2003.	Yes
Imprecision	The random dispersion of a set of replicate measurements and/or values expressed quantitatively by a statistic, such as standard deviation or coefficient of variation.	Clinical Laboratory Standards Institute EPO-05-A3, Vol. 34, No. 13. Evaluation of Precision of Quantitative Method Procedures; Approved Guideline—Third Edition. October 2014	Yes
Incurred Sample Reanalysis	A repeated measurement of analyte concentration from study samples to demonstrate reproducibility.	FDA Guidance for Industry, Bioanalytical Method Validation, Draft Guidance, U.S. Department of Health and Human Services Food and Drug Administration, September 2013. http://www.fda.gov/ucm/groups/fdagov-public/@fdagov-drugs-gen/documents/document/ucm368107.pdf	No
Interference (Matrix, Drug)	Matrix effect: The direct or indirect alteration or interference in response due to the presence of unintended analytes (for analysis) or other interfering substances in the sample.	FDA Guidance for Industry, Bioanalytical Method Validation, Draft Guidance, U.S. Department of Health and Human Services Food and Drug Administration, September 2013. http://www.fda.gov/ucm/groups/fdagov-public/@fdagov-	Yes

(continued)

Table 2 (continued)

Validation Parameter/Term	Definition	Reference	Achievable with Cell-Based Assays?
		drugs-gen/documents/ document/ucm368107.pdf	
Linearity	The ability (within a given range) to provide results that are directly proportional to the concentration (amount) of the analyte in the test sample. Linearity typically refers to overall system response (i.e., the final analytical answer rather than the raw instrument output. The linearity of a system is measured by testing levels of an analyte which are known by formulation or known relative to each other (not necessarily known absolutely); when the system results are plotted against these values, the degree to which the plotted curve conforms to a straight line is a measure of system linearity.	Clinical Laboratory Standards Institute EP6-A, Vol. 23 No. 16. Evaluation of the Linearity of Quantitative Measurement Procedures: A Statistical Approach; Approved Guideline, April 2003.	Instrument Linearity: Yes Assay Linearity is possible but without the accuracy aspect
Measurand	Quantity intended to be measured.	Clinical Laboratory Standards Institute EPO-05-A3, Vol. 34, No. 13. Evaluation of Precision of Quantitative Method Procedures; Approved Guideline–Third Edition. October 2014	Yes
Measurement	Set of operations having the objective of determining a value of a quantity.	International Standard ISO 15189:2007 (E). Medical laboratories – Particular requirements for quality and competence. Second Edition. 2007-4-15	Yes
Post-examination procedures / post-analytical phase	Processes following the examination, including systematic review, formatting and interpretation, authorization for release, reporting and transmission of the results, and storage of samples of the examinations.	International Standard ISO 15189:2007 (E). Medical laboratories – Particular requirements for quality and competence. Second Edition. 2007-4-15	Yes

(continued)

Table 2 (continued)

Validation Parameter/Term	Definition	Reference	Achievable with Cell-Based Assays?
Precision (Measurement)	Closeness of agreement between indications of measured values obtained by replicate measurement of the same or similar objects under specified conditions.	Clinical Laboratory Standards Institute EPO-05-A3, Vol. 34, No. 13. Evaluation of Precision of Quantitative Method Procedures; Approved Guideline–Third Edition. October 2014	Yes
Pre-examination procedures / pre-analytical phase	Steps starting, in chronological order, from the clinician's request and including the examination requisition, preparation of the patient, collection of the primary sample, and transportation to and within the laboratory, and ending when the analytical examination procedure begins.	International Standard ISO 15189:2007 (E). Medical laboratories – Particular requirements for quality and competence. Second Edition. 2007-4-15	Yes
Range of Quantification	The range of concentrations, including ULOQ and LLOQ, that can be reliably and reproducibly quantified with accuracy and precision through the use of a concentration-response relationship.	FDA Guidance for Industry, Bioanalytical Method Validation, Draft Guidance, U.S. Department of Health and Human Services Food and Drug Administration, September 2013. http://www.fda.gov/ucm/groups/fdagov-public/@fdagov-drugs-gen/documents/document/ucm368107.pdf	No
Reference Intervals	The interval between and including two reference limits. It is designated as the interval of values from the lower reference limit to the upper reference limit. In some cases only one reference limit is important, usually an upper limit.	Clinical Laboratory Standards Institute EP28-A3C: Defining, Establishing, and Verifying Reference Intervals in the Clinical Laboratory; Approved Guideline—Third Edition	Yes
Sensitivity (Limits of Detection)	Measured quantity value, obtained by a given measurement procedure, for which the probability of falsely claiming the absence of a measurand in a material is β, given a probability α of falsely claiming its presence.	Clinical Laboratory Standards Institute EP17-A2: Evaluation of Detection Capability for Clinical Laboratory Measurement Procedures; Approved Guideline—Second Edition	Yes

(continued)

Table 2 (continued)

Validation Parameter/Term	Definition	Reference	Achievable with Cell-Based Assays?
Selectivity / Specificity	The ability of the bioanalytical method to measure and differentiate the analytes in the presence of components that may be expected to be present. These could include metabolites, impurities, degradants, or matrix components.	FDA Guidance for Industry, Bioanalytical Method Validation, Draft Guidance, U.S. Department of Health and Human Services Food and Drug Administration, September 2013. http://www.fda.gov/ucm/groups/fdagov-public/@fdagov-drugs-gen/documents/document/ucm368107.pdf	Yes
Stability	The chemical stability of an analyte in a given matrix under specific conditions for given time intervals.	FDA Guidance for Industry, Bioanalytical Method Validation, Draft Guidance, U.S. Department of Health and Human Services Food and Drug Administration, September 2013. http://www.fda.gov/ucm/groups/fdagov-public/@fdagov-drugs-gen/documents/document/ucm368107.pdf	Yes
Standard Calibrators	A biological matrix to which a known amount of analyte has been added. Calibration standards are used to construct calibration curves from which the concentrations of analytes in quality control samples and in unknown study samples are determined.	FDA Guidance for Industry, Bioanalytical Method Validation, Draft Guidance, U.S. Department of Health and Human Services Food and Drug Administration, September 2013. http://www.fda.gov/ucm/groups/fdagov-public/@fdagov-drugs-gen/documents/document/ucm368107.pdf	No
Reference Materials	Reference materials with or without assigned quantity values can be used for measurement precision control, whereas only reference materials with assigned quantity values can be used for calibration or measurement trueness control.	Clinical Laboratory Standards Institute EP17-A2: Evaluation of Detection Capability for Clinical Laboratory Measurement Procedures; Approved Guideline—Second Edition	No
Trueness	The closeness of the agreement between the average value obtained from a large series of results of measurements and a true value.	International Standard ISO 15189:2007 (E). Medical laboratories- Particular requirements for quality and competence. Second Edition. 2007-4-15	No

Typically during method validation, accuracy, interference (matrix, drug), linearity, normal signal distribution, precision (robustness), prozone effect, the range of quantification, reference ranges, selectivity, sensitivity (limits of detection), specificity, and stability are characterized (Table 2).

The validation of cell-based assays is quite different from the validation of assays measuring soluble molecules such as cytokine immunoassays [16–21]. Because of the nature of the type of data generated and the complexity of cellular measurements, some validation parameters cannot be evaluated in cell-based assays. Specificity, precision, sensitivity, reference ranges, and stability are the parameters that can be validated with cell-based assays.

Depending on the intended use of the data generated from the assay and the regulatory requirements associated with that use, it may not be necessary to validate all parameters; a fit-for-purpose validation approach should always be applied [16]. In a research setting or drug-discovery environment, at a minimum, specificity and precision should be validated (Fig. 1). If an assay will be used, even only as an exploratory end-point, in pre-clinical or clinical settings, additional validation is recommended. If, for example, specimens will be assayed more than four hours post-collection, stability must also be included in the initial validation. For rare-event analysis and dimly expressed markers, sensitivity must also be included in the initial validation. Note that in the translational space, the intended use of the data may change. In such cases, an iterative approach should be applied wherein additional validation parameters are evaluated as needed (Figs. 2 and 3). It must be

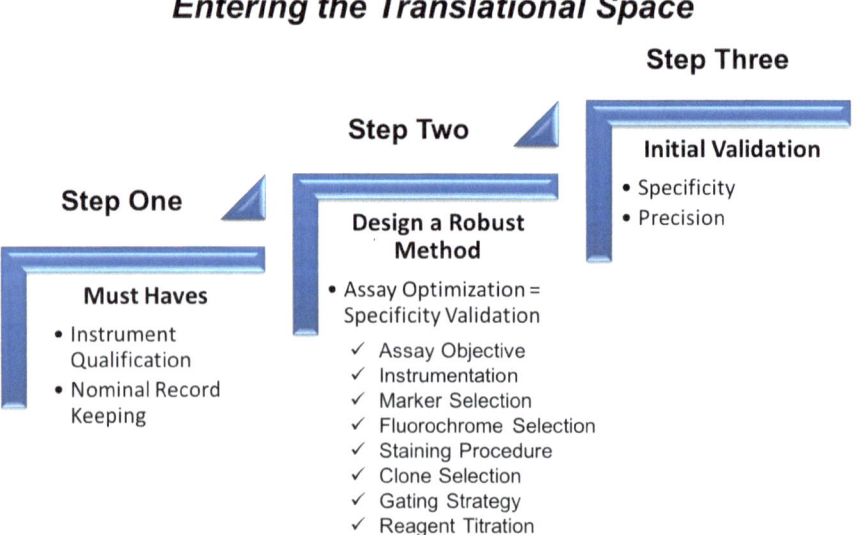

Fig. 1 Entering the translational space. Scientists working in the non-regulated space do not need to fully validate every assay. Requirements for entering the Translation Space are that high-quality cytometry is being conducted, meaning that instrumentation must be qualified and monitored and assays must be fully optimized according to current standards

The Iterative Validation Approach

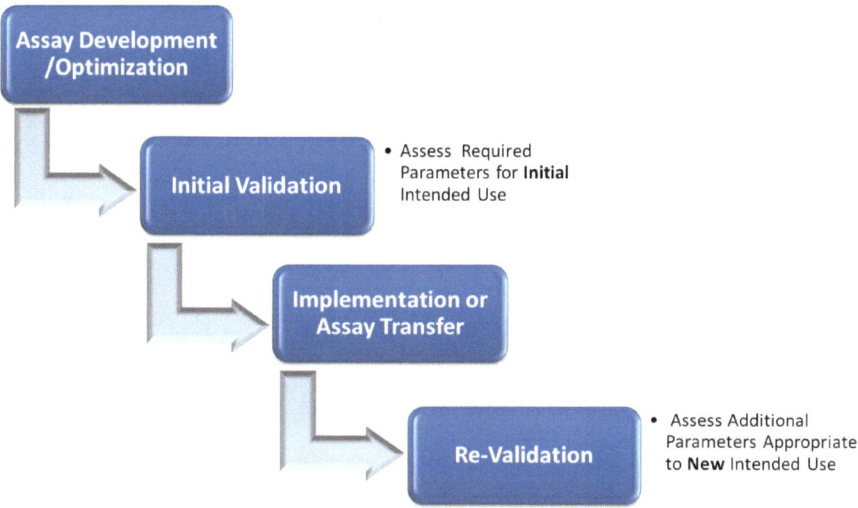

Fig. 2 **The iterative validation approach**. In the life cycle of a biomarker assay, the intended use of the data will change, as will the associated regulatory requirements. Additional validation parameters are evaluated as needed

emphasized that the iterative approach will be successful only if the initial assay optimization and characterization are robust.

Instrument qualification and monitoring must occur in every lab containing a flow cytometer. Erroneous results can easily be generated from an improperly installed and maintained instrument [22, 23]. Instruments must have a minimal level of characterization and qualification and must be maintained through a routine quality-control (QC) system. This is particularly important if antigen expression levels are included in the assay readout. If more than a single instrument will be used to perform the assay, a cross-instrument standardization method should also be implemented. Ensuring that the instruments are in optimal working condition and maintained over time is essential to delivering robust biomarker data.

(a) **Record Keeping**

In a laboratory operating under regulatory requirements, validation would follow a "Say It, Do It, Prove It" approach, meaning that a validation plan (Say It) would be prepared prior to conducting the validation experiments (Do It) and a final validation report (Prove It) would document the results and conclusions. The validation plan should include a detailed description of the method to be validated, including reagent sources and catalog numbers; for monoclonal antibodies, it is critical that the clone and fluorochrome conjugation be provided. In addition, the validation plan includes a detailed description of the samples that will be used in the validation, the number of replicates to be tested, and the number of analytical runs. Lastly, the validation plan includes the statistical analysis to be applied to the validation data and the acceptance criteria. The validation report then presents the

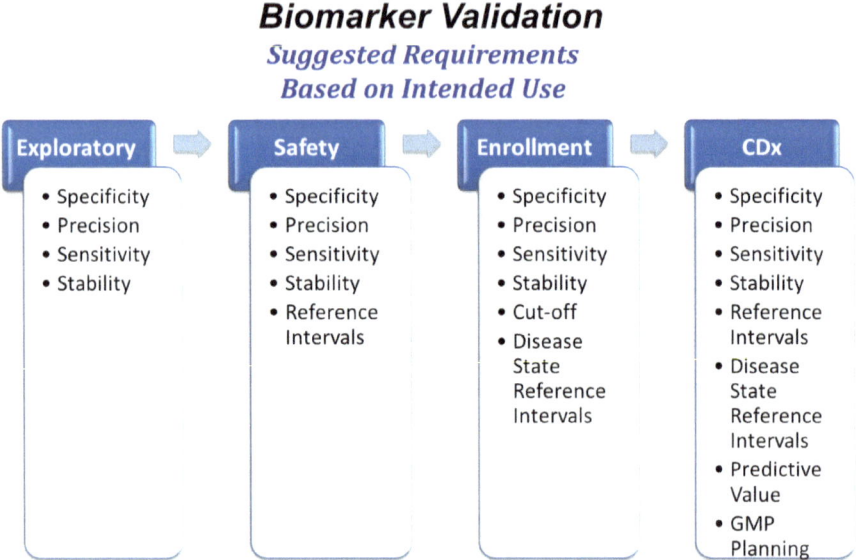

Fig. 3 Biomarker validation for intended use. Although the concept of fit-for-purpose validation is well accepted, there are no official guidance documents directly addressing which validation parameters are required for any given application of the data. Moreover, there are no official regulatory guidance documents regarding the validation of cytometric methods. This figure presents logical suggestions for validation requirements based on four scenarios of intended use of the data. These suggestions are the opinions of the authors and have not been endorsed by a regulatory agency. When biomarker data will be used for more than internal decision making, the best practice is to review the validation approach with the regulatory agency

data from the experimental phase as well as critical reagents (lot numbers and expiration dates), instruments (configuration and serial numbers), and software (version) used in the validation. The validation plan must answer the question "Is the assay performance acceptable for the intended use?" All data must be presented in the report; outliers may be excluded only when a recognized method for outlier evaluation has been applied to the data. Results that do not meet the acceptance criteria are discussed and next steps considered.

In a non-regulated environment, there are no requirements for formal documentation, but top-quality scientists typically do have a record-keeping system and maintain some type of laboratory notebook. Note that the International Society for the Advancement of Cytometry (ISAC)-sponsored publication of Optimized Multicolor Immunofluorescence Panels (OMIPs) requires that documentation of the assay optimization be submitted for the on-line component of the OMIP publication [24]. Many of the details required for the validation plan and validation reports, such as the results of monoclonal antibody (mAb) clone comparisons, fluorochrome pairings, and mAb titration, are mandatory aspects of the OMIP publications.

Record keeping is an essential aspect of generating translatable data. In particular, it is critical to keep track of reagent lot numbers and expiration dates. Lot-to-lot variability is a rare occurrence when reagents are purchased from reliable and established vendors, but it does happen. In the absence of lot-to-lot reagent qualification, a change in reagent could be misinterpreted as a real change in the biology. Furthermore, in the absence of precision validation, it is not possible to understand the extent of the variation introduced by the reagent change as compared to inherent assay variation.

(b) Assay Optimization and Specificity

Fortunately, top-quality scientists in basic research and non-regulated drug-discovery laboratories routinely address assay specificity validation, whether they are aware of this or not. Specificity is the ability of an assay to measure solely the intended measurand (Table 2). In a multicolor immunofluorescence panel, assay specificity is accomplished during panel design and optimization [24, 25]. A fully optimized cytometric assay will specifically measure the intended measurands (i.e., cellular population(s) of interest, intracellular antigens, nucleic acid) and not events present in a particular gate that are the result of compensation errors, conjugate degradation, or contamination by another cell type expressing shared antigens.

Best practices in multicolor panel design and optimization are beyond the scope of this chapter, but are well described elsewhere [24, 26, 27]. Briefly, for cytometric methods one must first consider the overall objective of the assay, the instrumentation, and the available reagents. Factors that influence assay specificity are the antigens used to define a particular cellular population of interest, the mAb clone, fluorochrome assignment, reagent titration, and the staining procedure. Furthermore, high-dimensional flow cytometry and mass cytometry enable increased assay specificity as they allow for the inclusion of markers for negative selection as well as multiple markers for positive selection that contribute to a more specific gating hierarchy.

(c) Data Category and Accuracy

The most challenging, and controversial, aspect of cell-based assay validation is the question of accuracy. In a landmark fit-for-purpose method validation paper, Jean Lee and colleagues from the American Association of Pharmaceutical Scientists (AAPS) discuss the four different categories of bioanalytical data (definitive quantitative, relative quantitative, quasi-quantitative, qualitative) (Table 3) [16]. They describe how the bioanalytical data category influences the method validation strategy as well as which validation parameters can be addressed for each data category.

Data generated in flow and mass cytometry could be categorized as either relative quantitative, quasi-quantitative, or qualitative. Most commonly, cytometry data fall into the quasi-quantitative bioanalytical data category, meaning that numerical results are generated that are reflective of the test sample but are not derived from a calibrator or reference material. For quasi-quantitative data, precision, sensitivity, specificity, and specimen stability can be validated, whereas

Table 3 Bioanalytical Data Categories

	Definitive Quantitative	Relative Quantitative	Quasi-Quantitative	Qualitative
Reference Standard	Well defined Representative of endogenous biomarker	Not well characterized Not fully representative of endogenous biomarker	None	None
Results	Continuous numeric units Generated from definitive standard curve	Continuous numeric units Generated from definitive standard curve	Continuous numeric units Characteristic of the test sample but not generated from a standard curve	Non-numeric results Categorical expressed in ordinal or nominal formats
Technology	Mass Spectrometry	Enzyme immunoassays Flow cytometry fluorescence intensity data expressed using fluorescence quantitation standards	Flow cytometry data expressed as relative percentage positive for a given subset or marker	Quantitative PCR Flow cytometry data expressed as phenotypic description
Biomarker Example	Insulin	Cytokines	Lymphocyte immunophenotyping	Single nucleotide polymorphism Leukemia/lymphoma characterization

Adapted from Lee et al., 2005

accuracy, dilutional linearity, and parallelism cannot be validated. An easily understandable illustration of quasi-quantitative data can be found when whole blood is stained for CD4 (Fig. 4). Three populations of cells are identified: CD4 negative cells, CD4 bright cells (CD4 T cells), and CD4 dim cells (monocytes). The bioanalytical results from this assay include the relative percentage of the total population of cells represented by each population and the fluorescence intensity of the CD4 expression on each of the three populations. Owing to the fact that there is no reference material or calibrator with a certified number of monocytes and CD4 T cells, accuracy cannot be established.

When fluorescence intensity units are calibrated by a variety of processes such as fluorescence quantitation beads, the resulting data would be considered relative quantitative [21, 28]. In this case, a calibrator is available in the form of the antigen-binding beads or beads with a pre-defined number of fluorescent molecules, but the calibrator is not representative of the test sample. In this case, only the fluorescence intensity results are relative quantitative, the other data remain quasi-quantitative.

Fig. 4 Quasi-quantitative data. Differential expression level of CD4 measured on human peripheral blood leukocytes. The side scatter versus CD4 bivariate representation of the data allows for the discrimination of three populations based on CD4 expression and cellular complexity: granulocytes (*dark grey*), monocytes (*red*), lymphocytes (*blue*). Data are considered quasi-quantitative given that numerical results are generated that are reflective of the test sample but not derived from a calibrator or reference material

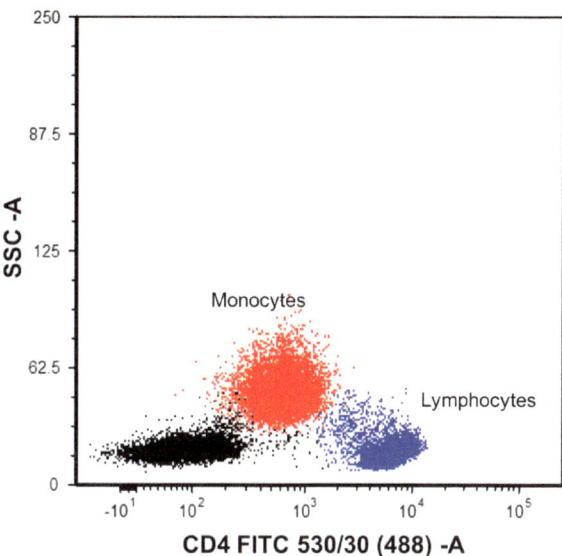

The classification of leukemia and lymphoma specimens based on immunophenotyping would be considered qualitative data.

(d) Precision

Precision is actually the most important and the easiest validation parameter to address. Understanding how close the results are when a same sample is tested repeatedly under the same conditions (intra-assay precision), is the foundation for understanding all other validation data. Intra-assay precision is as good as it gets and thus becomes the yardstick for assessing other factors that introduce variability and in determining the acceptance criteria for other validation parameters such as sensitivity and stability. It is critically important to establish assay acceptance criteria empirically for each and every result reported from the assay. Without an understanding of basic intra-assay precision, it is not possible to establish sample stability or evaluate lot-to-lot reagent differences.

Four to six samples are sufficient to evaluate intra-precision [18, 21, 29]. Ideally, each sample will have different levels of the cellular population of interest, or reportable result, but often it is not possible to obtain samples that fulfill this requirement [17, 30]. Each sample should be assayed (stained and acquired) in triplicate in a single analytical run. If the assay includes a fluorescence minus one (FMO) tube or similar gating control tube it is not mandatory to run that tube in triplicate; the value added of doing so is small but the expense and effort are large. The mean, SD, and %CV for each sample and each reportable result should be calculated. The overall intra-assay imprecision should be reported and the mean % CV (and range) achieved for all of the validation samples. The published acceptance criterion for cell-based assays is 10–30%CV [18, 21]. Obviously, the lower the better, but higher coefficients of variation are acceptable at the lower end of the reporting range, e.g., rare populations or dimly expressed antigens. The final intended use of the data must always be kept in mind when determining the acceptability of assay performance.

After intra-assay precision is established, other factors that contribute to variability can also be evaluated, such as inter-assay, inter-instrument, and inter-analyst variability.

(e) **Sensitivity**

Sensitivity is the reliability of measurements at the low end (frequency or fluorescence) and involves establishing the point at which the assay is no longer precise and for definitive quantitative and relative quantitative assays only, accurate. Depending on the expected results and the intended use of the data, it may not be necessary to evaluate assay sensitivity during the initial validation.

In flow cytometry, sensitivity relates to rare-event detection, which is becoming increasingly important in the clinical evaluation of cell-depleting immunotherapeutics and in monitoring minimal residual disease (MRD) in leukemia and lymphoma. Understanding and establishing the limits of fluorescence sensitivity can also be critical to the assay when modulation of the antigen is expected, such as measuring receptor occupancy of a dimly expressed antigen.

The main challenge in assessing sensitivity is to generate samples with low levels of the population of interest. Various approaches to creating samples for sensitivity have been described, and include depleting the target cell with immunomagnetic beads, admixing partially stained samples into fully stained samples, and for leukemia and lymphoma admixing disease-state samples into normal donor samples [21, 31]. Once the samples are created, sensitivity validation becomes a precision experiment. The acceptance criterion depends on the intended use of the data. Imprecision as high as 35% CV may be acceptable, or even higher if the expected changes in the population of interest will be great. Both the actual reportable result and the number of events in the gate should be considered when establishing the lower limit of quantitation (LLOQ) for the cell population of interest. A minimum of 20–50 events should be included the gate [32, 33].

(f) **Stability**

One aspect of assay performance and validation that is often overlooked is the impact of specimen and stained-sample stability. Antigen expression, cellular composition, and viability begin to change immediately after collection. Some cellular populations, such as T cells, are more stable than other populations, such as plasma cells. Stability assessment should be part of the experimental validation plan [31]. The design of the stability validation will depend on specimen type and collection procedure [31]. An acceptance criterion of less than a 20% difference, or 25% CV, between the baseline specimen value and the stored specimen value is the most commonly used [21, 31]. If samples are not assayed immediately after collection, specimen stability evaluation is mandatory to ensure that robust and translatable data are obtained.

5 Assay Implementation

As an assay travels in the translational space or along the drug-development pathway, the intended use of the data will change, as will the validation requirements. Challenges in implementing the assay under the new requirements will arise. For example, when designing a new cell-based method for use as a biomarker assay, it is critical to fully understand the clinical questions being addressed and what decisions the data will enable. This information must guide decisions regarding the matrix (whole blood, bone marrow, cerebrospinal fluid, bronchial alveolar lavage) selection, as well as the assay performance requirements for sensitivity and stability.

The inclusion of a cell-based biomarker assay in a clinical trial presents a myriad of new challenges, one of the most challenging being the logistics of flow and mass cytometry sample testing. The choices are either to run the samples at multiple local laboratories or to ship the samples to a central laboratory. Both options have advantages and disadvantages. When considering a limited specimen stability, using multiple local laboratories would appear to be the best option. On the other hand, when the high complexity of flow and mass cytometry is considered alongside the requirements for highly trained staff, then shipping the samples to a central laboratory would appear to be the best option. Additional concerns with local testing include the fact that the instrumentation may not be available at each investigative site; even if instrumentation is available, there is a high probability that each site will have different platforms and processes. As with other aspects of cell-based assay validation and implementation, there is no perfect solution. When choosing the option of a centralized laboratory for testing, the main limitations lie in addressing sample stability. Several solutions to extend sample stability exist, such as the use of cell-stabilization blood collection tubes or other partial processing at the investigative site.

6 Improving Translatability in Cytometry

The implementation of rigorous scientific procedures in the translational space will increase the reliability of data generated and increase the likelihood of accelerating the progression of scientific advances from the bench to the bedside.

Although the concept of analytical method validation is somewhat new to cytometry and not yet widely adapted, the concepts of quality and standardization are not. Publications on standardization in flow cytometry date back as early as 1983 [34]. Processes for daily instrument setup, cross-instrument standardization, and monitoring continue to improve [23, 35, 36]. In addition, Holden Macker, Phil McCoy, and Robert Nussenblatt have proposed a model for harmonizing flow cytometry in multisite clinical trials, and the EuroFlow Consortium has developed an extensive inter-laboratory harmonization program [37, 38].

As emphasized above, an important part of assay optimization (specificity validation) is the selection of the correct phenotype for the cell subset of interest.

While we will never have 100% consensus on the phenotype for every cell type, reaching consensus is an area of interest to several organizations. The Human Immunology Project (http://www.immuneprofiling.org/) has suggested optimized practices for the characterization of human immune-cell phenotypes. Several clinical groups, such as EuroFlow, the European Research Initiative in CLL (ERIC) [39–41], and the US–Canadian Consensus group, have recommended phenotyping for leukemia and lymphoma panels.

All this effort contributes to improving the quality of data and thereby the understanding of disease biology and the mechanism of action of novel therapies that will lead to more effective treatments of human disease.

References

1. Zerhouni EA (2005) Translational and clinical science–time for a new vision. N Engl J Med 353:1621–1623. doi:10.1056/NEJMsb053723
2. Reis SE, Berglund L, Bernard GR, Califf RM, Fitzgerald GA, Johnson PC, National C, and Translational Science Awards C (2010) Reengineering the national clinical and translational research enterprise: the strategic plan of the National Clinical and Translational Science Awards Consortium. Acad Med 85:463–469. doi:10.1097/ACM.0b013e3181ccc877
3. Dahlin JL, Inglese J, Walters MA (2015) Mitigating risk in academic preclinical drug discovery. Nat Rev Drug Discov 14:279–294. doi:10.1038/nrd4578
4. Gehr S, Garner CC (2016) Rescuing the lost in translation. Cell 165:765–770. doi:10.1016/j.cell.2016.04.043
5. Robinson JP, Rajwa B, Patsekin V, Davisson VJ (2012) Computational analysis of high-throughput flow cytometry data. Expert Opin Drug Discov 7:679–693. doi:10.1517/17460441.2012.693475
6. Sklar AL, Edwards SB (2011) HTS flow cytometry, small molecule discovery, and the NIH molecular libraries initiative, in flow cytometry in drug discover and development. In: Litwin V, Marder P (eds). Wiley, New Jersey
7. Yuan J, Hegde PS, Clynes R, Foukas PG, Harari A, Kleen TO, Kvistborg P, Maccalli C, Maecker HT, Page DB, Robins H, Song W, Stack EC, Wang E, Whiteside TL, Zhao Y, Zwierzina H, Butterfield LH, Fox BA (2016) Novel technologies and emerging biomarkers for personalized cancer immunotherapy. J Immunother Cancer 4:3. doi:10.1186/s40425-016-0107-3
8. Litwin V, Green C, Stewart JJ (2016) Receptor occupancy by flow cytometry. Cytometry B Clin Cytom 90:108–109. doi:10.1002/cyto.b.21364
9. Wehling M (2011) Drug development in the light of translational science: shine or shade? Drug Discov Today 16:1076–1083. doi:10.1016/j.drudis.2011.07.008
10. Begley CG, Ellis LM (2012) Drug development: raise standards for preclinical cancer research. Nature 483:531–533. doi:10.1038/483531a
11. Collins FS (2011) Reengineering translational science: the time is right. Sci Transl Med 3:90cm17. doi:10.1126/scitranslmed.3002747
12. Gergely P, Nuesslein-Hildesheim B, Guerini D, Brinkmann V, Traebert M, Bruns C, Pan S, Gray NS, Hinterding K, Cooke NG, Groenewegen A, Vitaliti A, Sing T, Luttringer O, Yang J, Gardin A, Wang N, Crumb WJ Jr, Saltzman M, Rosenberg M, Wallstrom E (2012) The selective sphingosine 1-phosphate receptor modulator BAF312 redirects lymphocyte distribution and has species-specific effects on heart rate. Br J Pharmacol 167:1035–1047. doi:10.1111/j.1476-5381.2012.02061.x

13. Kovarik JM, Stitah S, Slade A, Vitaliti A, Straube F, Grenet O, Winter S, Sfikas N, Seiberling M (2010) Sotrastaurin and tacrolimus coadministration: effects on pharmacokinetics and biomarker responses. J Clin Pharmacol 50:1260–1266. doi:10.1177/0091270009360534

14. Perl AE, Kasner MT, Shank D, Luger SM, Carroll M (2012) Single-cell pharmacodynamic monitoring of S6 ribosomal protein phosphorylation in AML blasts during a clinical trial combining the mTOR inhibitor sirolimus and intensive chemotherapy. Clin Cancer Res 18:1716–1725. doi:10.1158/1078-0432.CCR-11-2346

15. Pers JO, Devauchelle V, Daridon C, Bendaoud B, Le Berre R, Bordron A, Hutin P, Renaudineau Y, Dueymes M, Loisel S, Berthou C, Saraux A, Youinou P (2007) BAFF-modulated repopulation of B lymphocytes in the blood and salivary glands of rituximab-treated patients with Sjogren's syndrome. Arthritis Rheum 56:1464–1477. doi:10.1002/art.22603

16. Lee JW, Devanarayan V, Barrett YC, Weiner R, Allinson J, Fountain S, Keller S, Weinryb I, Green M, Duan L, Rogers JA, Millham R, O'Brien PJ, Sailstad J, Khan M, Ray C, Wagner JA (2006) Fit-for-purpose method development and validation for successful biomarker measurement. Pharm Res 23:312–328. doi:10.1007/s11095-005-9045-3

17. Litwin V, Green CL (2013) The role of biomarkers in clinical trials and the fit-for-purpose method validation approach. FDA Public Workshop-Clinical Flow Cytometry in Hematologic Malignancies, Silver Springs, MD, February 2013. http://www.fda.gov/MedicalDevices/NewsEvents/WorkshopsConferences/ucm334772.htm

18. O'Hara DM, Xu Y, Liang Z, Reddy MP, Wu DY, Litwin V (2011) Recommendations for the validation of flow cytometric testing during drug development: II assays. J Immunol Methods 363:120–134. doi:10.1016/j.jim.2010.09.036

19. Sommer U, Morales J, Groenewegen A, Muller A, Naab J, Woerly G, Kamphausen E, Marsot H, Bennett P, Kakkanaiah V, Vitaliti A (2015) Implementation of highly sophisticated flow cytometry assays in multicenter clinical studies: considerations and guidance. Bioanalysis 7:1299–1311. doi:10.4155/bio.15.61

20. Tanqri S, Vall H, Kaplan D, Hoffman B, Purvis N, Porwit A, Hunsberger B, Shankey TV, Group IIW (2013) Validation of cell-based fluorescence assays: practice guidelines from the ICSH and ICCS—part III—analytical issues. Cytometry B Clin Cytom 84:291–308. doi:10.1002/cyto.b.21106

21. Wood B, Jevremovic D, Bene MC, Yan M, Jacobs P, Litwin V, Group IIW (2013) Validation of cell-based fluorescence assays: practice guidelines from the ICSH and ICCS—part V—assay performance criteria. Cytometry B Clin Cytometry 84:315–23. doi:10.1002/cyto.b.21108

22. Green CL, Brown L, Stewart JJ, Xu Y, Litwin V, Mc Closkey TW (2011) Recommendations for the validation of flow cytometric testing during drug development: I instrumentation. J Immunol Methods 363:104–119. doi:10.1016/j.jim.2010.07.004

23. Perfetto SP, Ambrozak D, Nguyen R, Chattopadhyay PK, Roederer M (2012) Quality assurance for polychromatic flow cytometry using a suite of calibration beads. Nat Protoc 7:2067–2079. doi:10.1038/nprot.2012.126

24. Mahnke Y, Chattopadhyay P, Roederer M (2010) Publication of optimized multicolor immunofluorescence panels. Cytometry A 77:814–818. doi:10.1002/cyto.a.20916

25. Roederer M, Tarnok A (2010) OMIPs–Orchestrating multiplexity in polychromatic science. Cytometry A 77:811–812. doi:10.1002/cyto.a.20959

26. McLaughlin BE, Baumgarth N, Bigos M, Roederer M, De Rosa SC, Altman JD, Nixon DF, Ottinger J, Li J, Beckett L, Shacklett BL, Evans TG, Asmuth DM (2008) Nine-color flow cytometry for accurate measurement of T cell subsets and cytokine responses. Part II: Panel performance across different instrument platforms. Cytometry A 73:411–420. doi:10.1002/cyto.a.20556

27. McLaughlin BE, Baumgarth N, Bigos M, Roederer M, De Rosa SC, Altman JD, Nixon DF, Ottinger J, Oxford C, Evans TG, Asmuth DM (2008) Nine-color flow cytometry for accurate measurement of T cell subsets and cytokine responses. Part I: Panel design by an empiric approach. Cytometry A 73:400–410. doi:10.1002/cyto.a.20555

28. Hoffman RA, Wang L, Bigos M, Nolan JP (2012) NIST/ISAC standardization study: variability in assignment of intensity values to fluorescence standard beads and in cross calibration of standard beads to hard dyed beads. Cytometry A 81:785–796. doi:10.1002/cyto.a.22086
29. Davis BH, McLaren CE, Carcio AJ, Wong L, Hedley BD, Keeney M, Curtis A, Culp NB (2013) Determination of optimal replicate number for validation of imprecision using fluorescence cell-based assays: proposed practical method. Cytometry B Clin Cytometry 84:329–337. doi:10.1002/cyto.b.21116
30. Oldaker TA (2013) LDTs in flow cytometry: ICSH/ICCS guidelines for validation of fluorescent cell-based diagnostic testing. FDA Public Workshop-Clinical Flow Cytometry in Hematologic Malignancies, Silver Springs, MD, February 2013. http://www.fda.gov/MedicalDevices/NewsEvents/WorkshopsConferences/ucm334772.htm
31. Brown L, Green CL, Jones N, Stewart JJ, Fraser S, Howell K, Xu Y, Hill CG, Wiwi CA, White WI, O'Brien PJ, Litwin V (2015) Recommendations for the evaluation of specimen stability for flow cytometric testing during drug development. J Immunol Methods 418:1–8. doi:10.1016/j.jim.2015.01.008
32. Rawstron AC, Orfao A, Beksac M, Bezdickova L, Brooimans RA, Bumbea H, Dalva K, Fuhler G, Gratama J, Hose D, Kovarova L, Lioznov M, Mateo G, Morilla R, Mylin AK, Omede P, Pellat-Deceunynck C, Perez Andres M, Petrucci M, Ruggeri M, Rymkiewicz G, Schmitz A, Schreder M, Seynaeve C, Spacek M, de Tute RM, Van Valckenborgh E, Weston-Bell N, Owen RG, San Miguel JF, Sonneveld P, Johnsen HE, European Myeloma N (2008) Report of the European Myeloma Network on multiparametric flow cytometry in multiple myeloma and related disorders. Haematologica 93:431–438. doi:10.3324/haematol.11080
33. Roederer M (2008) How many events is enough? Are you positive? Cytometry A 73:384–385. doi:10.1002/cyto.a.20549
34. Jakobsen A (1983) The use of trout erythrocytes and human lymphocytes for standardization in flow cytometry. Cytometry 4:161–165. doi:10.1002/cyto.990040209
35. Hoffman RA (2001) Standardization and quantitation in flow cytometry. Methods Cell Biol 63:299–340
36. Maecker HT, Trotter J (2006) Flow cytometry controls, instrument setup, and the determination of positivity. Cytometry A 69:1037–1042. doi:10.1002/cyto.a.20333
37. Kalina T, Flores-Montero J, van der Velden VH, Martin-Ayuso M, Bottcher S, Ritgen M, Almeida J, Lhermitte L, Asnafi V, Mendonca A, de Tute R, Cullen M, Sedek L, Vidriales MB, Perez JJ, te Marvelde JG, Mejstrikova E, Hrusak O, Szczepanski T, van Dongen JJ, Orfao A, EuroFlow C (2012) EuroFlow standardization of flow cytometer instrument settings and immunophenotyping protocols. Leukemia 26:1986–2010. doi:10.1038/leu.2012.122
38. Maecker HT, McCoy JP, Nussenblatt R (2012) Standardizing immunophenotyping for the human immunology project. Nat Rev Immunol 12:191–200. doi:10.1038/nri3158
39. van Dongen JJ, Lhermitte L, Bottcher S, Almeida J, van der Velden VH, Flores-Montero J, Rawstron A, Asnafi V, Lecrevisse Q, Lucio P, Mejstrikova E, Szczepanski T, Kalina T, de Tute R, Bruggemann M, Sedek L, Cullen M, Langerak AW, Mendonca A, Macintyre E, Martin-Ayuso M, Hrusak O, Vidriales MB, Orfao A, EuroFlow C (2012) EuroFlow antibody panels for standardized n-dimensional flow cytometric immunophenotyping of normal, reactive and malignant leukocytes. Leukemia 26:1908–1975. doi:10.1038/leu.2012.120
40. Rawstron AC, Bottcher S, Letestu R, Villamor N, Fazi C, Kartsios H, de Tute RM, Shingles J, Ritgen M, Moreno C, Lin K, Pettitt AR, Kneba M, Montserrat E, Cymbalista F, Hallek M, Hillmen P, Ghia P, European Research Initiative in CLL (2013) Improving efficiency and sensitivity: European Research Initiative in CLL (ERIC) update on the international harmonised approach for flow cytometric residual disease monitoring in CLL. Leukemia 27:142–9. doi:10.1038/leu.2012.216
41. Stelzer GT, Marti G, Hurley A, McCoy P Jr, Lovett EJ, Schwartz A (1997) U.S.–Canadian consensus recommendations on the immunophenotypic analysis of hematologic neoplasia by flow cytometry: standardization and validation of laboratory procedures. Cytometry 30:214–230. doi:10.1002/(SICI)1097-0320(19971015)30:5<214:AID-CYTO2>3.0.CO;2-H

Flow Cytometry in Microbiology: The Reason and the Need

Cidália Pina-Vaz, Sofia Costa-de-Oliveira, Ana Silva-Dias,
Ana Pinto Silva, Rita Teixeira-Santos
and Acácio Gonçalves Rodrigues

Abstract The diagnosis of infection is based on methodologies that are, even in the 21st century, based on the study of the ability of microorganisms to grow in the presence of different substrates in the case of identification or in the presence of different antimicrobial drugs in the case of susceptibility evaluation. Despite the use of revolutionary techniques like molecular biology and more recently mass spectrometry, a lot of effort needs to be made to speed the results and to understand what happens to cells as individuals and not just as populations. Flow cytometry is an excellent tool still unexplored in microbiology; several applications are here described in the hope of contributing to the real use of flow cytometry in the clinical lab and to inspire future applications.

Keywords Flow cytometry · Identification of microorganisms · Susceptibility phenotype determination · Mechanisms of resistance · Drug monitoring · Microbial adhesion · Bacteria · Fungi · Parasites

C. Pina-Vaz (✉) · S. Costa-de-Oliveira · A. Silva-Dias · A.P. Silva · R. Teixeira-Santos · A.G. Rodrigues
Department of Microbiology, University of Porto and Faculty of Medicine,
Al. Hernâni Monteiro, 4200-319 Porto, Portugal
e-mail: cpinavaz@med.up.pt; cpinavaz@yahoo.com

S. Costa-de-Oliveira
e-mail: sqco@med.up.pt

A. Silva-Dias
e-mail: asilvadias@med.up.pt

A.P. Silva
e-mail: atsilva@med.up.pt

R. Teixeira-Santos
e-mail: ritadtsantos@med.up.pt

A.G. Rodrigues
e-mail: agr@med.up.pt

© Springer Nature Singapore Pte Ltd. 2017   153
J.P. Robinson and A. Cossarizza (eds.), *Single Cell Analysis*,
Series in BioEngineering, DOI 10.1007/978-981-10-4499-1_7

1 Introduction

Classic microbiological diagnosis is commonly based upon the study of microbial ability to use different substrates during the growth process for identification and on the measurement of growth inhibition in the presence of different antimicrobials for definition of the antimicrobial susceptibility phenotype. Such cornerstones are invariably time-consuming procedures since the microorganisms need time to replicate.

Developments in molecular biology and in immunology, namely monoclonal antibodies and more recently mass spectrometry, have provided new tools to microbiologists, particularly with respect to identification strategies, but not in the area of antimicrobial susceptibility profiles.

Flow cytometry (FC) has a strong potential to change the diagnostic paradigm in microbiology. Individual cells can be studied regarding their morphology or structure, or even more interesting, regarding functional aspects. Kinetic studies can be performed, evaluating cell events over time. The potential to sort different microbial cell populations represent a challenge almost completely unexplored. The ancestors of modern flow cytometers were apparatus used during World War II by the US Army in experiments for the detection of bacteria and spores [1]. Many technical improvements were achieved before publication of **Flow Cytometry in Microbiology** from which most microbiological cytometrists have learned [2]. In the late 1990s, the applications of FC in microbiology significantly increased [3–5], and **Practical Flow Cytometry** by Howard Shapiro became the "holy Bible" of the field [6]. Our research group has enthusiastically published research papers, protocols, and book chapters regarding applications of cytometry in microbiology; more recently U.S. patent WO 2012164547 A1, focusing on antimicrobial-susceptibility evaluation, was granted.

Several protocols and microbiological applications taking advantage of the potential of FC will be detailed, with the aim of inspiring the reader to devise additional applications and developments. This section is organized according to the main applications, namely identification, susceptibility to antimicrobials, mechanisms of action and resistance, assessment of drug combinations and post-antibiotic effects, and drug monitoring, as well as research applications. In addition, technical details will be also discussed in order to help future users.

2 Detection/Identification

The use of fluorochrome-labeled monoclonal antibodies targeting specific antigens is one of the most powerful approaches for microbial identification. The method is simple and fast; cells can be detected with high sensitivity and specificity. It could be performed directly in clinical products, avoiding the need for culture (saving time), or with isolates in pure cultures. The disadvantage is that it represents a "target method" that will find only what we are looking for. Conventional detection involves fluorescence microscopy, requiring an expert observer and time. Alternatively, with flow cytometry a larger sample can be analyzed in a few seconds, and the process automated after optimization of the protocol.

Several protocols have been published using fluorochrome-labeled antibodies for detection of *Pneumocystis jirovecii* [7], *Encephalitozoon intestinalis* [8], *Giardia lamblia* [9], and *Legionella pneumophila* [10]. Regarding *Giardia* and *Legionella*, it was possible not only to confirm presence but also to assess viability, which is of high relevance for environmental studies (Figs. 1 and 2). Only viable microorganisms are infectious for humans. Culture methods are the single approach to evaluate viability but take a long time, for example in the case of *Legionella*, or are impossible to perform in the case of unculturable microorganisms. As medical microbiologists we dedicated our efforts in particular to biological products such as

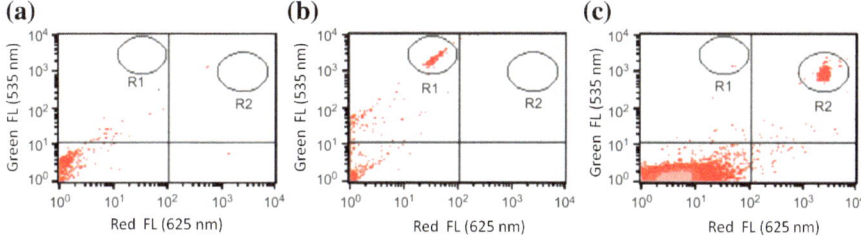

Fig. 1 A two-dimensional dot plot correlating green FL (*green* fluorescence—535 nm) with red FL (*red* fluorescence—625 nm) of: (**a**) *Giardia lamblia* cysts without staining (autofluorescence), (**b**) *G. lamblia* cysts stained with specific fluorescent antibody, and (**c**) *G. lamblia* cysts stained with specific antibody followed by propidium iodide (PI). The acquisition gate defined as R2 corresponds to *G. lamblia* cysts stained with both specific antibody and PI, and as R1 to *G. lamblia* cysts stained with specific antibody [9]

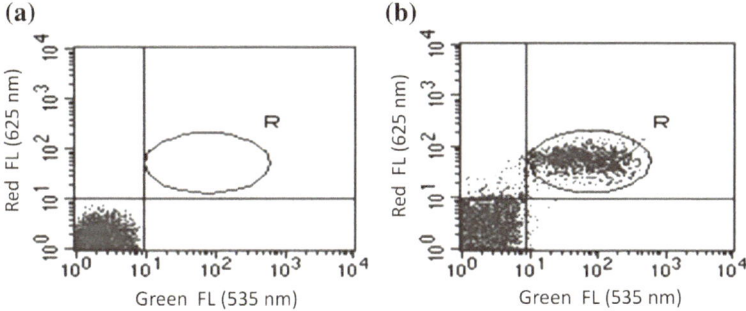

Fig. 2 Two-dimensional dot plot correlating FL1 (*green* fluorescence, 535 nm) and FL3 (*red* fluorescence, 620 nm): (**a**) Negative sample without *L. pneumophila* cells. (**b**) *L. pneumophila* ATCC 33152 cell suspension (10^4 bacteria/ml) labeled with 20 µl of specific monoclonal antibody MONOFLUO™ anti–*Legionella pneumophila* conjugated with fluorescein isothiocyanate (FITC) and 1 µg/ml propidium iodide (PI). Region R corresponds to the specific acquisition gate for labeled *L. pneumophila* cells [10]

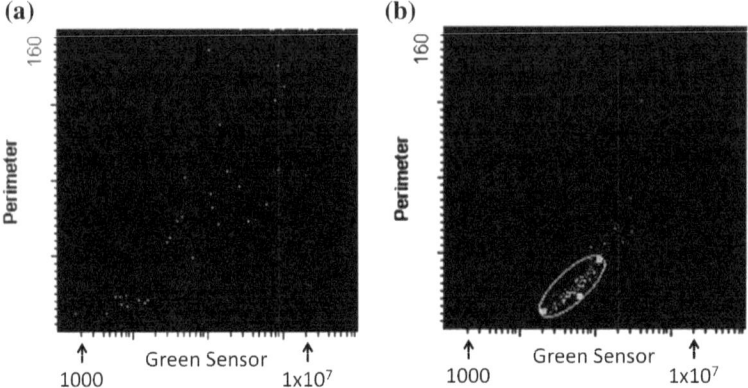

Fig. 3 Laser-scanned image of (**a**) negative and (**b**) positive smears stained by the PI-Auramine O method. The slide map shows perimeter on the *y* axis and *green* fluorescence on the *x* axis

stools and bronchial secretions; nevertheless, the protocols developed could have a special impact regarding water analysis.

In order to detect *Mycobacterium tuberculosis* in bronchial secretions we improved the standard auramine staining analyzed by florescence microscopy; instead, we proposed a protocol for a laser scanning cytometer [11]. Following the preparation of a smear it is possible to use an automated reading method, more sensitive than the traditional auramine staining (Fig. 3).

3   Antimicrobial Susceptibility Profile

According to the World Health Organization (WHO), "without urgent, coordinated action by many stakeholders, the world is headed for a post-antibiotic era, in which common infections and minor injuries which have been treatable for decades can once again kill." Effective antibiotics have been one of the pillars allowing longer and healthier lives. Unless we take significant actions to improve efforts to prevent infection and also change how we prescribe and use antibiotics, the global implications will be devastating. In fact, we are almost reaching such a post-antibiotic era, as we daily have to combat an increasing number of microorganisms known as "superbugs," microorganisms resistant to all drugs available for clinical use. Many professionals are accountable: clinicians who over-prescribe, pharmaceutical companies, who out of economic interests favor antimicrobial prescriptions, and farmers, who use and release massive amounts of antibacterial and antifungals in the environment. And what about microbiologists? If the susceptibility profile of a microorganism could be provided as fast as the value of hemoglobin or glucose, this would avoid the so common non-specific, blind antimicrobial prescription. Whenever urine or spinal-fluid samples arrive to the lab, only after culture and isolation of pure colonies (at least 24 h) is it possible to start antimicrobial-susceptibility profile evaluation. As these tests are traditionally based on study of the ability of the organism to grow in the presence of different antimicrobial drugs, they take considerable time. Standard conventional methods take at least 24 h and even the most recent automated methods take 8–24 h to provide a report. Whenever a blood culture is flagged as positive, a minimum of an additional 48 h is needed to obtain the antimicrobial susceptibility profile of the agent responsible for a very critical clinical situation in the patient. Thus, empiric therapy is most often initiated, frequently with more than a single antimicrobial, with a broad-spectrum profile; after the lab result is made available, de-escalation should be performed, but most often it is not. Knowing that increased resistance to an antimicrobial is directly related to its use, we should limit such use as much as possible. Only when we shorten the time to obtain results from susceptibility tests can we recommend a targeted and safer therapy sooner. Although several stewardship programs have been proposed, only with a quicker laboratory support could they be of real help in terms of antimicrobial guidelines to be implementated.

Molecular biology has the potential to provide information regarding some mechanisms of antimicrobial resistance; for example, when mecA and mecC genes are detected in a *Staphylococcus aureus* isolate, the microbiologist is aware that it is a methicillin-resistant strain (MRSA). However, as only a few resistance genes are yet known, a huge amount of prior knowledge would be needed in order to have a large genetic data base available. In addition, since resistance might be coded by non-genetic mechanisms, a molecular test will theoretically always lag behind a functional test. Remarkably, such a genetic test would say nothing about susceptibility. However, it has the potential to be used with colonies or directly in clinical samples, as, for example, for carrier studies like nasal or rectal swabs.

More recently a few mass spectrometry protocols have been described for detection of some mechanisms of antimicrobial resistance. Briefly, these have similar advantages (faster results) and disadvantages to molecular methods (the prior need to know all the profiles associated to resistance).

Our original proposal was the flow cytometric detection of cellular lesions produced by different antimicrobials, using the most adequate fluorescent probes. A flow cytometric antimicrobial susceptibility test (FAST) was optimized in order to provide the susceptibility phenotype of the most clinically relevant microorganisms to the recommended antimicrobial drugs. The optimization of the protocols involved assessment of the time of incubation, selection of the probe that more quickly detected distinct cell lesions, and selection of the cytometric settings. Cell morphology and intensity of fluorescence were recorded in order to develop an algorithm of analysis. Dedicated software was developed as well in order to provide an automated result with minimal technical interference.

Our initial studies involved yeasts of medical importance; at that time, only a few clinical labs performed susceptibility tests to antifungals, and all the existing methods had problems [12–15]. Fungi are eukaryotic cells and can be responsible

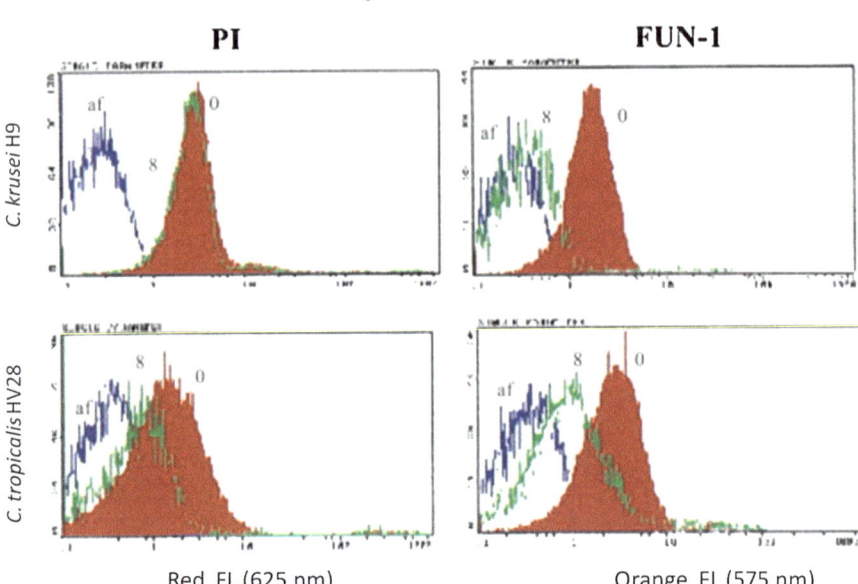

Fig. 4 Flow cytometry of strain *Candida krusei* H9 (an example of a strain with low MIC value to Am B) and *Candida tropicalis* HV28 (an example of a strain with high MIC value to Am B) after treatment with 0 and 8 mg/L of Am B (0 and 8 on the figure) for 1 h and stained with propidium iodide (PI) or FUN-1. The autofluorescence (af) of the cells is also represented [15]

for bloodstream infections, with high mortality. Three groups of antifungals belonging to different classes with different mechanisms of action were tested: azoles, echinocandins, and amphotericin B. Different probes such as propidium iodide (PI), $DiBAC_4(3)$, and FUN-1 were used to stain yeasts cells after treatment with different concentrations of the antifungals for different incubation times. Propidium iodide is a marker of cell death following severe lesion of the membrane. Nevertheless it was unable to stain yeasts exposed to fungicidal concentrations of amphotericin B (AmB); probably owing to its high size the AmB molecule makes the yeast cell impermeable to most stains, as is the case with PI and FUN-1 (Fig. 4), even after several hours of incubation [15]. With $DiBAC_4(3)$ it was possible to discriminate between susceptibility/resistance only after 60 min of incubation at 37 °C. Regarding the azoles (a fungistatic class of drugs interfering with synthesis of the cell membrane) and echinocandins (a class of drugs that act by impairing cell-wall synthesis), both could be studied using two fluorescent probes: FUN-1, a metabolic probe that is processed into red intracellular vacuoles by viable yeast cells, but displaying increasing yellow fluorescence when the cells are metabolically disturbed, although not yet dead, and $DiBAC_4(3)$, a membrane depolarization probe. Both probes are able to discriminate between susceptible and resistant strains after 60 min of exposure to azoles or echinocandins. A susceptibility kit is now ready for susceptibility evaluation of the most relevant antifungals in one hour versus 24–48 h required by other methods; it can be performed directly on colonies or on positive blood cultures.

Another microorganism requiring improvement in the speed of susceptibility profile evaluation is *Mycobacterium tuberculosis*; it usually takes at least 2 weeks to draw any conclusions about the susceptibility phenotype of a positive culture. We were confronted with two serious difficulties regarding this microorganism: *Mycobacterium* possesses a complex cell wall, almost impermeable to all drugs; because aerosols facilitate its propagation, we did not dare analyze a viable isolate in a flow cytometer. Our option was to incubate the isolate in MGIT vials (*Mycobacterium* growth indicator tube), with and without the main antituberculosis agents: streptomycin, isoniazid, rifampicin, and etambutol (SIRE) over a period of 3 days at 37 °C. Afterwards, MGIT tubes were autoclaved in order to kill *M. tuberculosis*; the nonspecific nucleic acid probe Syto 16 was added to count the number of cells and not the debris. An excellent correlation was found between the flow cytometric assay (time to result (TTR) 3 days) and the standard procedure (TTR 2 weeks) [16]. A 6–24 h antimicrobial assay was also published regarding nontuberculosis *Mycobacteria* strains [17] with considerable less virulence. Fluorescein diacetate (FDA), a viability indicator, was used to discriminate susceptible strains, which hydrolyzed less FDA than drug-free control.

Protocols addressing the most frequently isolated bacteria with regard to the most relevant medical antimicrobials were also developed by our team and made available as a diagnostic kit branded FASTinov®: a kit for Gram-negative and a kit for Gram-positive isolates. After 60 min of antimicrobial exposure, treated cells were compared with control cells (non-treated) (Fig. 5). Cut-off values were determined in order to discriminate susceptible from resistant phenotypes; excellent correlation with reference methods was found. In addition to application to isolates in pure culture, these kits can be used directly with positive blood cultures and with urine after using a microorganism extraction protocol.

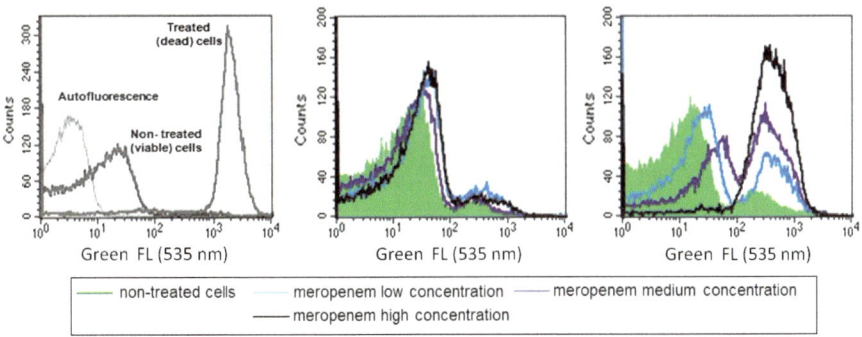

Fig. 5 *Escherichia coli* antimicrobial susceptibility to meropenem

With a TTR of 2 h, such kits provide important support to a targeted therapy and will allow the timely isolation of a patient with a "superbug" microorganism to avoid its spread in the hospital. At present, whenever a multiresistant bacterium is detected, the patient is admitted to the hospital; he is there for more than 2 or 3 days, often seen in more than one department, thus providing multiple opportunities to spread the organism.

Since in most instances we have used only breakpoint concentrations, which are the values used to classify the phenotype in susceptible (S), intermediate (I), or resistance (R), the kit is qualitative. However, for some drugs it might be important to determine the minimal inhibitory concentration (MIC), since this will help quantify the level of resistance and allow epidemiological surveillance. MIC determinations are also possible with flow cytometry, using a series of concentrations for each drug and not only the breakpoints. Flow cytometric protocols allowing MIC determinations have been developed for the most important drugs, such as vancomycin and more recently colistin.

One limitation for use of this kit in clinical labs might be its throughput (around 20–30 min are required for each strain). If we include more concentrations, as we do for MIC determinations, more time will be needed for analysis; this will represent a challenge for flow cytometer manufacturers. Another limitation of such tests is that like all other phenotypical tests, susceptibility profile assessment cannot be performed directly from mixed cultures.

4 Drug Associations

In clinical settings the combination of different antibacterial and/or antifungal drugs is quite common. However, these combinations are not routinely tested in vitro because that requires cumbersome procedures difficult both to perform and to interpret. Our team evaluated several drug combinations using flow cytometry, with promising results [18, 19]. The combination of antifungals required selection of a probe that could be used for the evaluation of both drugs; for example, $DiBAC_4(3)$ was used for the combination of echinocandin with an azole and for amphotericin B and an echinocandin (Fig. 6) [18]. As mentioned above, azole and echinocandin effects could be studied using FUN-1 or $DiBAC_4(3)$, although the amphotericin B effect was shown only with $DiBAC_4(3)$.

Synergic associations between compounds such as ibuprofen [20, 21] and local anesthetics [22] with antifungals were also studied through cytometric assays with great advantages. Regarding combinations of different antibacterial drugs, flow cytometry clearly demonstrated a synergic effect between carbapenems (such as ertapenem and imipenem or ertapenem and meropenem) acting on *Enterobacteriaceae* producing different types of carbapenemases [23]. Such observations were corroborated by classic susceptibility test results and by computational analysis (Fig. 7).

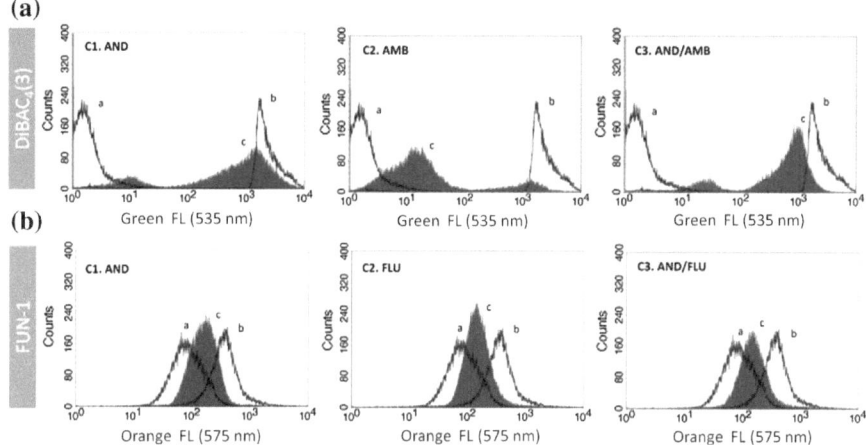

Fig. 6 Evaluation of antifungal combination effect using flow cytometry. (**a**) Flow cytometric analysis of the combined effect of anidulafungin and amphotericin B on *C. albicans*. Line a, fluorescence of untreated cells stained with $DiBAC_4(3)$; line b, fluorescence of cells treated with 70% ethanol and stained with $DiBAC_4(3)$; line c, fluorescence of cells treated with antifungal drugs and stained with $DiBAC_4(3)$; C1, cells treated with a subinhibitory concentration of anidulafungin (AND) (0.5 × MIC); C2, cells treated with a subinhibitory concentration of AMB (0.5 × MIC); and C3, cells treated with a subinhibitory concentrations of both antifungal drugs in association anidulafungin (AND 0.5 × MIC + AMB 0.5 × MIC). (**b**) Flow cytometric analysis of the combined effect of anidulafungin and fluconazole (FLU) on *C. albicans* OL196 strain. Line a, fluorescence of untreated cells stained with FUN-1; line b, fluorescence of cells treated with 70% ethanol and stained with FUN-1; line c, fluorescence of cells treated with antifungal drugs and stained with FUN-1; C1, cells treated with a subinhibitory concentration of AND (0.5 × MIC); C2, cells treated with a subinhibitory concentration of FLU (0.5 × MIC); and C3, cells treated with subinhibitory concentrations of both antifungal drugs in association (AND 0.5 × MIC + FLU 0.5 × MIC) [18]

5   Mechanisms of Drug Action

Very complex and diverse methodologies are used to study the mechanism of action of antimicrobial drugs. Flow cytometry was used by our team in order to understand how several compounds, not classical antimicrobial agents, exhibited antimicrobial activity.

5.1   Lesion of the Cytoplasm Membrane

In order to understand the antimicrobial mechanism of action of different compounds kinetic studies were performed. The microorganisms were incubated for varying lengths of time with different drug concentrations and with different

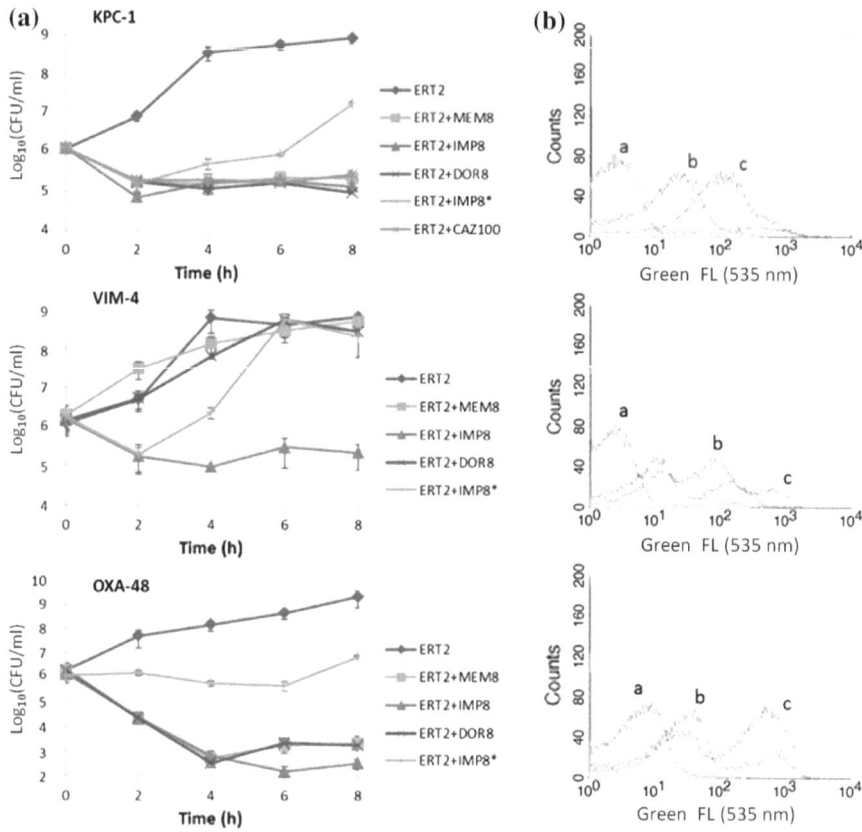

Fig. 7 a Time-kill curves of various antibiotic combinations against a strain of *Klebsiella pneumoniae* KPC-1, a strain of *Enterobacter cloacae* VIM-4, and a *K. pneumoniae* OXA-48. Each line represents ertapenem (ERT) either alone or associated with meropenem (MEM), imipinem (IMP), or doripenem (DOR) reinforced with the respective drugs each 2 h. Ertapenem plus imipenem (*) is shown as an example of the treatment without reinforcement. The association of ERT 2 mg/L with ceftazidime (CAZ) 100 mg/L against KPC-1 producing bacteria is also represented. **b** Overlay of histograms obtained by flow cytometry. For each strain, cells stained with DIBAC$_4$(3) are represented after 2 h treatment with ERT 2 mg/L (*a*), ERT 0.5 mg/L plus IMP 8 mg/L (*b*), and ERT 2 mg/L plus IMP 8 mg/L (*c*) [23]

fluorescent probes. When propidium iodide (PI) stained the cells after a short incubation time (5 min), it meant that the drug acted on the cell membrane, producing a fast and severe cell lesion. Ibuprofen at high concentrations [20], essential oils [24, 25], and some local anesthetics [22] produced early membrane lesions leading to cell death; all these studies were corroborated by other methodologies such as electron microscopy studies.

5.2   Efflux Pump Blockade

Resistance to antibacterials and antifungals often relates to efflux pumps, exporting drugs out of the cell; several modulators may block these pumps, reducing or eliminating such resistance.

Ibuprofen at low concentrations was initially shown to be synergic with azoles for an unknown reason [20]. FUN-1 showed to be a good marker of efflux in *Candida* and helped to understand the mechanism of action of ibuprofen. Following incubation of *Candida* strains having overexpression of efflux pumps with ibuprofen at sub-inhibitory concentrations, the increase of FUN-1 staining was parallel to the increase of the radioactive azole content. We concluded that ibuprofen at low concentrations blocks efflux pumps, increasing the amount of the azole inside the yeast cells; such efflux pumps are usually unspecific drug transporters also transporting FUN-1. Ibuprofen was described for the first time as an efflux pump blocker [21].

5.3   Decrease of Drug Input

Propofol is a sedative, often used in intensive care units, for patients receiving multiple antifungal drugs. An increase of MIC values for azole antifungals was observed when cells were concomitantly incubated with propofol at plasma concentrations. Flow cytometry showed that owing to the propofol lipidic vehicle the antifungals reduced access and/or permeabilization in the case of *Candida* and *Aspergillus* [26]. Fungal cells were stained with FUN-1 after treatment with sodium azide at low concentrations in order to block the efflux, and compared with cells treated with propofol; a non-stained population was evident, showing impermeability to the stain. Our hypothesis was that propofol would reduce the uptake of antifungals. Radioactive studies with (^{3}H) itraconazole demonstrated what flow cytometry analysis showed; reduced access and/or permeabilization of fungal cells to the antifungal was observed in the presence of propofol [26].

6   Mechanisms of Antimicrobial Resistance

The characterization of the strain phenotype in terms of resistance does not say anything about the underlying mechanisms of resistance. Special attention is devoted to this subject for several reasons: individual treatment; infection control and public health; the design of new therapeutic strategies. Antimicrobial resistance can be the result of different mechanisms: changes in or overexpression of the target; efflux pumps that extrude the drugs from the cells; enzymatic production, particularly extended-spectrum β-lactamases and carbapenemases. Genes on

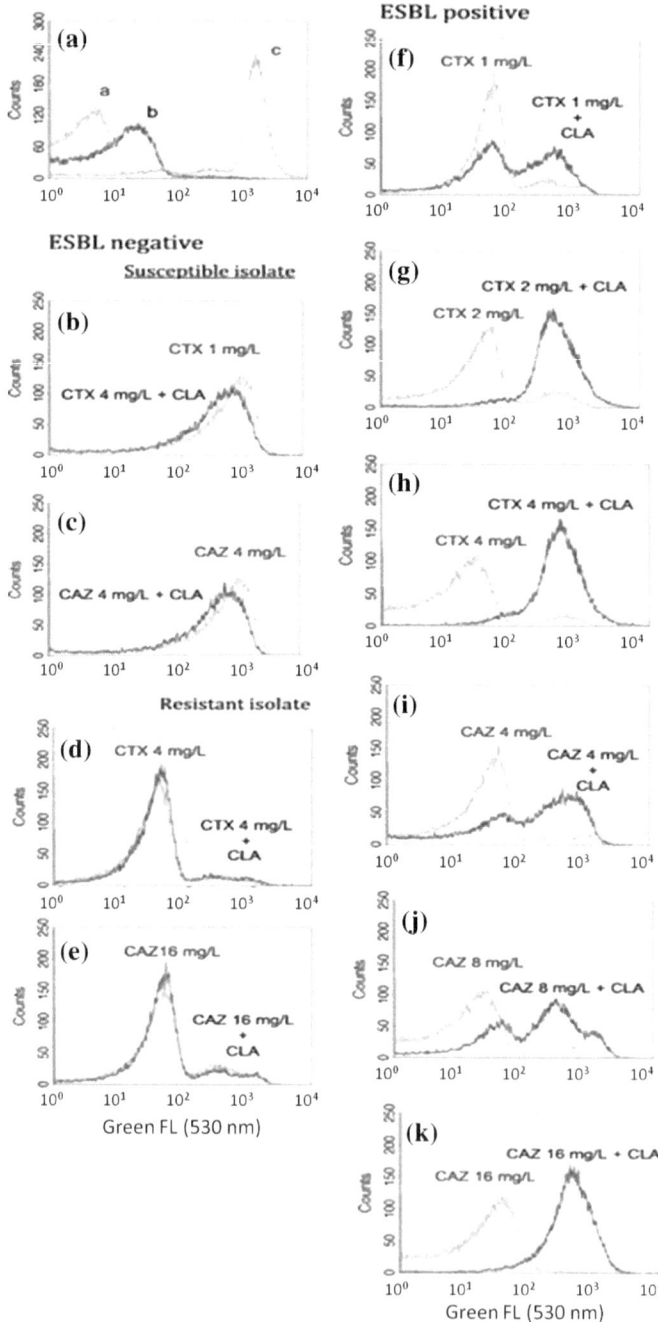

◄**Fig. 8** Flow cytometric histograms representing the emitted fluorescence at FL1 (*green* −530 nm) of the bacterial population. **a** Control − 10^6 bacterial cells per ml: (**a**) autofluorescence, **b** viable (non-treated) and **c** dead cells both stained with bis-(1,3-dibutylbarbituric acid) trimethine oxonol (DiBAC$_4$(3)). Typical example of extended-spectrum beta-lactamase (ESBL)-negative strains, susceptible (**b**, **d**) and resistant (**e**, **f**) to cefotaxime (CTX) and ceftazidime (CAZ), with and without clavulanic acid (CLA) at 4 mg/L. Typical example of ESBL-positive isolate after treatment with CTX at 1, 2, 4 mg/L (**f**, **g**, **h**), respectively, and CAZ at 4, 8, and 16 mg/L—(**i**, **j**, **k**), respectively, with and without CLA at 4 mg/L [28]

transposable elements located on plasmids, displaying the ability to spread quickly from bacterium to bacterium, from patient to patient, encode most such enzymes. The molecular characterization of these mechanisms could represent the ideal approach, considering the short TTR and the possibility of direct application in biological samples like nasal or rectal swabs. However, only very well character-ized genes/mechanisms could be searched; very often they are not known or there are several genes that could lead to the same resistance mechanism—for instance, several genes codify for β-lactamases—or they might be expressed in susceptible strains but overexpressed in resistant strains. Notably, as previously mentioned, molecular methods do not provide information about susceptibility, which is very relevant for patient treatment.

Classic phenotypic methods for characterization of resistance mechanisms are cumbersome and require a long TTR since they are based on the ability of microorganisms to grow, for example in the presence of different inhibitors [27].

Flow cytometry has the ability to provide susceptibility phenotypes and simul-taneously characterize some mechanisms of resistance, particularly enzymatic ones. Based on the most recent EUCAST guidelines for detection of resistance mecha-nisms, we have developed cytometric protocols that with a TTR of 2 h can detect an extended spectrum beta-lactamase (ESBL)- or carbapenemase-producing strain, helping determine therapeutic strategy and prevent further spread of infection.

In order to detect those mechanisms of resistance it is recommended to add different inhibitors (substrates of the distinct enzymes) to the antimicrobial drugs; whenever the phenotype reverts to a susceptible state we can conclude that those enzymes are present. Faria-Ramos et al. described a protocol for ESBL detection [28] and Silva et al. a protocol for carbapenemase detection [29]. Following one hour of incubation with and without different inhibitors we were able to determine the presence of an ESBL (Fig. 8) or a carbapenemase (Fig. 9); according to the synergic activity detected it is possible to classify the kind of carbapenemase present (KPC, metallocarbapenemase, or OXA-48 like).

Regarding antimicrobial heterorresistance, considered recently a putative cause of therapeutic failure, flow cytometry deserves to be explored, as it could be able to discriminate different cell populations much faster and better than other phenotypic methods like population analysis profiling. In fact, using flow cytometry analysis we were able to detect heteroresistance to vancomycin after exposing the cells to high concentrations of the drug for 4 h and redetermination MIC values. For

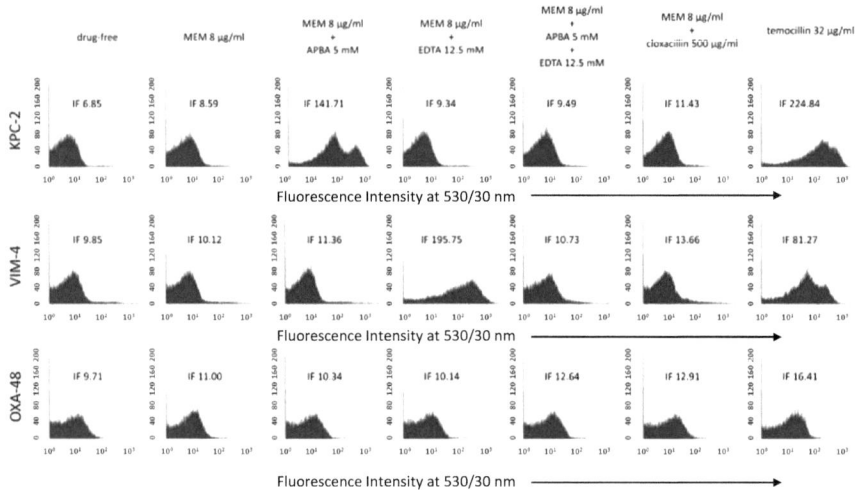

Fig. 9 Representative histograms obtained by flow cytometry for different types of carbapen-emases. For each strain (*Klebsiella pneumoniae* KPC-2, *Enterobacter cloacae* VIM-4, and *K. pneumoniae* OXA-48), cells not exposed to drugs, cells exposed to 8 μg/ml meropenem (MEM) and with the inhibitors 5 mM aminophenylboronic acid (APBA), 12.5 mM EDTA, 5 mM APBA plus 12.5 mM EDTA, or 500 μg/ml cloxacillin, and cells exposed to 32 μg/ml temocillin are shown. IF, intensity of fluorescence [29]

heteroresistant strains, the MIC value increases as the smaller population of more resistant cells increases [30].

Regarding yeasts, we were able to select a fluorescent probe, FUN-1, that can be used to detect resistance due to efflux; an excellent correlation between FUN-1 staining and the quantification of the expression of efflux pump–codifying genes was observed [21, 31]. Susceptible strains also harbor such genes, but with lower expression, resulting in less fluorescence after FUN-1 staining.

7 Post-antibiotic Effect and Drug Monitoring

The postantibiotic effect (PAE) corresponds to the period of bacterial recovery from the toxic effects of antimicrobial treatment. It depends on the bacteria, the type and concentration of antibiotic, and exposure time.

Compared to other methodologies, flow cytometry provides the advantage of facilitating the study of PAEs by analyzing simultaneously many characteristics of individual bacteria in a short time. The PAE of methicillin in *S. aureus* in terms of respiratory activity and membrane potential was studied by flow cytometry [32]; novel information not provided by colony forming unit (CFU) counts was obtained, as well as detection of several subpopulations.

The evaluation of patient response to antimicrobial treatment is usually based upon clinical findings. Unfortunately, in vitro results do not invariably correlate with clinical response since several often unpredictable factors may interfere and are very difficult to estimate. The immunological status of the patient, the possibility of selection of a subset of resistant population (heteroresistance, as referred above), or even the difficulty to achieve active and effective antibacterial levels at the infection site could lead to the lack of correlation between in vivo and in vitro results. In addition, antibacterial therapy regimens are standardized for general patients, but there are specific patients in whom drugs could have totally different pharmacokinetics (PK) and/or pharmacodynamics (PD), requiring adjustments of therapy. In critical-care patients, with several co-morbidities such as kidney and hepatic dysfunctions, increase of extravascular volume, antibacterial levels could vary between 1000 and 10,000-fold at the site of infection. Using flow cytometry, our team was able to quantify the antimicrobial drug level present at the site of infection (blood, urine, or cerebrospinal fluid). At present, only the plasma level of drugs such as vancomycin and aminoglycosides are quantified in patients by chemical determination. We are able to perform drug monitoring of all the main antimicrobial drugs for which we can evaluate the antimicrobial activity using flow cytometry. Such a test represents a functional evaluation using a microbiological approach. In theory, if the drug is present but chemically inactive, it will be detected when using chemical methods but not when using a microbiological method such as the flow cytometric assay hereby proposed.

8 Adhesion Studies and Other Research Applications

Microorganisms such as bacteria and fungi often adopt a community-based and sedentary lifestyle, usually binding to biotic or abiotic surfaces. There is no infection without microbial adhesion, often followed by biofilm formation, turning its detection and whenever possible, its prevention into a considerable challenge. After adhesion, microorganisms are much less exposed to host defense mechanisms and much more resistant to antimicrobial drugs. Several methodologies to study adhesion and biofilm formation have been developed, all cumbersome and very subjective. Our research team has developed a quantitative flow cytometric assay to measure yeast-cell adhesion [33]. The main concept involves the use of highly green-fluorescent carboxylated polystyrene microspheres and their incubation with unstained microspheres. After a short incubation time (30 min), the percentage of cells with yeast attached and the adhesion pattern were characterized by flow cytometry. We were able to distinguish non-adherent (non-fluorescent) from adherent (fluorescent) yeast cells. Many potential studies are now possible, since such beads could be coated with a wide range of molecules, like host constituents, therefore allowing measuring and comparing adhesion to distinct substrates. Yeast cells expressing specific adhesins encoded by the ALS gene family were studied. The microspheres were coated with fibronectin, gelatin [33], and other host proteins, and the results correlated quiet well with classic 48-h incubation methodology. Flow cytometric adhesion assays can provide broader information following shorter incubation times, with considerable less laboratory manipulation.

The quantification of chitin content in the fungal cell wall was another flow cytometric assay developed by our team. Chitin is a β-1,4 homopolymer of N-acetylglucosamine present in most fungi; together with β-1,3-glucan, it plays an important role in maintaining fungal cell integrity and rigidity. An important group of antifungals, the echinocandins, reduce the amount of glucan, which can lead to a compensatory increase of the synthesis of chitin. The amount of chitin could be, according to some authors [34], correlated with echinocandin resistance, but its determination is very laborious and time-consuming by current methods. Methods based on glucosamine release through acid hydrolysis are equally very laborious and time-consuming. Using calcofluor white, chitin content could be evaluated by flow cytometry [35]. No relationship between chitin content and echinocandin resistance was found, but data recently provided evidence that more than a single mechanism is needed to confer antifungal resistance, i.e. concomitance of point mutations. A novel, simple, and reliable method to estimate cell-wall chitin content based on flow cytometry technology is now available [35].

9 Technical Issues

As a principle, antimicrobial drugs and fluorochromes should be tested sequentially and not simultaneously as there could be interference between them. Before adding drugs we should perform experiments to exclude that possibility. It is important to note that according to the mechanism of action of a specific drug (bacteriostatic or fungistatic), we should incubate the drug before adding the fluorochrome, especially if it is a metabolic fluorescent probe.

Classically, in microbiological studies the unstained control (autofluorescence) is adjusted to the first logarithmic decade (depending on the flow cytometer dynamic range). Nevertheless, when it is important to see differences between different strains within the same species or even find a pattern between different species it is useful to employ other strategies. We can use microspheres with well-established fluorescence and size, and adjust the voltage such that the microsphere peak falls in a specific logarithmic decade, ideally 10^3 or 10^5 or 10^6, depending on the flow cytometer.

10 Concluding Remarks

New tools for microbial detection and more importantly for antimicrobial susceptibility assays are available based on flow cytometry, giving a quicker and much more informative susceptibility phenotype. By experience, we all know that a new method is seldom able to replace all those pre-existing. Owing to its great potential, flow cytometry deserves to be much further explored in clinical microbiology and in research. We agree with Michael Ormerod[1] *"the limitation is our imagination."*

References

1. Gucker Jr. FT, O'Konski CT et al (1947) A photoelectronic counter for colloidal particles. J Am Chem Soc 69(10):2422–2431
2. Lloyd D (1993) Flow cytometry in microbiology, 1st edn. Springer, London
3. Alvarez-Barrientos A et al (2000) Applications of flow cytometry to clinical microbiology. Clin Microbiol Rev 13(2):167–195
4. Betz JW, Aretz W, Hartel W (1984) Use of flow cytometry in industrial microbiology for strain improvement programs. Cytometry 5(2):145–150
5. Jansson JK, Prosser JI (1997) Quantification of the presence and activity of specific microorganisms in nature. Mol Biotechnol 7(2):103–120
6. Shapiro HM (1995) Practical flow cytometry, 3rd edn. Wiley-Liss
7. Barbosa J et al (2010) A new method for the detection of *Pneumocystis jirovecii* using flow cytometry. Eur J Clin Microbiol Infect Dis 29(9):1147–1152
8. Barbosa J, Rodrigues AG, Pina-Vaz C (2009) Cytometric approach for detection of *Encephalitozoon intestinalis*, an emergent agent. Clin Vaccine Immunol 16(7):1021–1024
9. Barbosa J et al (2008) Optimization of a flow cytometry protocol for detection and viability assessment of *Giardia lamblia*. Travel Med Infect Dis 6(4):234–239
10. Faria-Ramos I et al (2012) Detection of *Legionella pneumophila* on clinical samples and susceptibility assessment by flow cytometry. Eur J Clin Microbiol Infect Dis 31(12):3351–3357
11. Pina-Vaz C et al (2004) Novel method using a laser scanning cytometer for detection of *Mycobacteria* in clinical samples. J Clin Microbiol 42(2):906–908
12. Pina-Vaz C, Rodrigues AG (2010) Evaluation of antifungal susceptibility using flow cytometry. Methods Mol Biol 638:281–289
13. Pina-Vaz C et al (2005) Comparison of two probes for testing susceptibilities of pathogenic yeasts to voriconazole, itraconazole, and caspofungin by flow cytometry. J Clin Microbiol 43(9):4674–4679
14. Pina-Vaz C et al (2001) Susceptibility to fluconazole of *Candida* clinical isolates determined by FUN-1 staining with flow cytometry and epifluorescence microscopy. J Med Microbiol 50(4):375–382
15. Pina-Vaz C et al (2001) Cytometric approach for a rapid evaluation of susceptibility of *Candida* strains to antifungals. Clin Microbiol Infect 7(11):609–618
16. Pina-Vaz C, Costa-de-Oliveira S, Rodrigues AG (2005) Safe susceptibility testing of *Mycobacterium tuberculosis* by flow cytometry with the fluorescent nucleic acid stain SYTO 16. J Med Microbiol 54(Pt 1):77–81
17. Bownds SE et al (1996) Rapid susceptibility testing for nontuberculosis *Mycobacteria* using flow cytometry. J Clin Microbiol 34(6):1386–1390
18. Teixeira-Santos R et al (2012) Novel method for evaluating in vitro activity of anidulafungin in combination with amphotericin B or azoles. J Clin Microbiol 50(8):2748–2754
19. Teixeira-Santos R et al (2015) New insights regarding yeast survival following exposure to liposomal amphotericin B. Antimicrob Agents Chemother 59(10):6181–6187

---

[1]Personal Communication.

20. Pina-Vaz C et al (2000) Antifungal activity of ibuprofen alone and in combination with fluconazole against *Candida* species. J Med Microbiol 49(9):831–840
21. Pina-Vaz C et al (2005) Potent synergic effect between ibuprofen and azoles on *Candida* resulting from blockade of efflux pumps as determined by FUN-1 staining and flow cytometry. J Antimicrob Chemother 56(4):678–685
22. Pina-Vaz C et al (2000) Antifungal activity of local anesthetics against *Candida* species. Infect Dis Obstet Gynecol 8(3–4):124–137
23. Pina-Vaz C et al (2016) A flow cytometric and computational approaches to carbapenems affinity to the different types of carbapenemases. Front Microbiol 7:1259
24. Pina-Vaz C et al (2004) Antifungal activity of Thymus oils and their major compounds. J Eur Acad Dermatol Venereol 18(1):73–78
25. Pinto E et al (2006) Antifungal activity of the essential oil of Thymus pulegioides on *Candida*, *Aspergillus* and dermatophyte species. J Med Microbiol 55(Pt 10):1367–1373
26. Costa-de-Oliveira S et al (2008) Propofol lipidic infusion promotes resistance to antifungals by reducing drug input into the fungal cell. BMC Microbiol 8:9
27. Testing, E.C.O.A.S (2013) EUCAST guidelines for detection of resistance mechanisms and specific resistances of clinical and/or epidemiological importance (version 1.0)
28. Faria-Ramos I et al (2013) A novel flow cytometric assay for rapid detection of extended-spectrum beta-lactamases. Clin Microbiol Infect 19(1):E8–E15
29. Silva AP et al (2016) Rapid flow cytometry test for identification of different carbapenemases in Enterobacteriaceae. Antimicrob Agents Chemother 60(6):3824–3826
30. Pinto-Silva A, Da S, Teixeira-Santos R, Costa de-Oliveira S, Rodrigues AG, Pina-Vaz C (2016) Determination of vancomycin susceptibility for *Staphylococcus aureus* by flow cytometry. ASM Microbe. Boston, USA
31. Ricardo E et al (2009) Ibuprofen reverts antifungal resistance on *Candida* albicans showing overexpression of CDR genes. FEMS Yeast Res 9(4):618–625
32. Suller MT, Lloyd D (1998) Flow cytometric assessment of the postantibiotic effect of methicillin on *Staphylococcus aureus*. Antimicrob Agents Chemother 42(5):1195–1199
33. Silva-Dias A et al (2012) A novel flow cytometric protocol for assessment of yeast cell adhesion. Cytometry A 81(3):265–270
34. Munro CA (2013) Chitin and glucan, the yin and yang of the fungal cell wall, implications for antifungal drug discovery and therapy. Adv Appl Microbiol 83:145–172
35. Costa-de-Oliveira S et al (2013) Determination of chitin content in fungal cell wall: an alternative flow cytometric method. Cytometry A 83(3):324–328

Flow Cytometer Performance Characterization, Standardization, and Control

Lili Wang and Robert A. Hoffman

Abstract Flow cytometry is a widely used technique for the analysis of single cells and particles. It is an essential tool for immunological research, drug and device development, clinical trials, disease diagnosis, and therapy monitoring. However, measurements made on different instrument platforms are often inconsistent, leading to variable results for the same sample on different instruments and impeding advances in biomedical research. This chapter describes methodologies to obtain key parameters for characterizing flow cytometer performance, including precision, sensitivity, background, electronic noise, and linearity. Further, various fluorescent beads, hard dyed and surface labeled, are illustrated for use in quality control, calibration, and standardization of flow cytometers. To compare instrument characteristics, fluorescence intensity units have to be standardized to mean equivalent soluble fluorochrome (MESF) or equivalent reference fluorophore (ERF) units that are traceable to the existing primary fluorophore solution standards. With suitable biological controls or orthogonal method, users will be able to quantitatively measure DNA and RNA content per cell or biomarker expression in antibodies bound per cell. Comparable, reproducible, and quantitative measurements using flow cytometers can be accomplished only upon instrument standardization through performance characterization and calibration, and use of proper biological controls.

Keywords Standard · Calibrate · Quality control · Fluorescence · Sensitivity · Linearity · Flow cytometer · MESF · ERF · ABC

L. Wang (✉)
Biosystems and Biomaterials Division, National Institute of Standards and Technology, 100 Bureau Drive, Galthersburg, MD 20899, USA
e-mail: llli.wang@nist.gov

R.A. Hoffman
Livermore, CA, USA

© Springer Nature Singapore Pte Ltd. 2017
J.P. Robinson and A. Cossarizza (eds.), *Single Cell Analysis*,
Series in BioEngineering, DOI 10.1007/978-981-10-4499-1_8

1 Introduction

1.1 Why Is It Important

This chapter covers three distinct but related activities that insure that the results from a flow cytometer will be as comparable, reliable, and accurate as possible. No two flow cytometers are exactly alike. Every instrument and instrument subsystem is made to within defined specifications, but every specification has a tolerance. Flow cytometers in current use have been developed over a period of more than 20 years. Differences among instruments are greater if they are different models or if they are made with newly available technology rather than with older technology from past decades. There are significant differences among instrument models in the linearity over their multi-decade measurement range. This affects spectral compensation accuracy, the ability to resolve dimly fluorescent particles, and the ability to resolve particle populations (especially submicron ones) by light scatter.

It is helpful and often necessary to standardize the settings on a flow cytometer by adjusting the detector gains to place signals from stable particles at specified levels. This allows results from an application to be compared to previous results on that instrument. In many cases the same particles can be used for quality control of some aspects of the instrument performance. The detector gain or PMT voltage that must be used to reach required signal levels as displayed in a dot plot or histogram can be recorded daily to monitor drifts or sudden changes that alert the user it is time to troubleshoot a problem. The CV of a bead population measured on a detector channel can show whether the sample stream is adequately aligned to the laser beams and detection optics. With the proper stable particles, it is possible to standardize groups of instruments so that populations from a biological sample would be displayed in the same locations on histograms or dot plots from all instruments in the group. But having data displayed the same on all instruments in a group does not insure that the results from each instrument would be the same. Dim populations may be resolved on some instruments but not on others. Submicron particles may be detected above background on some instruments but not on others. Compensated fluorescence plots can have artefactual positive or negative populations if the signals from the electronics are not in an adequately linear range. Information about the key performance characteristics of the instrument will help to interpret results as being truly biologically meaningful within a performance limit of the instrument.

This chapter builds on previous work and publications on standardization and flow cytometer performance characterization [1–3]. The critical issues that should be considered when using beads to standardize, calibrate, and control are discussed along with fluorescence intensity units and methods used to assign intensity units to beads. Practical approaches for characterizing instrument performance are discussed, with examples for linearity and the factors determining fluorescence sensitivity. Standards for DNA and RNA measurements are reviewed. Different approaches to convert measured fluorescence intensity to antibodies bound per cell (ABC) are described. The chapter concludes with some thoughts on the future.

1.2 What Do Instrument Manufacturers Provide?

Most instrument manufacturers provide beads and instructions or automated software to set up instrument gains for typical applications. The beads can be used to monitor instrument performance—particularly optical alignment and for regular checks to determine whether the instrument response has stayed within an acceptable range and to alert the user when performance has changed so much that troubleshooting or service is required. A few manufacturers, e.g., BD Biosciences CS&T System, provide additional characterization of instrument performance, including measuring the range of linear response, electronic noise level, optical background noise, and detection response. If the manufacturer also sells clinical applications, there will be specific application setup conditions—sometimes with application-specific beads and software.

However, instrument manufacturers cannot anticipate every application that users will develop or every experimental condition that will be tried. So it is a good idea to know what alternatives are available for setting up instruments and evaluating and characterizing performance. This will be particularly important when instruments in a laboratory or group study are from multiple manufacturers or consist of several different models. Materials and methods used to get consistent measurement scales over a variety of different instruments may require creating an alternative set of beads and setup procedures not available from any of the instrument manufacturers.

2 Beads as Standards

2.1 Bead Characteristics

Most beads (also called microparticles) used for standardization and applications are made from polymers. Some specialty beads are made of silica and have an optical refractive index closer to that of cells. In either case, beads are available in a wide size range covering submicron to tens of microns. There are two basic approaches for making fluorescent beads. The first approach embeds fluorescent molecules within the bead, which keeps the fluorophore from contact with the suspension buffer and greatly improves the stability of the fluorophore. These beads are often referred to as "hard dyed" and have the advantage of long shelf life without loss of fluorescence. The disadvantage of hard-dyed beads is that the fluorophores used to stain cells are water soluble and not generally compatible with the hard-dye manufacturing process. As a result, the spectral response of hard-dyed beads almost never matches well with that of fluorescently stained cells. In addition, fluorescence from hard-dyed beads and fluorophores on cells can behave differently (photobleaching, emission saturation, etc.) with respect to excitation intensity. Since the spectral responses of flow cytometers vary to some extent even among the same model, hard-dyed beads cannot be used to set up all instruments to respond exactly the same when stained biological samples are analyzed.

A second type of fluorescent bead is stained on the surface with the actual fluorophore used to stain cells. The fluorophore is in essentially the same environment as in or on a cell. In particular, fluorescent beads used to best standardize instruments for immunofluorescence are surface-stained. Unfortunately, surface-stained beads are less stable over time and can be more expensive to make. The most stable surface-stained beads are freeze-dried, which adds to the expense. So flow cytometrists need to be aware of when it is appropriate to use surface-stained, fluorophore-specific beads and when the use of hard-dyed beads will be adequate. This decision will be determined by the application and the degree to which the individual instrument needs to compare to other instruments. Figure 1 shows emission spectra from a commonly used hard-dyed bead and spectra from two common fluorophores used for immunofluorescence. It is clear from comparing the spectra that using filters with different pass bands for FITC or PE will change the relative amount of fluorescence detected from beads and the fluorophores.

To have some objective criteria for deciding when hard-dyed beads are appropriate fluorescent standards, a study was conducted on 133 instruments among 28 laboratories and instrument manufacturers [4]. Ten different instrument models were included in the study. Each instrument was first set up with stable, freeze-dried surface-stained beads, and then a variety of hard-dyed beads were analyzed at the same settings. The ratio of mean fluorescence of the hard-dyed bead to the surface-stained beads was then compared for all instruments in the study. If the hard-dyed beads gave the same fluorescence scale as the surface-stained beads (and the same mean fluorescence for stained cells), there would be no variation of this

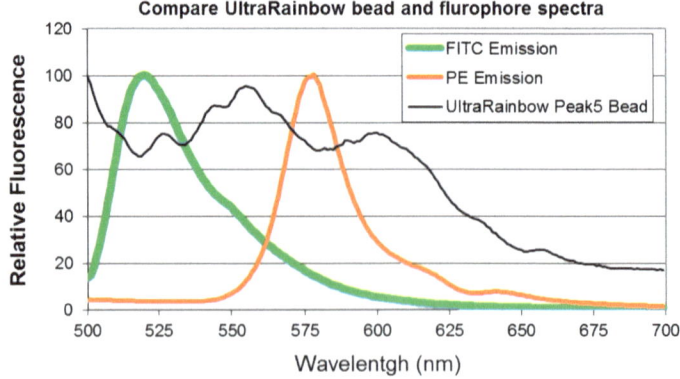

Fig. 1 Emission spectra of Spherotech Ultra Rainbow beads, FITC and PE fluorophores

ratio among different instruments. The results of the study showed just the opposite: there was considerable variation on the fluorescence scales with all the hard-dyed beads. Figure 2 shows results of the study for the PE channel.

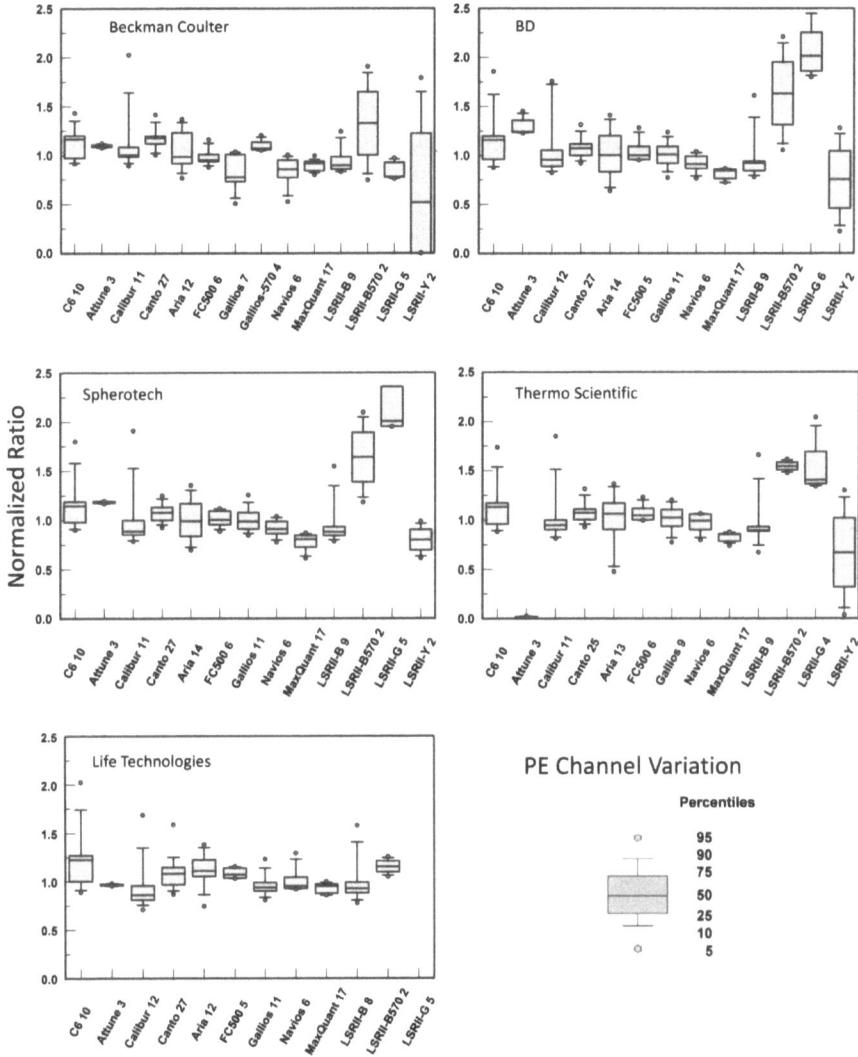

Fig. 2 Box-and-whisker plots of the normalized ratio of the MFI of the indicated hard-dyed beads to the MFI of the PE-stained fluorophore-specific standard bead for 10 different flow cytometer models. The box shows the 25–75th percentiles, and the line in the box indicates the median value. Horizontal bars outside the box indicate 10 and 90th percentiles and the circles indicate 5th and 95th percentiles. The percentile markers indicate the percentage of instruments for which the cross-calibration was within the indicated normalized range. The number of instruments represented for each instrument model is noted after the model name on the X-axis of each plot (this figure is from reference [4] Cytometry Part A, 81A, 785)

Typical variation using hard-dyed bead fluorescence standardization even among the same instrument model was 20% or more using robust standard deviation as a measure. But the hard-dyed calibration range of 90% of the instruments varied by factors of 1.5–2 or more. If an assay, such as some clinical assays, requires a mean fluorescence to be measured within 10% accuracy, none of the hard-dyed beads would be suitable calibrators. Indeed, clinical assays that require mean fluorescence measurements with 10% accuracy use fluorophore-specific surface-stained beads for calibration. Hard-dyed beads can be a good standard to set up the fluorescence scale and verify linearity and dynamic range of the instrument among a group of study instruments. If a factor-of-2 variation in the mean fluorescence from cells can be tolerated, hard-dyed beads can be used as a standard.

2.2   Fluorescence Intensity Units Used in Flow Cytometry

MESF stands for molecules of equivalent soluble fluorochrome, and ERF refers to equivalent number of reference fluorophore. In both cases, the assigned number is the equivalent number of fluorophore molecules in solution that produce the same fluorescence intensity as the bead. MESF assignments use solutions of the same fluorophores used to label antibodies. MESF assignments are in units of fluorescein, PE, APC, etc. In the case of ERF unit assignment, however, the fluorophore reference solution may not be one that is used for antibody labeling. The only requirement for an ERF reference solution is that it can be excited with the same excitation wavelength and fluoresce in the wavelength range overlapping significantly with the fluorochrome associated with beads. For example, a calibration bead stained with PE can have ERF assignments in units of Nile Red. The additional requirement for an ERF assignment is that the excitation wavelength and emission wavelength range must also be specified. A complete ERF assignment for a bead labeled with PE, for example, could be equivalent to 45,000 molecules per bead of Nile Red excited with 488 nm and in the emission range 560–590 nm. In essence, MESF is a special case of ERF; both are a measure of particle fluorescence that is equivalent to the fluorescence signal from a known number of reference fluorophores in solution. The advantage to the ERF unit is that a small number of reference fluorophores can provide assignments to an unlimited number of different fluorophores used to tag antibodies, including fluorophores developed in the future. And it is practical for an authoritative body such as NIST to provide those few fluorophores as traceable Standard Reference Material. It would not be practical for such a body to provide Standard Reference Material fluorophores for all the different fluorophores used as antibody conjugates.

There are a few fluorescence intensity units defined and used by bead manufacturers for quality control of their beads. A unit of fluorescence specific for BD Biosciences is the assigned BD unit (ABD). A fluorescence intensity unit was needed for the cytometer setup and tracking (CS&T) system developed for instrument performance characterization and QC. Intensity values in ABD units were assigned to many more detection channels than had calibrators available. In essence the ABD values for CS&T beads are tied to a gold-standard bead lot to which the initial ABD values were assigned through correlation with (not calibration to) human lymphocytes stained with CD4 conjugates tagged with a wide variety of fluorophores. Like the ERF intensity unit, the ABD unit for a particular detection channel is defined with a specific laser excitation wavelength and emission filter (emission spectral range). A fluorescent bead that is calibrated in ABD can be cross-calibrated to ERF units.

2.3 Bead Fluorescence Assignments Vary Among Manufacturers

Although the basic approach to assigning MESF or ERF values to beads is followed by all bead manufacturers, there seem to be differences in detail that produce differences in the assigned values. A simple comparison of commercially available beads with assigned MESF values was performed by one of the authors (RAH). With no change in the flow cytometer, calibration beads for FITC and PE from several manufacturers were run. Using the MESF values assigned by the manufacturers, the FITC and PE channels were calibrated in MESF per channel. Results are shown in Table 1. In this small sample, it appears the ratio values of MESF and MFI for FITC calibration beads are consistent within either bead type, surface-labeled beads or hard-dyed beads. There is a factor of five difference between the hard dyed and surface labeled FITC standards. However, there are large discrepancies in the ratio values for PE beads. The variations in the ratios of PE beads might likely be due to the absence of a common PE primary solution standard for bead manufacturers performing the fluorescence intensity value assignment.

Table 1 Different beads with assigned MESF values give varying fluorescence calibration in MESF/MFI

	Bead product	Bead type	MESF/MFI
FITC standards			
	1	Surface labeled	14.68
	2	Surface labeled	13.19
	3	Hard dyed	2.87
	4	Hard dyed	2.80
PE standards			
	1	Surface labeled	2.97
	2	Surface labeled	0.85
	3	Surface labeled	0.70
	4	Hard dyed	1.21
	5	Hard dyed	0.86

Table 2 ERF values assigned to the four surface-labeled microsphere reference standards by four manufacturers in addition to NIST

ERF$_{major}$					
Microsphere	NIST	Vendor A	Vendor B	Vendor C	Vendor D
FITC	7.74×10^4	3.08×10^4	2.19×10^7	1.33×10^7	3.11×10^5
PE	7.94×10^5	5.01×10^4	1.89×10^{10}	1.81×10^7	1.58×10^6
APC	3.21×10^4	6.12×10^3	1.93×10^8	3.62×10^7	not done*
PB	1.59×10^6	3.36×10^4	4.12×10^9	8.00×10^6	7.12×10^6

Table reproduced from Hoffman et al. [4]

To evaluate what variation might occur when different bead manufacturers assign MESF or ERF values to beads, four manufacturers and NIST used the same surface-labeled beads, reference fluorophore solutions, and protocol to assign ERF values using their own equipment and personnel. Results from this study (4) are shown in Table 2.

The study showed large differences among the different manufacturers and compared to NIST, which was considered as the reference laboratory. Partly owing to this result, NIST and ISAC organized a series of workshops that culminated in an agreement to establish an ERF assignment service at NIST available to members of a consortium described in the next section.

2.4 Authoritative, Traceable Fluorescence Intensity Assignments (NIST)

NIST has published a series of reports detailing the fundamental scientific basis and reference methods for assigning MESF or ERF values to fluorescently labeled microparticles [5–10]. Most recently, NIST has produced a primary fluorophore solution kit, Standard Reference Material 1934, that includes fluorescein, Nile Red, coumarin 30, and allophycocyanin for ERF value assignment following its published standard operating procedure [10]. NIST uses a specially designed and calibrated spectrofluorometer equipped with laser excitation and a CCD detector to perform ERF value assignment of calibration microparticles. Laser wavelengths can be selected from any commonly used in flow cytometry. This ERF value assignment service is provided to the participating members of the newly formed flow cytometry quantitation consortium [11]. The use of SRM 1934 establishes the traceability of the ERF value assignment and ultimately enables the standardization of the fluorescence intensity scale of flow cytometers in quantitative ERF units.

2.5 Considerations Using Beads as Cell Analogs for Light Scatter

The most important particle factors that affect light scatter are size and refractive index. While size can be well controlled in microparticle production to correspond to various cell sizes, the refractive index of all polymer particles is significantly higher than that of cells. Silica particles are closer to most cells in refractive index, but are not a true analog. The use of beads to standardize light scatter is further complicated by the fact that different instrument models measure different ranges of scatter angles. Cells are also not homogeneous structures. The nucleus and other substructures have refractive indexes different from that of the more homogeneous cytoplasm. So while homogenous beads cannot reproduce the light scatter from cells, they do provide a useful standard on a particular instrument model for setting up the instrument so cells are displayed in a predetermined location on the scatter scales. Because of the difference in light scatter from the cells, beads are produced and used as an internal counting standard for measuring biological cell concentrations. The relative position of beads and cells can vary quite a lot among different instrument manufacturers and models, but is reasonably consistent for a particular instrument model.

Hydrogels are new materials that are being used to make particles that could be light-scatter standards for flow cytometry. The material allows control over refractive index in the same range as cells and also offers the possibility of heterogeneous structure more similar to nucleated cells [12].

3 Standardization, Calibration, and Quality Control/QC

3.1 How Standardization, Calibration, and Control Differ

As a generally understood term in flow cytometry, standardization is the process that assures that the response of an instrument will be set up to produce expected results when an application is run. This essentially means assuring that cell populations will appear at expected locations on the data scales such as histograms and dot plots. Hard-dyed beads are most often used to set gains or check that gains are set appropriately for the application. But for both fluorescence and light scatter there are limits to how reproducible the setup will be on different instruments, as discussed in Sect. 2.

The best standard particle for setting up a particular fluorophore channel will have the same excitation and emission spectrum as the fluorophore that will be measured in that channel. This assures that all instruments will be set up the same regardless of differences in their spectral response. If the particles have intensity units such as ERF assigned, the fluorescence scale will be calibrated. In that case, the fluorescence from cells can be reported in quantitative units rather than arbitrary mean fluorescence intensity (MFI). Expressing fluorescence intensity measurements in calibrated units is essential in order to quantitatively compare results from different labs and over time—perhaps over decades.

Quality control of a flow cytometer requires regular monitoring of at least the stability of the detection system and alignment of the sample stream. Stability of the detection system can be monitored either by measuring the PMT voltage or detector gain required to put the scatter and fluorescence signals from stable particles at the same level each time or by measuring the signal levels at fixed PMT voltage or gain settings. The CV of a bright, uniformly fluorescent bead is used to monitor alignment. When the day-to-day change is beyond a predetermined amount, it is time to do maintenance or troubleshooting. Some instrument models or QC software such as BD's CS&T system provide additional QC tracking information based on measurements of hard-dyed beads.

3.2   Control/QC

Two ways that hard-dyed beads are particularly useful are for secondary standards and quality control. Unless components such as filters or lasers in a particular flow cytometer are changed (or change with time), one can use a fluorophore-specific surface-stained bead as the initial primary standard or calibrator and cross-calibrate a hard-dyed bead to it. This is easy to do by simply running the primary and secondary standards at the same instrument settings, preferably as successive samples. Thus occasional cross-calibration of a hard-dyed bead standard to a fluorophore-specific standard allows the hard-dyed bead to be used on a routine or daily basis owing to its superior stability. When used for quality control, the hard-dyed beads are run daily, and the instrument response is monitored for short-term and long-term change in response. For example, the beads can be used to adjust the detector gains so the bead fluorescence mean channel is the same each day and to monitor the gain required to accomplish this. When the detector gain change is more than a prescribed amount, this can alert the user to troubleshoot for a problem. If the problem requires changing an optical component or detector, the primary fluorophore-specific standard should be used to cross-calibrate the hard-dyed bead again.

Stained and fixed cells could be used as fluorophore-specific standard particles for some situations. For example, a study among a group of laboratories might send such stained, fixed cells to each lab in the study. Each lab could cross-calibrate the fluorophore-specific standard bead or cell sample to hard-dyed beads on each instrument in the study and use the hard-dyed beads as secondary standards over an extended time. Although there is no traceable fluorescence value assignment to the cells, their use would assure that all instruments in the study group were set up with identical fluorescence scales.

3.3 Standardization and Calibration

If only one flow cytometer is providing all the data and it is only important that semi-quantitative results be reported, then using a hard-dyed bead without reliable assigned intensity values to standardize the instrument setup can be sufficient. It is necessary to cross-calibrate a new lot of beads to the lot currently being used in order to maintain consistency in instrument setup.

But if fluorescence intensity results need to be compared quantitatively across labs and over time, beads for standardization should be more carefully chosen. If possible, fluorophore-specific primary standards with assigned intensity units should be used. If this is not possible, then complete description of the filters, lasers, and laser power used with the beads should be disclosed. This would allow at least the possibility to quantitatively compare fluorescence results from other instruments.

4 Standardizing and Calibrating DNA and RNA Content Per Cell

4.1 Total DNA Content

Total DNA content is one of the earliest measurements made in flow cytometry [13]. Fluorescent dyes such as propidium iodide that bind stoichiometrically to DNA are used to measure the relative amount of DNA in cells. With some sample preparations, RNA is removed by enzymes so it does not interfere. Since DNA per cell is highly controlled and conserved, the measurement of total DNA per cell requires the highest precision of any flow cytometer application—preferably with less than 2% CV in measurements of non-replicating cells. Sample preparation is critical for quantitative DNA measurements [14, 15].

As this is one of the first applications of flow cytometry, standardization and controls are well developed [16]. Several types of cells are used as standards. Chicken erythrocytes, rainbow trout erythrocytes, and human lymphocytes are well-characterized standard cell types [17, 18]. These cells may be either used as separately stained samples, or if the DNA content is sufficiently different from the test sample, mixed in and stained together with the test sample. Chicken or rainbow trout erythrocytes can be used as internal stain controls with human samples. With careful sample preparation and appropriate standards, DNA content of cells can be expressed in pg of DNA per cell. Tiersch et al. determined the DNA content of a wide variety of vertebrate cells using female human lymphocytes with 7.0 pg DNA per cell as the reference calibration [18]. In studying abnormal DNA content in malignancies, one can use normal lymphocytes from the patient or a healthy individual as an internal control with the test sample and lymphocytes prepared in the identical manner [16, 17].

4.2 DNA and RNA Measurements Using Molecular Biology Techniques

Researchers studying genetic profiles of different cell subsets by sequencing and PCR-based methods have two different technology options available to them: affinity bead-based separation and cell sorting. Both methods give more precise information than bulk analysis methods do, but still suffer from major limitations that have thus far limited clinical, therapeutic, and diagnostic advancements. With the advent of more quantitative technologies to measure isolated genomic material, improved microscopy functionality, and more powerful flow cytometry instrumentation, we are just beginning to break the barriers that previously limited us in quantitative genomic measurements. Flow cytometry allows an investigator to decisively measure genomic material within intact cells while simultaneously cross-referencing these measurements to specific cellular subsets.

Studies of simultaneous, single-cell measurement of RNA and cell-associated proteins have recently been reported [19–21]. He et al. combined florescence in situ hybridization (FISH) with flow cytometry and correlated the intracellular microRNA (miRNA) expression measurements by digital PCR from purified cell-associated miRNA [22]. Significant advancement can be further made to FISH-flow cytometry for quantitative measure of miRNA expression in terms of copy number in specific blood-cell subsets. Quantitative FISH-flow has many advantages over traditional quantitative nucleic-acid measurement techniques. Most notably it allows one to measure cell subtype–specific miRNA expression instead of averaged expression from all cell types and avoids creating artifacts introduced during RNA purification processes. The method correlation transitions the FISH-flow technology into a quantitative, single-cell measurement system.

5 Standardizing and Calibrating Antibodies Bound Per Cell

Cytometrists often use the term "ABC" to stand for "antibodies bound per cell." This term may not always imply a saturating staining condition, which is a requirement for "antibody binding capacity," partly due to interference caused by simultaneous staining of many different kinds of antibodies on the same cell population. An ultimate goal of immunofluorescence standardization and calibration is to express cytometry measurement results of biomarkers in terms of ABC. Four approaches have been used to estimate ABC. Each approach has different critical technical requirements and potential sources of error. Although not, strictly speaking, a source of error, it must be kept in mind that different antibody clones with the same cluster designation (CD) can have different binding affinity and avidity. Particular examples of clone variability have been noted for CD4 (Davis et al. [31]) and CD34 [23]. Therefore, if all approaches to quantitative ABC are to be compared, they should be compared with the same clone or with clones that are demonstrated to give the same ABC. In addition, the sample preparation method can affect the antibody binding and must be taken into consideration [23, 24].

One of the essential qualifications for antibody selection is the antibody binding affinity that is assessed by the affinity binding dissociation constant, K_d. However, it is challenging to understand and model the binding titration curves performed using a test antibody and cells carrying the antigen/receptor. Complications arise due to dual surface-binding interactions, cooperative effects associated with multivalent binding, and cell-surface roughness [25]. Figure 3 shows cooperative binding between anti-CD4 FITC (SK3 clone) and cryopreserved PBMCs, which is dominated by divalent binding. Presently, K_d can be estimated comparatively by fitting the linear portion of the binding titration curves [26]. The use of high-binding-affinity monoclonal antibodies, e.g., in the sub-nanomolar range, would minimize non-specific cell staining. For the same antibody clone, the values of K_d can be used for assessing the effect of fluorophore labeling to the antibody clone. Another important parameter in antibody selection is the staining index of the fluorescently labeled antibody, defined as fluorescence signal difference between positive and negative cell populations divided by 2 standard deviation of the negative population [27, 28]. The larger the staining index, the more sensitive the antigen detection would be. This parameter is extremely valuable for choosing the brightest fluorophore-conjugated antibody for the sensitive detection of dimly expressed biomarkers, in particular, in the case of multicolor antibody panel design. In essence, staining index allows the evaluation of the brightness of fluorophore-conjugated

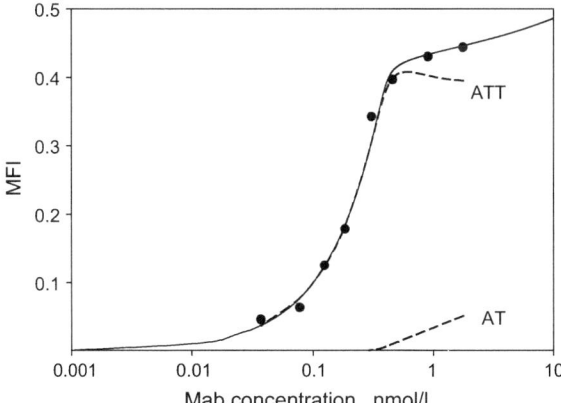

Fig. 3 Mean fluorescence intensity (MFI) measured for peripheral blood mononuclear cells (PBMCs) stained with CD4 antibodies conjugated with FITC fluorophore. The horizontal axis gives the concentration of CD4 antibodies (Mabs) used in the staining of PBMCs. The solid circles are measured values of MFI obtained from the antibody titration. The solid trace is the predicted response assuming both monovalent (trace AT) and divalent binding (trace ATT) of the CD4 antibody to CD4 receptors on the cell surface. The result suggests that CD4 antibody undergoes cooperative binding to the CD4 receptor. The binding of the first site of the CD4 antibody enhances the likelihood of the binding of the second site to another CD4 receptor

antibodies as well as non-specific cell staining. It is expected that an antibody clone with a large value of K_d should have a large value of staining index. However, owing to differences in the process of fluorophore antibody conjugation performed by different manufacturers, it is possible that antibodies with the same clone have similar K_d, but somewhat different values of staining index. It is likely the differences in the staining index are due to difference in fluorescence yield of individual antibody molecules characterized by the number of effective fluorophores per antibody molecule (effective F/P). Therefore, it is important to characterize changes in fluorescence yield induced by fluorophore conjugated to the antibody and further binding of the labeled antibody to the receptor on the cell.

The first two approaches for estimating ABC, quantitative indirect immunofluorescence (QIFI) and Quantum Simply Cellular (QSC), have recently been illustrated in detail [29]. A third method uses antibody conjugates that have been prepared with a known MESF/antibody ratio and a flow cytometer that has been calibrated in MESF. Phycoerythrin is an attractive fluorochrome for this approach since antibody conjugates can be prepared with exactly one PE molecule per antibody. Because the fluorescence-emitting unit of the PE molecule is insulated within the protein [30], it is expected that the fluorescence yield of a single PE molecule is the same as the yield of a unimolar antibody-PE conjugate, meaning the effective F/P is equal to 1. Successful initial experiments [31, 32] ultimately led to the development and production of the Quantibrite products that include purified 1:1 PE-antibody conjugates and freeze-dried beads surface-stained with known numbers of PE molecules per bead. The Quantibrite method provides a great example of quantifying antigen expression levels in the PE channel of flow cytometers. However, the availability of unimolar PE-antibody conjugates is an issue. And although unimolar PE-antibody conjugates provide a known F/P, the effective F/P is not yet available for antibodies labeled with other fluorophores.

The QIFI and Quantibrite methods have been found to be generally comparable [23, 33] for ABC quantitation, but the QSC method frequently gives significantly different results from the other methods [23, 33]. Since the amount of CD4, CD45, and many other molecules on normal human lymphocytes is generally reproducible [34–36], these cell-surface markers may be useful as biological calibrators with a relatively small variability and uncertainty. The use of biological calibrators has become the latest method for quantifying unknowns in ABC.

A detailed protocol of quantitative flow cytometry measurements in ABC based on the human CD4 reference marker has recently been developed jointly by NIST and the FDA [37]. The reference marker, CD4 receptor protein on human T helper cells, can come from either whole blood of normal healthy individuals or Cyto-Trol^{TM} control cells, a commercially available peripheral blood mononuclear cell (PBMC) preparation, depending on the preference of users and the accessibility of normal individual whole-blood samples. The CD4 expression levels in ABC are approximately 45,000 for fixed normal whole-blood samples and approximately 40,000 for Cyto-Trol cells, respectively [26, 38]. These CD4 expression levels have been verified by orthogonal measurement methods, quantitative flow cytometry, and mass cytometry using a well-characterized anti-human CD4 monoclonal antibody (SK3 clone from BD Biosciences) as well as quantitative mass spectrometry using an isotope-labeled, full-length recombinant CD4 receptor protein as the internal quantification standard. The known reference CD4 expression enables the translation of a linear fluorescence intensity scale to the ABC scale that ultimately ensures quantitative measure of target antigen expression levels independent of flow cytometers used. This approach is illustrated in Fig. 4 for determination of CD20 expression.

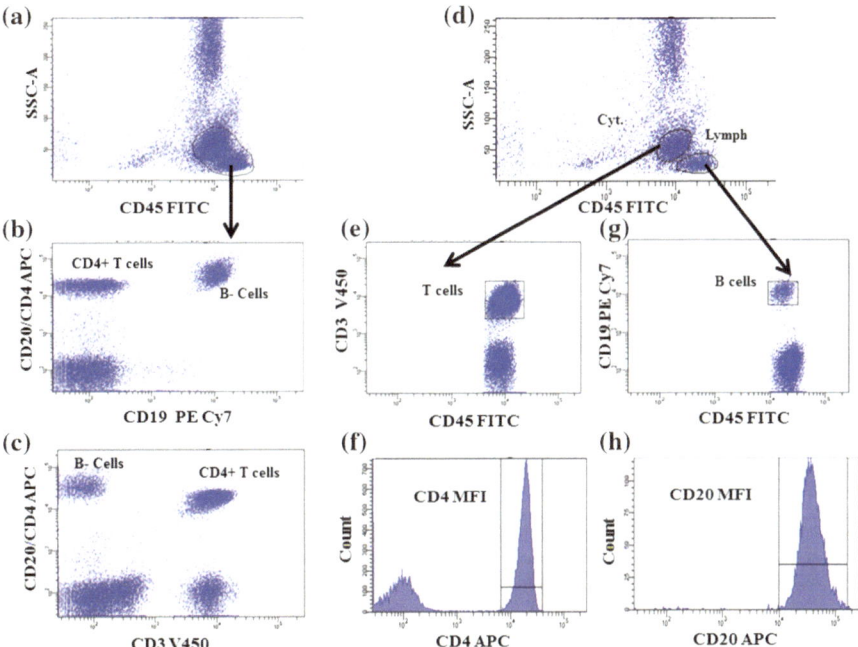

Fig. 4 Quantifying CD20 expression level in ABC units based on a known CD4 expression level on T helper cells from Cyto-Trol control cells, both stained in APC. The unknown whole-blood sample was stained with CD45 FITC, CD19 PE-Cy7, and CD20 APC, and Cyto-Trol was stained with CD45 FITC, CD3 V450, and CD4 APC, in two separate sample tubes. After staining and washing, the two samples were combined in a single tube and run on a linearity-calibrated flow cytometer. Two different gating strategies are shown. Gating strategy I: **a** a large lymphocyte gate (CD45+ and low SSC) was drawn in CD45 FITC versus SSC-A; **b** gated on lymphocytes, CD4+ T cells and CD19+CD20+ B cells were identified in a dot plot of CD19 PE-Cy7 versus CD20/CD4 APC; **c** alternatively, CD4+ T cells and CD20+ B cells can also be identified in a dot plot of CD3 V450 versus CD20/CD4 APC. The MFI values of CD20 and CD4 can then be obtained from a CD20/CD4 histogram under the respective CD20+ B-cell gate and CD4+ T-cell gate. Gating strategy II: **d** two individual lymphocyte gates (CD45+ and low SSC) were drawn as 'Cyt' for Cyto-Trol cells and 'Lymph' for unknown whole blood sample in CD45 FITC versus SSC-A; **e** gated on 'Cyt,' T cells were identified in a dot plot of CD45 FITC versus CD3 V450; **f** under T-cell gate, CD4 histogram shows the positive CD4+ gate, which was used to obtain the respective MFI value of CD4; **g** gated on 'Lymph,' B cells were identified in a dot plot of CD45 FITC versus CD19 PE-Cy7; **h** gated on B cells, CD20 histogram shows the positive CD20+ gate that was used to obtain the MFI value of CD20. With measured MFI values of CD20 and CD4, CD20 expression in ABC can be determined on the basis of the CD4 expression level from Cyto-Trol

Biological-cell reference materials have been gaining momentum as phenotypic benchmarks for quantitative and reproducible measure of patient characteristics in longitudinal studies and/or across locations. High-quality measurement data generated for patients on drug treatments will fill the gap between drug/therapy treatment and clinical treatment outcome. Currently, three different dried or lyophilized human PBMCs are commercially available: FACSCyto PBMC from BD Biosciences, Cyto-Trol Control Cells from Beckman Coulter, and Veri-Cells PBMC from Biolegend. Proper characterization of these cell reference materials would enhance their utility in clinical trials, disease diagnosis, immune-cell manufacturing, and therapy monitoring, drug, and device development.

The biological reference approach relies on antibody conjugates with a particular fluorophore having essentially the same fluorescence intensity per antibody independent of the antibody specificity. One approach to determine the relative fluorescence per antibody relies on measuring the fluorescence from beads that capture antibody. If different antibodies are captured identically at saturation staining levels, then the relative fluorescence per antibody can be determined from the mean fluorescence of the beads. This approach has been problematic, however, since various factors can affect the binding of antibodies to capture beads and affect the degree of fluorescence quenching at near saturated staining levels. Kantor et al. [39] propose an improved approach to determine the relative fluorescence per antibody molecule that does not depend on the saturated staining level. Instead, the approach measures the fluorescence from two antibodies, conjugated to two different fluorophores, which together saturate the binding sites of an antibody capture bead. The antibody conjugated to a first fluorophore (the Test antibody) is used in several dilutions to load the capture beads with a range of antibody levels. After washing the Test-stained beads, the second (Fill antibody) conjugated to a second fluorophore is added to the Test samples in adequate amount to fill the remaining capture sites on the beads. If staining were ideal, the relationship between fluorescence of the Fill and Test antibodies would be linear, with decreasing fluorescence of the Fill antibody as the beads captured more of the Test antibody. To account for possible non-linear behavior near saturation, the method by Kantor et al. fits the data with a quadratic function and uses the linear term of the fit to estimate relative brightness at low antibody density. Unless the relationship between Test and Fill reagents is highly non-linear, this approach gives quantitative measures of the relative brightness among different antibodies conjugated to the same fluorophore. If the relationship is highly non-linear, the Test reagent is considered unsuitable for quantitative measurements. If a fluorophore conjugate with a known quantitative relationship between fluorescence and antibodies bound is used, the system can be calibrated to give fluorescence per antibody conjugate of any fluorophore. Kantor et al. use antibody conjugated to exactly 1 PE molecule and beads with known numbers of PE molecules per bead to make this quantitative step.

An ideal simulated situation is illustrated in Fig. 5, where the Fill antibody is conjugated to FITC and Test antibodies are conjugated to either CY5 or PE, with the PE conjugate highly purified with exactly 1 PE molecule per antibody. Panel A illustrates how the relative brightness of two different CY5 antibody conjugates is

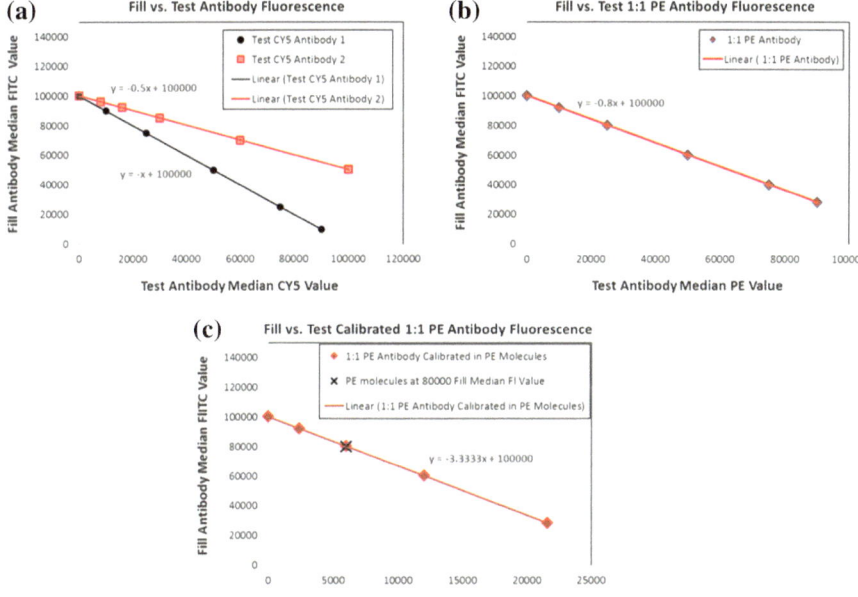

Fig. 5 Simulated example of the Test-Fill method to compare (panel A) and calibrate (panels B and C) fluorescent antibody conjugates. **a** Antibody capture beads are first stained with various amounts of a Test antibody conjugated to one fluorophore (CY5 in this example) and then stained with sufficient Fill antibody conjugated to a different fluorophore (FITC in this example) to saturate all the antibody binding sites on the beads. If fluorescence is proportional to the amount of antibody on the bead, there is a linear relationship of capture bead fluorescence between the two fluorophores. The greater the fluorescence of the Test reagent at a particular level of Fill reagent fluorescence, the brighter the Test reagent. Two different CY5-labeled Test antibody conjugates are compared. **b** A 1:1 PE conjugate is used as Test reagent. **c** If the PE scale is calibrated in PE molecules, equivalent to antibody molecules for a 1:1 conjugate, the Fill axis is calibrated in fluorescence intensity per antibody molecule

determined. The relative amount of test antibody is indicated by the reduction in Fill antibody from the saturation level (zero Test antibody added). In this example, antibody conjugate 2 is brighter because it has a smaller slope, indicating less Test antibody is on the beads at any level of Test antibody fluorescence. In this case CY5 antibody 2 is twice as bright as CY5 antibody 1. Panel B illustrates the relationship when a highly purified PE conjugate with exactly 1 PE molecule per antibody (such as BD Quantibrite reagents) is used. If the PE fluorescence axis is calibrated in PE molecules (for example with BD Quantibrite PE beads), then the relationship between reduced Fill fluorescence and the number of PE molecules is obtained.

Once the Fill fluorescence scale has been calibrated in antibody molecules per fluorescent unit for a particular Fill reagent, the relative relationship between antibody brightnesses can be translated to absolute fluorescence per antibody for any Test reagent conjugated to a fluorophore other than the one used for the Fill reagent. In the illustration of Panel C, 6000 PE molecules or equivalently 6000 antibody molecules conjugated to exactly 1 PE molecule cause a reduction of 20,000 units of fluorescence in the Fill reagent. Or equivalently 20,000 units of fluorescence from the Fill reagent is equal to 6000 antibody molecules or 6000/20,000 ABC/FITC FlUnit, or 0.3 ABC/FITC FlUnit. With this additional information the CY5 fluorescence scale can be translated to ABC for each of the CY5 conjugates. With CY5 Antibody 1, which has a slope of 0.5, the scale translates to (0.3 ABC/FITC FlUnit)x(0.5 FITC FlUnit/CY5 FlUnit) = 0.15 ABC/CY5 FlUnit. With CY5 Antibody 2, which has slope of 1, the CY5 scale for this antibody would translate to 0.3 ABC/CY5 FlUnit.

6 Fluorescence Performance Characterization

When controls are run regularly and quality control is practiced, a flow cytometer will provide reproducible results. But this does not guarantee that the results will be adequate for all applications. The performance of flow cytometers varies among different instrument models. Even different instruments of the same model will have different levels of performance, particularly regarding fluorescence. Performance can degrade over time as well. It is best if a flow cytometrist has objective and measurable criteria for instrument performance. This is particularly important when data from multiple instruments are used in a study.

A sample of multilevel beads such as the Spherotech Rainbow beads shown in Fig. 6 tells much about instrument performance. Such mixtures of beads stained at different levels are made from the same batch of unstained beads and all have nearly the same intrinsic CV. The brightest beads in the mixture are used to assess optical

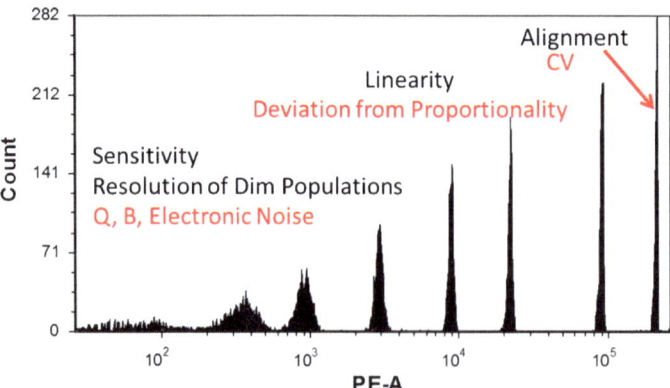

Fig. 6 Histogram of Spherotech 8-peak Rainbow beads (catalog number RCP-30-5A). MFIs and CVs of the seven stained populations can be applied in different ways to characterize performance of the fluorescence detection system

alignment—the smaller the CV, the better the alignment of the sample stream to the focused laser beam. Comparing the measured mean fluorescence intensity (MFI) to the intensity value of each bead population provided by the manufacturer gives information about the linearity of the instrument response. The broadening of the populations as the beads have less fluorescence is not due primarily to the beads themselves but rather to the decreasing number of photoelectrons generated in the detector (usually a PMT) and the effect of background light (such as Raman scatter from water) and the fixed level of electronic noise that is present in the amplifier and digitizing electronics. For most practical purposes the CV of the brightest bead can usually be treated as having the same intrinsic CV as all the other beads, and broadening of the dimmer bead populations is due to the other factors detailed later in this chapter. In some very sensitive instruments, the dimmest stained bead in the Rainbow bead set has a small but measurable increase in intrinsic CV compared to the brighter beads, but the dominant contribution to broadening of the populations are instrument related.

6.1 Linearity

Before measuring the contributions to population broadening of dim particles, however, it is important to know the range over which the measurements are linear [40]. An underappreciated effect of nonlinearity is the significant error that can be introduced into the calculation of spectral overlap compensation, which assumes that the measured signal is strictly proportional to the input optical signal. Under some conditions, nonlinearity of a few percent at the top of the scale in one fluorescence channel can cause an order-of-magnitude error in compensated values of a double-stained population at the low end of the scale in another channel. For clear data interpretation and quantitative measurements, a maximum deviation from linearity of 2% or less is recommended. Significant nonlinearity at the low end of the scale will cause errors in measured CVs that affect characterizing detection of dim fluorescence.

A set of multi-intensity beads such as the Spherotech Rainbow beads shown in Fig. 6 can provide a limited test of linearity using the manufacturer's assigned intensity values for each population. Figure 7 shows the result of such a test, where the MFI is plotted versus the assigned intensity units (MEF) for the FITC channel on a flow cytometer. Data are plotted on a log-log scale and fitted with a linear function of slope 1, which assumes that the MFI is proportional to the assigned MEF. The visual plot indicates a good fit, but the result shows deviations from proportionality of up to 4% at some parts of the scale. This instrument was also tested for linearity by an alternative method described next.

A better way to test for proportionality (strict linearity) is to compare the measured ratio of two output signals whose relative input values (ratio) are known. If the electronics are strictly linear, the ratio of the two measured signals will be the same as the ratio of the two input signals. The standard manufacturer's specification

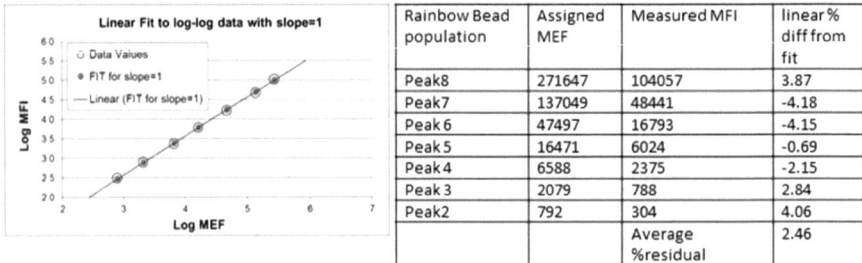

Rainbow Bead population	Assigned MEF	Measured MFI	linear% diff from fit
Peak8	271647	104057	3.87
Peak7	137049	48441	-4.18
Peak 6	47497	16793	-4.15
Peak 5	16471	6024	-0.69
Peak 4	6588	2375	-2.15
Peak 3	2079	788	2.84
Peak2	792	304	4.06
		Average %residual	2.46

Fig. 7 Evaluation of linearity in the FITC channel of a flow cytometer using comparison of MFI to manufacturer-assigned fluorescein intensities per bead (MEF) of Spherotech 8-peak Rainbow beads. Data were acquired on the same instrument used for data shown in Fig. 8

for flow cytometer linearity (if it is specified at all), is that the ratio of the MFIs of doublet and singlet chicken erythrocyte nuclei stained with a DNA dye will be 2.00 ± tolerance. For example, the doublet to singlet ratio will be 1.95–2.05. While this ratio approach is useful at one point on the scale, it does not give any information about other points on the scale, which can range over four to seven orders of magnitude.

The reference method for testing the linearity of an optical detection system exposes the detector to flashes of light from a light-emitting diode (LED), with alternating flashes of light at two different but consistent levels. While the electrical drive to the LED is not changed, the amount of light reaching the detector (e.g., PMT) is varied by positioning the LED closer to or further from the detector or by using neutral density filters to reduce the intensity. If the detector is linear (i.e., output proportional to the input light) the ratio of the two output signals will be constant no matter how much of the LED light reaches the detector. Deviation of the output ratio from the expected value is an indication and measure of non-linearity. This approach is easy to do at an engineering level but is not usually practical for routine use in most flow cytometer labs. An alternative by Bagwell et al. [41] used the ratio of florescence intensities of two different beads to evaluate the linearity of detector system electronics by varying the PMT voltage to cause the signals to the electronics to cover the entire measurement scale. One of the authors (RAH) extensively compared this approach using PMT voltage to vary the input signal to the reference LED ratio method during the development of the BD CS&T system at BD Biosciences. The two approaches gave equivalent measures of detector system linearity, and the PMT voltage variation approach was integrated into the CS&T system to measure linearity.

Table 3 shows the results of this ratio method from the same instrument used for the data in Fig. 7. Two of the Rainbow bead populations were used and the ratio of their MFI determined over the entire measurement scale by varying the PMT voltage for the FITC channel. The ratio method indicated a much higher degree of linearity than suggested by the comparison with manufacturer-assigned intensity values. This method to evaluate linearity is easy to do and takes only a short time.

Table 3 Example of electronics linearity test using the ratio of means of two bead populations as PMT voltage is varied to place beads along a histogram scale of 0–262544

PMT voltage	Bead 1 median	Bead 2 median	Bead2/Bead1	% deviation from average ratio
300	26	65	2.50	−1.75
350	77	200	2.60	2.07
400	203	514	2.53	−0.50
500	1033	2636	2.55	0.28
600	4040	10309	2.55	0.28
700	12730	32474	2.55	0.25
800	36441	92438	2.54	−0.31
900	98468	245714	2.50	−1.94

The same instrument was used for the linearity test shown in Fig. 7

6.2   Noise Contributions Broaden Measured Populations

In simplest terms, the CV or variance of a population is the sum of the CVs or variances intrinsic to the sample itself and the added variance from the measurement process in the instrument. The contributions to measurement variation from the instrument are due to a constant level of electronic noise in the electronics, optical background light, statistical variation in the number of photoelectrons generated by a light pulse, excitation variation (or laser noise), and variation in how uniformly each particle is illuminated and the fluorescence collected on the detector. The total instrument contribution to the standard deviation (SD) is calculated from the squares of individual contributions. SD^2 is also called the variance.

$$SD^2_{Instrument} = SD^2_{Photoelectron} + SD^2_{Backgnd} + SD^2_{LaserNoise} + SD^2_{Position} + SD^2_{ElectronicNoise}$$

$$(1)$$

For bright signals, the variability of particle illumination and detection based on particle position in the sample stream (grouped in the contribution $SD_{Position}$) and laser noise are dominant, but for lower signals, the statistical nature of the photon detection process adds variance along with variance from added non-signal photoelectrons from background light. Variance due to the limited number of signal photoelectrons is determined by the detection efficiency, Q, which is described more fully in Sect. 6.3 below. Conceptually, Q is the equivalent number of photoelectrons generated in the detector by a fluorophore molecule passing through the laser beam. At the low end of the measurement scale (independent of PMT voltage), a contribution from electronic noise can be expected.

Quantitative relationships for the various factors are:

$$SD^2_{Photoelecton} = \frac{F}{Q},\text{ where F is fluorochrome per particle measured in intensity units}$$

$$SD^2_{Backgnd} = \frac{B}{Q},\text{ where B is equivalent background fluorochrome}$$

$$SD^2_{LaserNoise} = [n*(Signal + Background)]^2,\text{ where n is fractional laser noise}$$

$$SD^2_{Position} = (Signal* CV_{particle})^2$$

$$SD^2_{ElectronicNoise} = Constant.$$

Electronic noise does not change with PMT gain and can be measured in several ways. If accurate measurements around zero signal can be made, as in most recent BD flow cytometers other than FACSCalibur, the SD due to electronic noise can be measured by turning the PMT voltage to zero and measuring the SD of the resulting noise signal. Alternatively, one can monitor the SD of a bead with relatively bright, uniform fluorescence as the PMT voltage is reduced to successively lower values. The distribution on the histogram will broaden as electronic noise becomes a significant factor of the total variance, and the SD will tend toward a stable number no matter how bright the initial bead fluorescence. An example is shown in Fig. 8.

The SD approached by all the beads at low signal levels is the electronic noise, which is always present but becomes insignificant at sufficiently higher signal levels. For best resolution of dim signals, the gain should be set so the CV of the unstained cell population is not significantly broadened by electronic noise. For

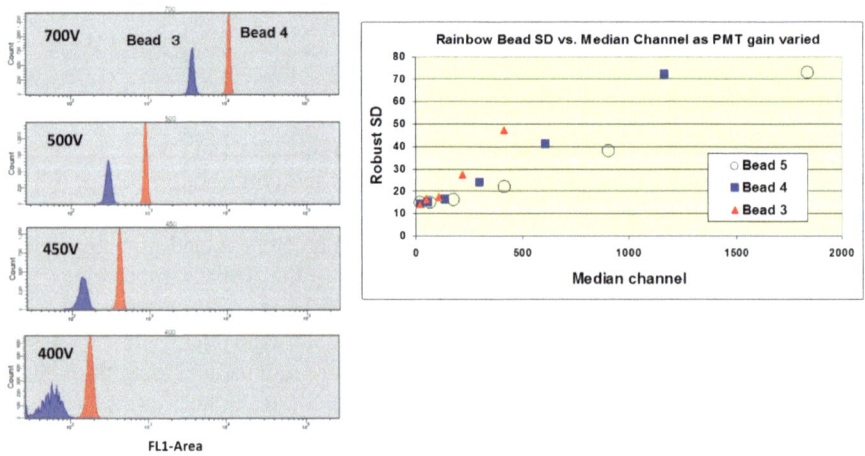

Fig. 8 Histogram of gated populations of Spherotech Rainbow beads at various PMT gains and robust standard deviation of the bead populations versus median fluorescence intensity at different PMT gains

example, if the electronic noise SD is 15, and the median channel of a population is 300, the electronic noise contribution to the measured CV is only 5%.

Since all the contributions to measured CV except $SD^2_{ElectonicNoise}$ do not vary with PMT voltage, one can also measure electronic noise by plotting measured fluorescence CV^2 versus $1/Mean^2$ of a particle over a range of PMT voltages. The slope of the plot is $SD^2_{ElectronicNoise}$, which is expected from the following relationships from dividing the SD^2 factors by $Mean^2$ to put the relationships in terms of CV rather than SD.

$$CV^2_{Instrument} = CV^2_{Photoelectron} + CV^2_{Backgnd} + CV^2_{LaserNoise} + CV^2_{Position} + \frac{1}{Mean^2} * SD^2_{ElectronicNoise}$$

(2)

$$CV^2_{Instrument} = Constant + \frac{1}{Mean^2} * SD^2_{ElectronicNoise}$$

(3)

6.3 Detection Efficiency, Q, and Background Light

If both signal and background light contributions are considered together, the variance in photoelectron contribution is the sum of both variances. Background, B, is expressed as the amount of fluorophore units that would produce the background light. When measured under conditions where signals are detected well above electronic noise and with flashes from an LED, one has [42]

$$SD^2_{TotalPhotoelectron} = f \cdot \frac{1}{Q} + \frac{B}{Q}$$

(4)

f = calibrated particle signal intensity in fluorescence units, Q = statistical photoelectrons per fluorescence unit, B = background in fluorescence units.

The best way to measure the instrument contributions to variance is to use light flashes from a light-emitting diode (LED) to simulate signals from a sample with zero intrinsic CV [43]. To make this performance characterization broadly available, Chase and Hoffman showed that sets of beads stained at varying levels could adequately replace LED flashes when the intrinsic CV and instrument broadening of the brightest bead in the set are taken into account [44]. They proposed the term Q as the measure of photoelectrons generated per particle fluorescence unit (e.g., MESF or ERF) and B as the constant background light always present when particles are measured. The variation due to the statistical nature of photon conversion to photoelectrons is increased slightly in a PMT owing to the amplification process. If there were no added noise in amplification, the SD of photoelectrons would be the square root of the average number created by repetition of identical light pulses. The concept of statistical photoelectrons is a measure of that variance and has been given the symbol S_{pe} [45]. So Q is more properly described as S_{pe} per particle fluorescence unit. B is also measured in the same fluorescence units as Q.

The approach used by Chase and Hoffman [44] estimated B at zero signal level and separately measured Q at a sufficiently high signal so that electronic noise and background light were negligible. The measured CVs were corrected for the CV of the brightest bead in the set. Hoffman and Wood [46] used a linear fit to Eq. 4 to determine Q and B, where the slope is 1/Q and intercept of the fit is B/Q. Again in this case measured CVs were corrected for the CV of the brightest bead in the set to calculate the SD due to photoelectron statistics. Figure 9 is an example of a spread sheet using this approach. The data for the Q and B measurement should be obtained using linear rather than logarithmic amplifiers for instruments such as BD FACSCalibur where both options are available.

Rather than estimate the intrinsic measurement CV from the CV of the brightest bead in a set, Parks et al. [45] improved the fitting for Q and B determination by adding to Eq. 4 a term $C*f^2$ that includes the intrinsic measurement variance. The data are then fit with a quadratic function that gives best estimates for 1/Q, B/Q and the "intrinsic" CV of the measurement when the particle is so bright that factors other than photoelectron statistics are dominant. The quadratic fitting approach was applied to both LED flashes and multi-level bead sets and found to work well. Beads used in the study and LED flashes generally gave equivalent results on an instrument. The exceptions were when the instrument used log amplifiers, which affected accuracy of the measurements, and when instruments had particularly high

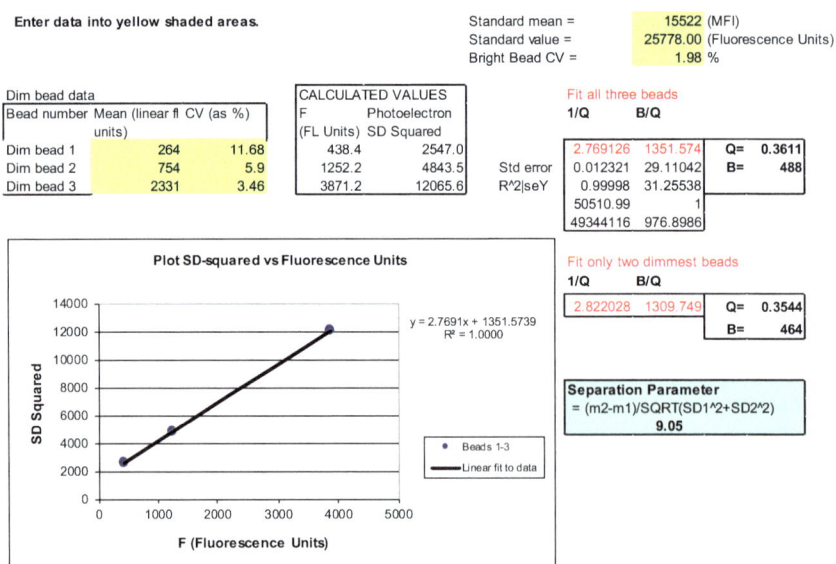

Fig. 9 Example of Q and B determination for the PE channel on a flow cytometer. BD Quantibrite beads were used as the standard and Spherotech 8-peak Rainbow beads as the test sample. See Fig. 6 for an example of a histogram of these beads

sensitivity (high Q), in which cases the assumption that all beads in the set had the same intrinsic CV seemed to not apply to the dimmest beads in the set.

All the performance characteristics—optical alignment, linearity, electronic noise, background light, and detection efficiency—are straightforward to assess. BD Biosciences' CS&T system conveniently performs these tests automatically with proprietary beads and software. In whatever way the performance is objectively measured, the results can be used to predict performance of biological applications. See Chase and Hoffman [44] for a simple example. In the future, this may be the most valuable use of this information. At least it can be used to set the minimum instrument performance requirements for important research or clinical assays. Furthermore, when these critical measures of instrument performance, Q and B, are standardized with traceable fluorescence units of MESF and ERF, users can take into account the difference in the performance of various cytometer platforms and design the most sensitive assays in the multisite studies.

6.4 Buyer Beware

Historically, simpler approaches to characterizing fluorescence sensitivity have been proposed that reduce this performance measure to a single number. Unfortunately, a simple but non-informative method has become the industry standard used in marketing literature and technical specifications. It is easy to show how any attempt to do this (e.g., detection threshold or delta channel) allows two instruments with the identical single-number "sensitivity" measure to have significantly different ability to resolve dim populations [47]. It is disappointing that as this chapter is being written, all instrument manufacturers still use the extrapolated intercept of a plot of Spherotech-assigned intensity values versus MFI of Rainbow or Ultra Rainbow beads to advertise "sensitivity" in terms of "molecules of equivalent soluble fluorochrome." This is no longer a scientifically justifiable measure and in fact has been refuted by presentations at international cytometry meetings. BD Biosciences continues to use the non-informative "molecules of equivalent soluble fluorochrome" specification in marketing literature for instruments that provide users with BD's rigorous CS&T performance characterization system. This measure of "sensitivity" has been around so long that a brochure or technical specification sheet apparently must have "molecules of equivalent soluble fluorochrome" in order to show how the instrument compares to the competition. Perhaps if enough customers ask serious questions about fluorescence sensitivity, technical data sheets will eventually have scientifically meaningful specifications for Q, B, and electronic noise.

7   Future Possibilities

Flow cytometry has had rapid growth since the middle of 1980. It has moved from a technology platform that only a few hundred "initiated" experts understood and could use to become an essential immunological tool for research, drug and device development, clinical trials, disease diagnostics, and immune-cell manufacturing and therapy monitoring. Flow cytometry is essential for accurate measurement of CD4+ cell counts for ensuring that patients receive the appropriate antiretroviral treatment for HIV/AIDS monitoring. A validated reference standard has been developed for quality control of clinical CD4+ cell counting following the World Health Organization's call for establishing an external quality assessment program [48]. At present, multiplexed flow cytometry assays are routinely used in clinics for disease diagnosis and therapies [49–51]. Moreover, flow cytometry has also become an essential clearance tool for the production of protein and cell therapeutics [52, 53]. All these applications essentially require that comparable and reproducible results can be generated using different flow cytometer platforms at different locations and times.

Consistency of flow cytometry measurements can only be accomplished using proper controls and standards, e.g., particles for instrument standardization and calibration and biological cell reference materials in the measurement process. Without proper use of these process controls, the value of this information-rich instrument will not be realized, nor will further advancement be made into new biological and clinical applications.

In an ideal world, a flow cytometrist would be able to compare flow cytometer performance requirements that have been previously determined and recorded in a newly published journal article, check those requirements against the performance of the necessary light scatter and fluorescence channels on the instrument in their lab, and know in advance whether the new application should run successfully. If the lab's instrument is capable of successfully performing the application, the flow cytometrist would run controls that check whether the instrument is still performing as well as previously. If the controls indicate performance is still good enough for the application, beads would be run to standardize or "set up" the instrument for the application by setting the detector gains so populations of cells will be in the expected range on the detector channels. If it is not automated by the software, the user may have to run separate samples to set spectral compensation.

The only part of this idealized scenario that is not yet commonly done is the publication of instrument performance requirements that are necessary to assure that an application will give adequate results. It is now possible to predict and model multicolor flow cytometry data once the fluorescence characteristics of the sample and critical performance characteristics of the flow cytometer are known. If engineering support is available, it is also possible to intentionally detune and degrade aspects of instrument performance until an assay is just still giving reliable results. With either approach it is possible to define the minimum performance required of critical instrument characteristics such as those affecting fluorescence sensitivity. The tools are available to do this. Now they need to be used.

Acknowledgements The authors would like to thank Adolfas Gaigalas of NIST for his pioneering work and continued engagement in developing traceable fluorescence standards. They would also like to thank David Parks of Stanford for reviewing and clarifying the chapter section on the TEST-FILL method.

Disclaimer Certain commercial equipment, instruments, and materials are identified in this paper to specify adequately the experimental procedure. In no case does such identification imply recommendation or endorsement by the National Institute of Standards and Technology, nor does it imply that the materials or equipment are necessarily the best available for the purpose.

References

1. Hoffman RA (2000) Standardization and quantitation in flow cytometry. In: Darzynkiewicz Z, Crissman HA, Robinson JP (eds) Methods in Cell Biol 63, Cytometry, 3rd edn. Part A:300–340
2. Tanqri S, Vall H, Kaplan D, Hoffman B, Purvis N, Porwit A, Hunsberger B, Shankey TV, on behalf of ICSH/ICCS working group (2013) Validation of cell-based fluorescence assays: practice guidelines from the ICSH and ICCS–Part III–analytical issues. Cytometry Part B 84B:291–308
3. Wang L, Hoffman RA (2016) Standardization, calibration, and control in flow cytometry. Curr Protoc Cytom 79:1.3.1–1.3.27
4. Hoffman RA, Wang L, Bigos M, Nolan JP (2012) NIST/ISAC standardization study: variability in assignment of intensity values to fluorescence standard beads and in cross calibration of standard beads to hard dyed beads. Cytometry A 81A:785–796
5. Gaigalas AK, Li L, Henderson O, Vogt RF, Barr J, Marti GE, Weaver J, Schwartz A (2001) The development of fluorescence intensity standards. J Res Natl Inst Stand Technol 106:381–389
6. Schwartz A, Wang L, Early E, Gaigalas AK, Zhang Y-Z, Marti GE, Vogt RF (2002) Quantitating fluorescence intensity from fluorophores: the definition of MESF assignment. J Res Natl Inst Stand Technol 107:83–91
7. Wang L, Gaigalas AK, Abbasi F, Marti GE, Vogt RF, Schwartz A (2002) Quantitating fluorescence intensity from fluorophores: practical use of MESF values. J Res Natl Inst Stand Technol 107:339–353
8. Wang L, Gaigalas AK, Marti GE, Abbasi F, Hoffman RA (2008) Toward quantitative fluorescence measurements with multicolor flow cytometry. Cytometry 73A:279–288
9. Wang L, Gaigalas AK (2011) Development of multicolor flow cytometry standards: assignment of ERF units. J Res Natl Inst Stand Technol 116:671–683
10. Wang L, DeRose P, Gaigalas AK (2016) Assignment of the number of equivalent reference fluorophores to dyed microspheres. J Res Natl Inst Stand Technol 121:269–286
11. Flow Cytometry Quantitation Consortium. 81 Federal Register 136(15 Jul 2016): 46054–46055
12. http://www.slingshotbio.com
13. Dittrich W, Göhde W (1969) Impulsfluorometrie bei Einzelzellen in Suspensionen. Z Naturforsch 24b:360–361
14. Darzynkiewicz Z, Traganos F, Kapuscinski J, Staiano-Coico L, Melamed MR (1984) Accessibility of DNA in situ to various fluorochromes: relationship to chromatin changes during erythroid differentiation of Friend leukemia cells. Cytometry 5:355–363

15. Myc A, Traganos F, Lara J, Melamed MR, Darzynkiewicz Z (1992) DNA stainability in aneuploid breast tumors: comparison of four DNA fluorochromes differing in binding properties. Cytometry 13:389–394
16. Dressler LG, Seamer LC (1994) Controls, standards, and histogram interpretation in DNA flow cytometry. Methods Cell Biol 41:241–262
17. Jakobsen A (1983) The use of trout erythrocytes and human lymphocytes for standardization in flow cytometry. Cytometry 4:161–165
18. Tiersch TR, Chandler RW, Wachtel SS, Elias S (1989) Reference standards for flow cytometry and application in comparative studies of nuclear DNA content. Cytometry 10:706–710
19. Van Hoof D, Lomas W, Hanley MB, Park E (2014) Simultaneous flow cytometric analysis of IFN-γ and CD4 mRNA and protein expression kinetics in human peripheral blood mononuclear cells during activation. Cytometry A 85:894–900
20. Soh KT, Tario JD Jr, Colligan S, Maguire O, Pan D, Minderman H, Wallace PK (2016) Simultaneous, single-cell measurement of messenger RNA, cell surface proteins, and intracellular proteins. Curr Protoc Cytom 75:7.45.1–7.45.33
21. Frei AP, Bava F-A, Zunder ER, Hsieh EWY, Chen S-Y, Nolan GP, Gherardini PF (2016) Highly multiplexed simultaneous detection of RNAs and proteins in single cells. Nat Methods 13:269–275
22. He H-J, Almeida JL, Hill CR, Lund S, Choquette S, Cole KD (2016} Development of a NIST standard reference material (SRM) 2373:genomic DNA standard for HER2 measurements. BDQ 8:1–8
23. Serke S, van Lessen A, Huhn D (1998) Quantitative fluorescence flow cytometry: a comparison of the three techniques for direct and indirect immunofluorescence. Cytometry 33:179–187
24. Islam D, Lindberg AA, Christensson B (1995) Peripheral blood cell preparation influences the level of expression of leukocyte cell surface markers as assessed with quantitative multicolor flow cytometry. Cytometry 22:128–134
25. Sengers BG, McGinty S, Nouri FZ, Argungu M, Hawkins E, Hadji A, Weber A, Taylor A, Sepp A (2016) Modeling bispecific monoclonal antibody interaction with two cell membrane targets indicates the importance of surface diffusion. MABS 8:905–915
26. Wang M, Misakian M, He H-J, Abbasi F, Davis JM, Cole KD, Turko IV, Wang L (2014) Quantifying CD4 receptor protein in two human CD4 + lymphocyte preparations for quantitative flow cytometry. Clin Proteomics 11:43
27. Maecker HT, Frey T, Nomura LE, Trotter J (2004) Selecting fluorochrome conjugates for maximum sensitivity. Cytometry A 62:169–173
28. Maecker HT, Trotter J (2008) Application notes: selecting reagents for multicolor flow cytometry with BD™ LSR II and BD FACSCanto™ sytems. Nat Methods 5:6–7
29. D'hautcourt J-L (2002) Quantitative flow cytometric analysis of membrane antigen expression. Curr Protoc Cytom 22:6.12.1–6.12.22
30. Gaigalas AK, Gallagher T, Cole KD. Singh T, Wang L, Zhang Y-Z (2006) A multistate model for the fluorescence response of R-phycoerythrin. Photochem Photobiol 82:635–644
31. Davis KA, Abrams B, Iyer SB, Hoffman RA, Bishop JE (1998) Determination of CD4 antigen density on cells: Role of antibody valency, avidity, clones and conjugation. Cytometry 33:197–205
32. Iyer SJ, Hultin LE, Zawadzki JA, Davis KA, Giorgi JV (1998) Quantitation of CD38 expression using QuantiBRITE beads. Cytometry 33:206–212
33. Lenkei R, Gratama JW, Rothe G, Schmitz G, D'hautcourt JL, Arekrans A, Mandy F, Marti G (1998) Performance of calibration standards for antigen quantitation with flow cytometry. Cytometry 33:188–196
34. Brown MC, Hoffman RA, Kirchanski S (1986) Controls for flow cytometers in hematology and cellular immunology. Ann NY Acad Sci 468:93–103
35. Bikoue A, George F, Poncelet P, Mutin M, Janossy G, Sampol J (1996) Quantitative analysis of leukocyte membrane antigen expression: normal adult values. Cytometry 26:137–147

36. Poncelet P, Bikoue A, Lavabre T, Poinas G, Parant M, Duperray O, Sampol J (1991) Quantitative expression of human lymphocyte membrane antigens: definition of normal densities measured in immunocytometry with the QIFI assay. Cytometry 5(Suppl):82–83

37. Wang L, Degheidy H, Abbasi F, Mostowski H, Marti G, Bauer S, Hoffman RA, Gaigalas AK (2016) Quantitative flow cytometry measurements in antibodies bound per cell based on a CD4 reference. Curr Protoc Cyto 75:1.29.1–1.29.14

38. Wang L, Abbasi F, Ornatsk O, Cole KD, Misakian M, Gaigalas AK, He HJ, Marti GE, Tanner S, Stebbings R (2012) Human CD4+ lymphocytes for antigen quantification: characterization using conventional flow cytometry and mass cytometry. Cytometry 81A:567–575

39. Kantor AB, Moore WA, Meehan S, Parks DR (2016) A quantitative method for comparing the brightness of antibody-dye reagents and estimating antibodies bound per cell. Curr Protoc Cytom 77:1.30.1–1.30.23

40. Wood JCS (2009) Establishing and maintaining system linearity. Curr Protoc Cytom 47:1.4.1–1.4.14

41. Bagwell CB, Baker D, Whetston S, Munson M, Hitchco S, Ault KA, Lovett EJ (1989) A simple and rapid method for determining the linearity of a flow cytometer amplification system. Cytometry 10:689–694

42. Wood JCS (1998) Fundamental flow cytometer properties governing sensitivity and resolution. Cytometry 33:260–266

43. Steen HB (1992) Noise, sensitivity, and resolution of flow cytometers. Cytometry 13:822–830

44. Chase ES, Hoffman RA (1998) Resolution of dimly fluorescent particles: a practical measure of fluorescence sensitivity. Cytometry 33:267–279

45. Parks DR, 1 Khettabi FE, Chase E, Hoffman RA, Perfetto SP, Spidlen J, Wood JCS, Moore WA, Brinkman RR (2017) Evaluating flow cytometer performance with weighted quadratic least squares analysis of LED and multi-level bead data. Cytometry A in Press

46. Hoffman RA, Wood JCS (2007) Characterization of flow cytometer instrument sensitivity. Curr Protoc Cytom 40:1.20.1–1.20.18

47. Wood JCS, Hoffman RA (1998) Evaluating fluorescence sensitivity on flow cytometers: an overview. Cytometry 33:256–259

48. Stebbings R, Wang L, Sutherland J, Kammel M, Gaigalas AK, John M, Roemer B, Kuhne M, Schneider RJ, Braun M, Leclere N, Dikshit D, Abbasi F. Marti GE, Porcedda P, Sassi M, Revel L, Kim SK, Marshall D, Whitby L, Jing W, Ost V, Vonski M, Neukammer J (2015) Determination of CD4 + cell count per µL in reconstituted lyophilized human PBMC pre-labelled with anti-CD4 FITC antibody. Cytometry 87A:244–253

49. Kalina T, Flores-Montero J, van der Velden VHJ, Martin-Ayuso M, Böttcher S, Ritgen M, Almeida J, Lhermitte L, Asnafi V, Mendonça A, De Tute R, Cullen M, Sedek L, Vidriales MB, Pérez JJ, te Marvelde JG, Mejstrikova E, Hrusak O, Szczepański T, van Donge, JJM, Orfao A, on behalf of the EuroFlow Consortium (2012) EuroFlow standardization of flow cytometer instrument settings and immunophenotyping protocols. Leukemia 26:1986–2010

50. Wood BL (2016) Principles of minimal residual disease detection for hematopoietic neoplasms by flow cytometry. Cytometry B 90:47–53

51. Lin YC, Winokur P, Blake A, Wu T, Manischewitz J, King LR, Romm E, Golding H, Bielekova B (2016) Patients with MS under daclizumab therapy mount normal immune responses to influenza vaccination. Neurol Neuroimmunol Neuroinflamm 3:e196

52. Dorvignit D, Palacios JL, Merino M, Hernandez T, Sosa K, Casaco A, Lopez-Requena A, Mateo de Acosta C (2012) Expression and biological characterization of an anti-CD20 biosimilar candidate antibody: a case study. Mabs 4(4):488–496

53. Fesnak AD, June CH, Levine BL (2016) Engineered T cells: the promise and challenges of cancer immunotherapy. Nat Rev Cancer 16(9):566–581

54. Hoffman et al. (2012) Cytometry Part A 81A:785

Alternative Approaches for Analysis of Complex Data Sets in Flow Cytometry

Carmen Gondhalekar

Abstract The field of flow cytometry is rapidly evolving as a result of advancements in sample preparation technology. Automated machinery, multi-channel PMTs, multiplexed assay formats, and mass cytometry introduce new opportunities in fields utilizing flow cytometry by creating very rich, albeit complex, data sets. Advances in cytometric data collection have always been accompanied by advances in analytical techniques. As high-content multi-parametric data sets are generated through new flow cytometry techniques and technologies, it is important to simultaneously develop computers and software with the ability to execute six major functions: manipulate an unlimited number of parameters, compare all data parameters in many different combinations, create and customize a data analysis strategy, see real-time changes in all parameters, offer advanced statistical tools, and rapidly report figures and graphs with endpoint conclusions. This chapter investigates methods of achieving each of these six functions by addressing data-analysis challenges in drug-screen assays, multiplexed cytokine assays, hyperspectral cytometry, and mass cytometry.

Keywords Data analysis · Software · Mass cytometry · Hyperspectral cytometry · Drug screens · Multiplexed assays · PlateAnalyzer · Cytospec

1 Introduction

For three decades, flow cytometry data analysis was focused on one to three parameters, then advanced to five to ten and subsequently up to twenty parameters. As the technology improved, data processing also increased in complexity to the point that we have shifted into a new age of research where many parameters can be

C. Gondhalekar (✉)
Weldon School of Biomedical Engineering and College of Veterinary Medicine, Room G221, Lynn Hall, 625 Harrison Street, West Lafayette, IN 47907, USA
e-mail: cgondhal@purdue.edu

© Springer Nature Singapore Pte Ltd. 2017
J.P. Robinson and A. Cossarizza (eds.), *Single Cell Analysis*,
Series in BioEngineering, DOI 10.1007/978-981-10-4499-1_9

201

measured on millions of cells in a very short time period. The computational cost of such complexity increases exponentially when mass cytometry is brought into the picture; the total number of parameters can be as high as a hundred for several million cells. Traditional flow-cytometry analytical techniques were not originally designed for very high-content and high-throughput data, prompting us to consider what the next generation of flow-cytometry data analysis should look like.

If we examine the kind of software that is commercially available today, we can identify a number of popular commercial packages, including FCS express, Flow Jo, FACSDiva, Kalooza, and Winlist. While the degree varies among packages, each hosts a more-or-less friendly interface to allow users to navigate quickly between options for loading data, saving protocols, exporting statistics, and visualizing scatter plots and histograms. For file sizes ranging from a couple of thousand events to a couple of million, data loading time can take from a few seconds to thirty seconds. In fact, one of the fundamental changes made in version 2.0 of the FCS file format [1] was the capacity for the collection and analysis of multiple data sets within a listmode file; this was followed by FCS3.0, which accommodated instruments that could collect more than 100 million events [2]—something previously not attainable. This set the scene for more complex and much larger file sizes, which had previously been restricted by both the file standard and the capacity of disc drives to accommodate such large files.

Manipulating data after they have loaded is also dependent on the number of events and parameters, but very large files can cause sluggish execution of the tasks being demanded of conventional software. The way in which the software reads and stores data files largely determines computational speed. In conventional flow cytometry software, a comprehensive read of every FCS file was practical because users typically needed to analyze one FCS file at a time. However, with large quantities of multiparametric data, and the demand to analyze multiple FCS files simultaneously, new approaches to data management are needed. For example, demands in the area of high-throughput cytometry and the emergence of mass cytometry have created new burdens on data processing. Common data-analysis challenges faced by researchers in these fields will be described in the upcoming sections along with a number of software solutions.

Aside from faster computing speed, what else does conventional data analysis need in order to adapt to advancements and diversification in flow cytometry? We identified six aspects of cytometry analysis approaches that we considered critical to the next generation of cellular analysis.

1. Ability to manipulate an unlimited number of parameters
2. Ability to compare all data parameters in many different combinations
3. Ability to create and customize a data analysis strategy
4. Ability to see real-time changes in all parameters
5. Ability to integrate advanced statistical tools
6. Rapid reporting toolsets (figures and graphs) with endpoint conclusions

Why are the above aspects important in cytometry? The answer lies in adaptation to the changing experimental approaches driven by interdisciplinary science. As technology advances impact the field of cytometry, it is necessary to modify the analytical toolsets. For example, with increased parameter space driven by mass cytometry, performing simultaneous analysis of 50–100 parameters is an absolute need [3]. Another example involves the introduction of hyperspectral flow cytometry [4, 5], where there emerged a need to change the way fluorescence signals are analyzed: instead of the signal intensity within a narrow wavelength range being the primary feature as in traditional flow cytometry, the spectrum itself became a parameter. However, full-spectrum analysis could not be handled with traditional software. Third, a need for change in traditional software also emerged hand-in-hand with automated data collection. Along with robotically driven flow cytometry, the ability to collect thousands of samples in an hour changed both file-structure [6] and file-management principles [7]. Fourth, issues are encountered with increasingly complex assays such as multiplex-bead assays, where separate merchant-specific software is required for data analysis. The following discussion addresses the issues of data management, data display, and computing capability in each of the four cases above through two alternative, free data-analysis packages: *CytoSpec* and PlateAnalyzer. Other practical tools include methods of re-naming parameters in FCS files and extracting raw data from FCS files. These two points are addressed through two additional free software packages: FCS Rename and LData.

2 PlateAnalyzer: Drug-Screen Assays

Automation of sample prep and flow-cytometry instrumentation is one of the main drivers for new techniques in data analysis because data collected in processes such as automated drug screens are stored in ever larger and more complex FCS files [8–11]. One example is provided in a 2012 study conducted by Robinson et al. to examine the effects of a panel of drugs on cell toxicity [12]. A 384-well plate was used to test the toxicity of 32 drugs in a 10-step dilution series using three indicators (dyes) of HL-60 functional integrity. This assay design had to take into consideration 960 samples ($32 \times 10 \times 3$; drug, dosage, dye), with each sample representing 5000 cells, a total of 4.8 million events! This is one of many similar studies regularly used in the field of drug development and can be challenging to analyze using traditional analytical software because of the file size and complexity.

Before the use of automated sample-preparation machinery, flow cytometry data analysis generally focused on single-file analysis—samples were prepped and run one at a time in test tubes. Even when using microtiter plates, analytical techniques focused on reproducing the same traditional analysis model of single-sample analysis. The fundamental problem observed with current commercial analysis software is the degree of difficulty in reading 384 listmode files, bringing all the

(a)

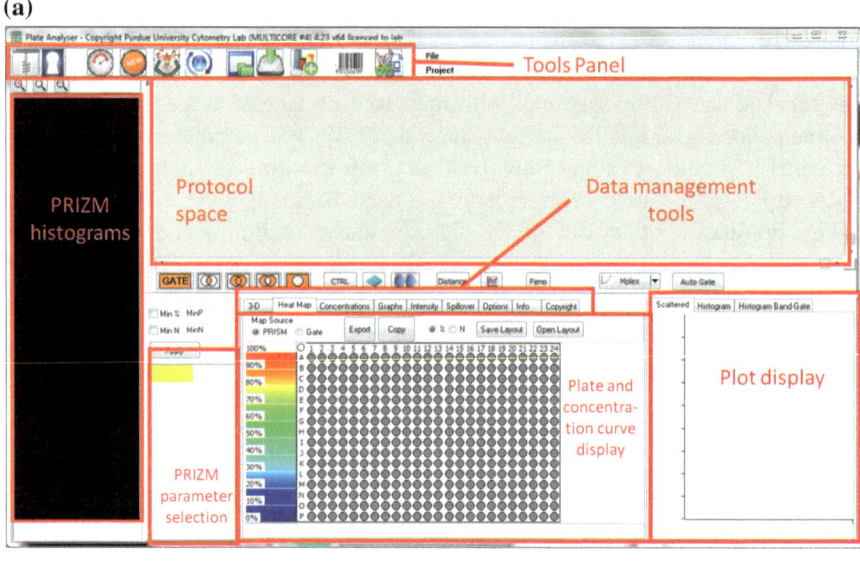

(b)

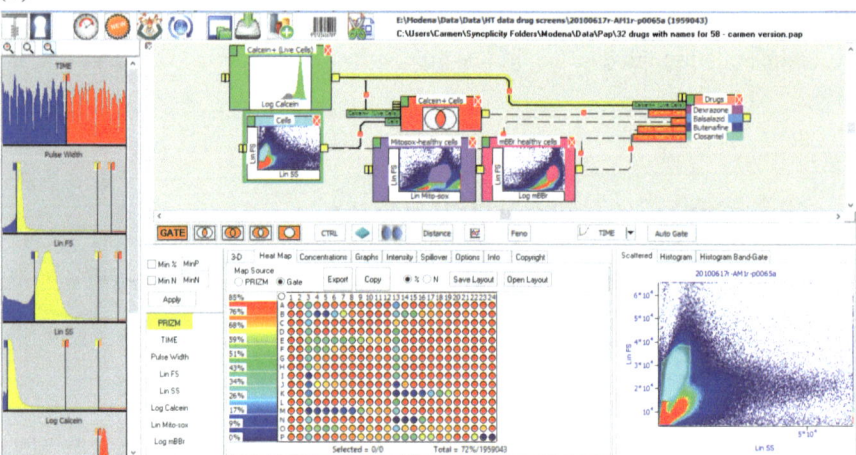

Fig. 1 To accommodate the analysis of complex data sets, PA utilizes methods of opening and closing files similar to those of most analysis software. However, it uploads all FCS files instead of just one. The upload of these files can be visualized in the plate and concentration-curve display. The microtiter-plate map is interactive, allowing the user to select the wells that will be plotted on the plot display. The workspace (**a**) allows the user to design a protocol that applies the gates to FCS files selected through the microtiter-plate display (**b**)—the key to rapid analysis of large data sets. Data provided by Robinson et al. [12]

data together, and reducing the results to dose-response curves (a standard tool for interpreting drug screens). In the example of Robinson's data, the starting process would require gating of light scatter (thus 384 plots), followed by an analytical gate for each of the three assays per well (1152 plots), followed by generation of dose-response curves for each drug in each of the three assays (another 96 plots). As seen in the example above, the bottleneck of the drug-screening process is sample analysis, assuming that one has a mechanism for fast sample preparation (robotics) and collection (also robotics). No commercial software is capable of processing cytometric data in addition to generating IC50 curves. Since the goal of Robinson's assay was to generate dose-response curves, it was necessary to develop a technology that would rapidly perform the required calculations.

For data such as that generated by Robinson et al. in 2012, hours if not days are required to analyze and compare each well of a 384-well plate. The PlateAnalyzer (PA) software was developed specifically to analyze large volumes of complex data quickly and efficiently and fits perfectly into the drug-screening paradigm. The software is not commercial, but is robust and designed to answer the demands of large data set–based flow cytometry.

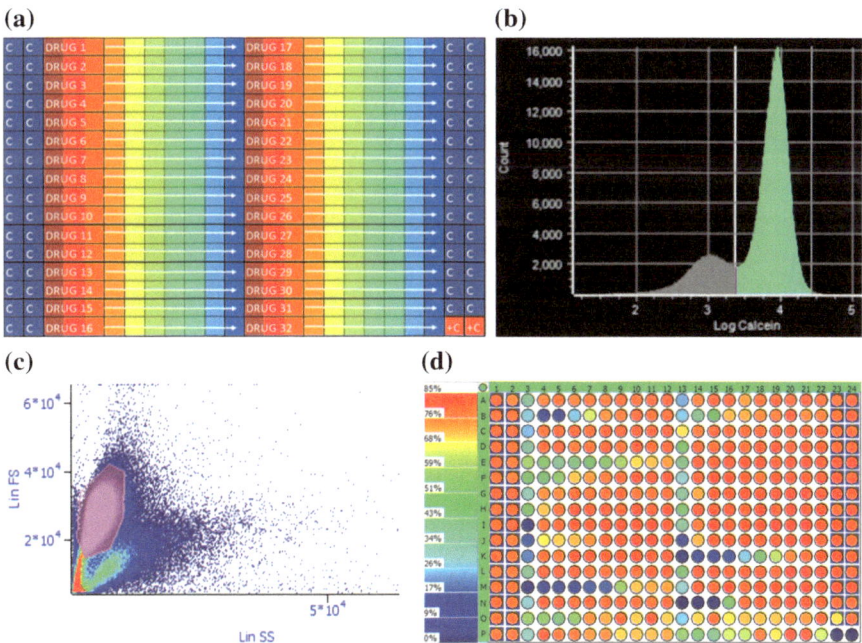

Fig. 2 **a** Drug-screen layout on a 384-well plate (C indicates control, C+ indicates positive control of cells treated with vancomycin); **b** gating live cells can be achieved by selecting cells positive for calcein (one of the three dyes used in the assay); **c** a gate (*pink*, indicating cells) is created on a forward-scatter (FS) versus side-scatter (SS) plot; **d** the number of events in the gate is displayed in the diagram of the 384-well plate as a heat map (*red* indicating a high number of cells). Data provided by Robinson et al. [12]

As previously mentioned, several criteria drive the development of solutions to analysis of high-throughput data sets, including the ability to read large amounts of data very quickly. While this appears to be a simple task, it is actually quite complex. PA utilizes several unique algorithms to increase the speed of data uploading and utilizes data layering to increase the speed of synthesizing large volumes of data. For a data set containing millions of data points, the applied algorithms and data layering shorten the computation time of uploading and sorting data during analysis. A typical set of 384 listmode files can be read in about 10 s, many times faster than pretty much any other software. Not only is the initial read speed faster, but the subsequent re-reading of the data set is almost instantaneous so that recalculations can be made in real time, even for very complex analyses. This feature allows the user to quickly test "what if" scenarios in PA before deciding on a plan for analysis—something often too difficult or time-consuming with traditional analytical approaches. Rapid sifting through data is a critical aspect of software design when dealing with very large data sets; one of the goals of PA is to reduce the total *time to result*.

The layout of PA is relatively fixed, as shown in Fig. 1. This layout shows the user all aspects of the analysis in a single view. Once data are uploaded to PA (Fig. 1b), they can be viewed through a number of interactive displays and graphs. PA has more gating options than most data analysis software, allowing users to make simple to very complex networks of gated populations that permit easy navigation of multi-parametric data. Like most analytical software, PA displays density scatter plots of the data. But unlike all current software packages, PA allows the display of only a single dot plot or histogram at a time (Fig. 2); these two plots are all that is needed, since all the FCS files of the microtiter plate are readily available to be analyzed simultaneously, individually, or in various combinations.

2.1 Alternative Data Displays

Each of the data display options can also be used for gating during the creation of a "logic map" (Fig. 3a). A logic map is a step-by-step pathway of analysis designed by the user. Whereas most analytical software limits the user to deduce the analytical pathway based on a simple display of scatter plots, histograms, and gates, PA clearly shows the analytical pathway as a network. The advantage of having a network is that there is a visible connection between gated plots. The network can be modified so the steps used to gate a certain population can be shuffled. This way, an analytical pathway is not only spatially re-organized, but the order in which populations are gated is changed (thereby changing the output). The ability to visually display the gating logic behind an output is particularly useful in multi-parameter data sets, where selecting for a unique population means selecting according to an organized series of scatter-plot gates. Therefore, logic maps range from being relatively simple (Fig. 3a), to very complex (Fig. 7e)—as is the case in many CyTOF experiments.

(a)

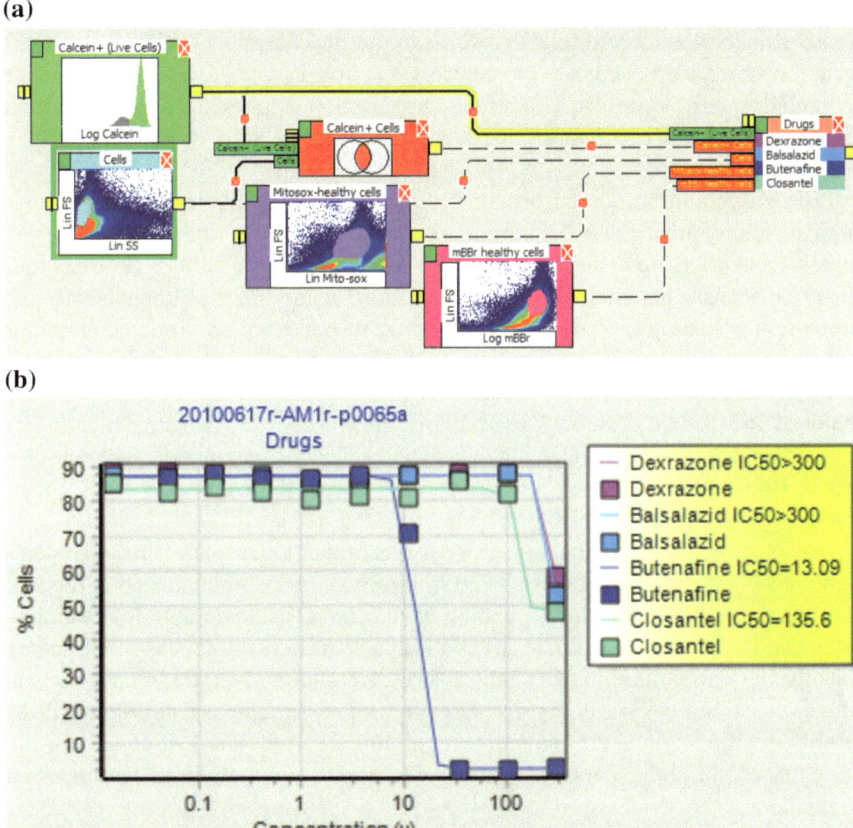

(b)

Fig. 3 **a** In the logic map, a number of different gates were made to investigate the effects of drugs on the dyes displayed in the table on HL-60 cells. One gate identifies calcein-positive cells. The calcein histogram gate can be networked with a forward-scatter (FS) versus side-scatter (SS) gate on HL-60 cells to have tighter criteria for selection. The process can be mimicked with gates for Mito-sox- and mBBr-positive populations. **b** A highlighted connection prompts PA to generate IC50 curves of the relationship between drug dosage and calcein dye intensity. Data provided by Robinson et al. [12]

2.2 Fast Gating of Multiple Parameters

There are essentially no restrictions in gating in PA, as this can be achieved in five ways. The first two are standard for cytometry software—gating on the histogram or the density scatter plot. The next three are less conventional, but highly practical from the standpoint of analyzing large quantities of data distributed across a multi-well plate. One option is to create gates using an automated gating algorithm that identifies dense clusters of data in a scatter plot and automatically applies a gate. This is particularly useful when many populations exist, as will be seen in an example of multiplexed bead analysis for cytokines discussed later in the chapter. Another method is to gate selected wells on the microtiter plate. For example, if a section of the plate is dedicated to controls, the user can create a gate that includes or excludes only controls. The fifth gating option applies a concept originally termed PRISM by Dr. James Wood when it was invented at the beginning of 1985 [13].

PRISM takes information from multiple parameters simultaneously and arranges them in histograms. Each histogram is associated with a linear gate separating negative events (below the threshold) and positive events (above the threshold). Analysis of each parameter involves selecting for the spectral "fingerprint" of a population, meaning the population's specific response to the group of antibodies being used as probes. Therefore, the PRISM function can be used as a gating criterion for selecting cells with different "fingerprints." The selection process is achieved in a single window by arranging the linear gates in automatically generated histograms, and identifying whether the desired population falls above or below the threshold for each histogram. This avoids the complex gating schemes commonly used to accomplish the same objective [13].

Gating populations is the first step in creating a logic map. The next step is taking each gated population displayed in the work-space and connecting the gates to refine the definition of the population of interest. Using available Boolean logic, gates can participate in multiple different analytical networks whose logic path is visually easy to follow. This offers a large advantage over traditional software where a user can easily get lost in the gating logic if multiple gates are needed to select for certain populations among many different samples. In PA, complex gating strategies can be applied and changed for all samples and viewed in real time for all populations.

For Robinson's drug screen, logic maps for three different assays are created for entire 384-well plates in just a few minutes. In conventional software, the final product would be a scatter plot or histogram and a couple of simple statistics regarding the percentage or intensity values of cells that fall inside or outside the gate. However, for the purpose of rapid drug-screening PA takes analysis one step further by making IC50 curves, and providing IC50 values for each drug identified (Fig. 3b). IC50 curves are a baseline evaluation tool for drug development, making its incorporation into flow cytometry analytical software particularly helpful. Within the PA software it is also possible to select any of four algorithms for IC50 generation, and it is also possible to switch off IC50 calculations to reduce demand on computing capacity during the process of designing the analytical logic map. In addition to generating IC50 curves, PA also exports into an excel sheet population data from scatter plots, histograms, and individual microtiter-plate wells.

Using Robinson et al.'s data as an example, we highlight the areas of modern flow cytometry techniques that are not compatible with the traditional analytical software, and resolve a number of these issues through a custom software specifically designed for rapid analysis of high-content, high-throughput data.

3 PlateAnalyzer: Multiplexed Cytokine Assays

Multiplexed bead assays are an example of an efficient way to analyze multiple analytes using very small volumes of sample. These types of assays have been performed for many years and were originally commercialized under the Luminex^{TM} technology in the mid-1990s [14, 15]. Many companies offer kits to perform multiplexed bead assays ranging from single analytes to more than 30. The concept behind these assays is to provide a powerful tool for researchers to study multiple co-occurring factors in a single assay. This clever technique bypasses the obstacle faced for years when multiplexing was limited by fluorescence overlap. It does so by using sets of differentially sized and stained beads. The differences in size and staining of each set of beads act as an ID tag (Fig. 4). Each bead set is conjugated to a different anti-target antibody. The sets are combined to form a heterogeneous mixture that along with a detection antibody against each target is

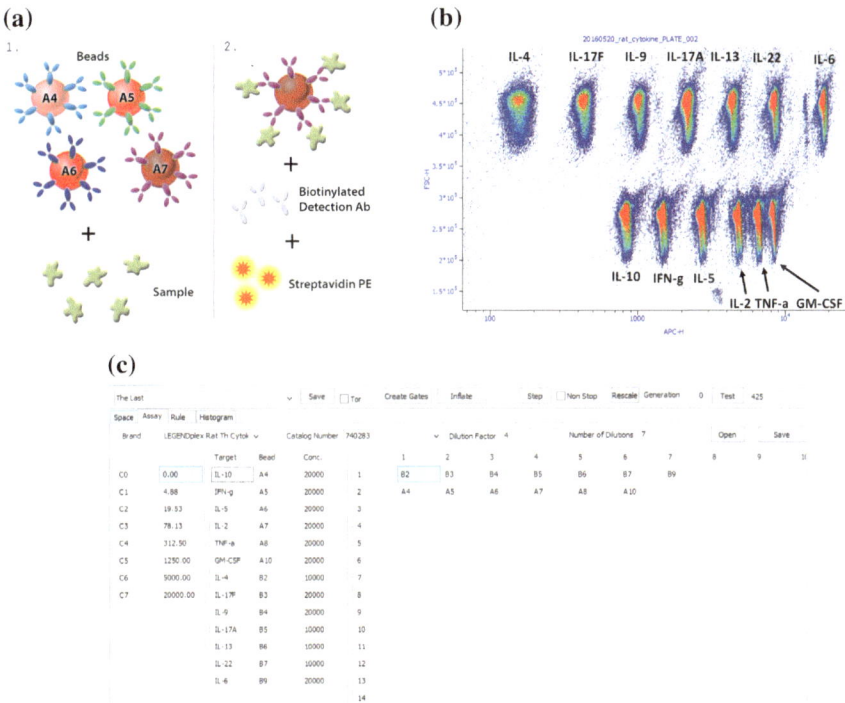

Fig. 4 **a** Biolegend kit (Catalog Number 740402) supplies differentially stained APC beads of two sizes conjugated to thirteen anti-cytokine antibodies. Once beads are introduced to the sample, an antibody conjugated to PE is used to detect the presence of the antigen on the bead. Part of the assay kit constitutes a standard dilution series in which the cytokine concentration is known [16]. **b** The differences between bead sets are observed in a FS versus APC scatter plot, where each bead set is a clear population distributed along the axes. **c** The information on the standard curve can be uploaded to PA through an interactive table

added to a sample containing unknown concentrations of the targets. During analysis, determination of the concentration of each target can be achieved by analyzing the correct bead, identified by its ID tag [16]. The samples can be in either a single test tube or a multi-well plate. The advantages of using a multi-well plate for multi-sample assays were emphasized in the previous section—being more efficient for experiments testing multiple different conditions.

An example of a multiplexed bead assay is demonstrated in an ongoing study where we describe a BioLegend 13-plex bead assay used to determine the cytokine response of plasma from rats experiencing vagus-nerve stimulation and/or treatments of saline or lipopolysaccharide (LPS) injection. In addition to the four treatments, samples are collected from each rat at six time points. A 96-well plate is used: columns 2–11 are loaded with rat plasma samples, while columns 1 and 12 are used to host a known 8-point dilution series of cytokines (these can be placed anywhere on the plate, but are described as being in columns 1 and 12 only for example), beads, and detection antibodies that will later be used to generate standard curves (Fig. 5).

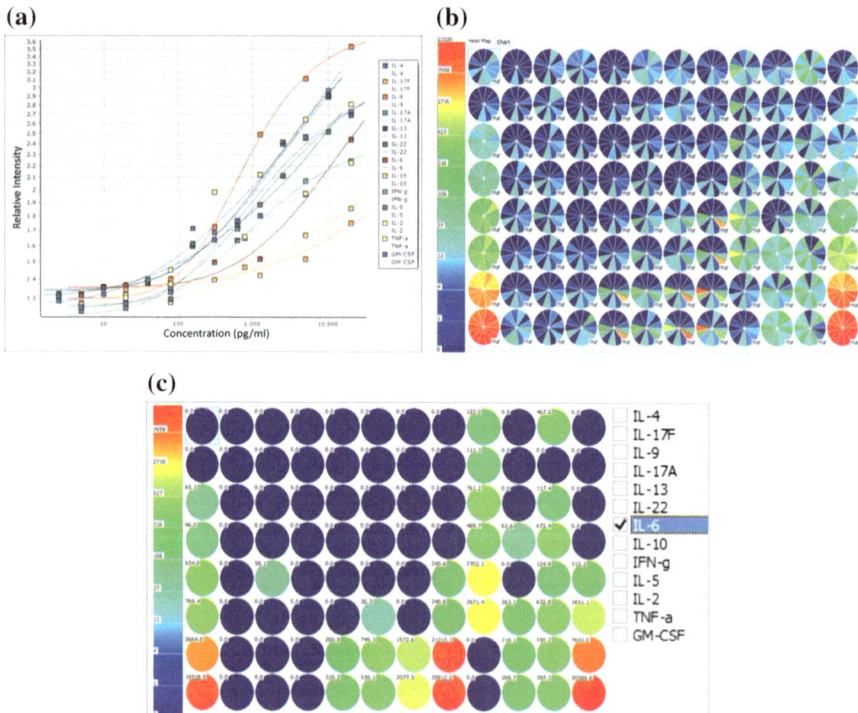

Fig. 5 Once information regarding the standard dilution series is input to the database, calibration curves are automatically generated by PA (**a**) and calculations are performed to determine the concentration (pg/ml) of each cytokine in each well. The calculated results can be qualitatively determined using a colorimetric chart where each pie slice represents a cytokine (**b**), or qualitatively determined by selecting the cytokine of interest (**c**). Data provided by Dr. Joseph Paul Robinson's lab group (unpublished)

3.1 Conventional Versus Alternative Approaches

How would conventional software be used to determine the concentration of each analyte for each well in a 96-well plate? Considering that each well is associated with one FCS file, the user would first upload 96 FCS files. The next step would be to make a protocol, gating each of 13 bead sets, and applying the protocol to all 96 files. The data on gated populations would then be exported (one well at a time) to another program so analyte concentration values could be obtained from a calibration curve that the user must also build. This would equate to 1248 calculations (96 × 13) that would need to be performed using a secondary software package after the relevant data were extracted from the listmode files of each well.

It is evident that multiplexed bead assays on microtiter plates are very challenging to analyze using standard software. To address this obstacle, manufacturers of multiplexed bead assay kits also sell data-analysis software specifically for those assays. However, the software sold by one company is not necessarily compatible with the multiplexed bead assay kit sold by another company, or with an assay designed by the user. Though differences between kits make them incompatible with mismatched software, bead kits have a number of common descriptors. One regards characteristics of the kit: the number of bead sets, how beads are identified, and the concentration and dilution steps of the standard curves. Another characteristic is directly related to the user: arrangement of experimental and standard samples on a microtiter plate and alterations of the kit's intended protocol. Either information regarding each characterizing feature of the kit must be embedded into the software, or the software must allow the user to easily input this information. To explore methods of resolving the latter of the two tasks, PA software uses an interactive assay-plate display (Fig. 5). The user can tell the program which wells were used for the calibration standards by highlighting them and selecting an option that opens an empty table dedicated to the input of assay information (Fig. 4c). The table is simplistic, prompting the user to enter the number of dilution steps, dilution factor, and initial/highest concentration. In the center of the table, the user can indicate the spatial arrangement and ID of the beads according to their location on a scatter plot. For the BioLegend 13-plex bead assay kit indicated previously, this entire process of detailing the kit characteristics and saving it to the PA kit database takes at maximum 5 min. The information input to the table can be saved, thereby embedding information about the kit in the software. The process of loading a saved kit and applying it to another experiment takes seconds. Using these tools, a user can not only build a database of commercial assay kits, but also develop and save his own custom assay protocols.

PA features other characteristics to facilitate multiplexed bead analysis, compared to both standard flow-cytometry software and software specific for multiplexing. One of these is the method of gating. A time-consuming aspect of multiplexed bead assays is gating each of the multiple bead sets. The number of bead sets in a single assay is increasing as more sophisticated ways to multiplex with flow cytometry emerge. Consequently, as more bead sets are added to a single kit, it will take more and more time to gate each population. In addition, between experiments users must gate each bead set based on the same criteria for gate size—though this process is up to the discretion of the user and susceptible to human error. Instead of requiring hand-made gates, PA can automatically identify dense clusters of data points and assign size-adjustable gates to each cluster—a process called auto-gating. This approach not only defines gate size according to a single algorithm applied to every experiment, eliminating human variation in gate creation, but also bypasses the need for manual gating—though this is still an option for an event where a combination of auto and manual gating is required.

3.2   Data Output Possibilities

Once data are appropriately gated, it is convenient to expedite the process of exporting the data regarding each gate. This is particularly important in a multiplexed bead assay, where there may be up to 30 gates. The next generation of flow-cytometry software should not only be capable of exporting data, but also make graphs and charts that are appropriate for the type of study being conducted. In the case of a multiplexed bead assay, such graphs and charts include calibration curves (Fig. 5a), automatic calculation of analyte concentration, and automatic application of those calculations to an easily understood visual display (Fig. 5b, c). Based on this need, PA has an integrated function that reads data from the wells that the user has identified as containing the calibration standards and automatically performs all the concentration calculations. The numerical data can be not only produced, but also displayed in an easily interpretable concentration heat map of all analytes. PA achieves this by arranging data in colorimetric pie charts (each slice indicating a different analyte) arranged in the same format as the microtiter plate.

Multiplexed bead assay kits are a major step forward in the field of flow cytometry because they permit simultaneous analysis of a multitude of variables. This feature is particularly powerful when conducting experiments in which variables of interest form a complex interconnected network. Software to examine such assays are available, but a single software compatible with kits from any manufacturer or even custom-built assays is not available. We have identified a number of unique features important to make this software of the future an effective tool: ability to generate calibration curves, rapid concentration calculations, automatic gating, easily interpretable graphics, and simultaneous analysis and visualization of all wells in the plate. These features as well as those common to most flow cytometry software were incorporated into PA, significantly simplifying calculations and abbreviating the time it takes to analyze a 13-plex bead assay to no more than a minute for an entire plate.

4   PlateAnalyzer: Mass Cytometry

Thus far in the chapter, we have noted two major advancements in the field of flow cytometry—automated high-throughput flow cytometry on microtiter plates and multiplexed bead assays. In both cases, traditional flow cytometry approaches are not designed, or are unable to offer the tools needed, for rapid analysis of very large and complex data sets directly from listmode files. The upcoming generation of software must feature methods by which to expedite sample processing by rapidly loading data containing multiple parameters that can be visualized and manipulated simultaneously. This feature is particularly important in the rapidly growing field of time-of-flight mass cytometry (CyTOF). As mentioned in the introduction, mass cytometry can involve the analysis of a few million events, each described by up to 100 parameters. One mechanism that allows for the use of so many parameters is "bar-coding" cells [17] in each treatment with unique combinations of metals/metal isotopes. If a user is conducting the assay in a microtiter plate where each well is experiencing a different experimental condition, bar-coding allows the user to identify which cell originates from which well if all the wells are combined into a single test tube before being processed by CyTOF. The strategy of condensing all experimental treatments to a single tube reduces the time it takes to run a microtiter plate and avoids washing and clearing the instrument between samples, a somewhat time-consuming process.

To keep pace with CyTOF's emerging barcoding technology [18], providers of barcoding kits such as Fluidigm also offer de-barcoding software as part of the instrument's collection and analysis routines. However, it is quite easy to perform

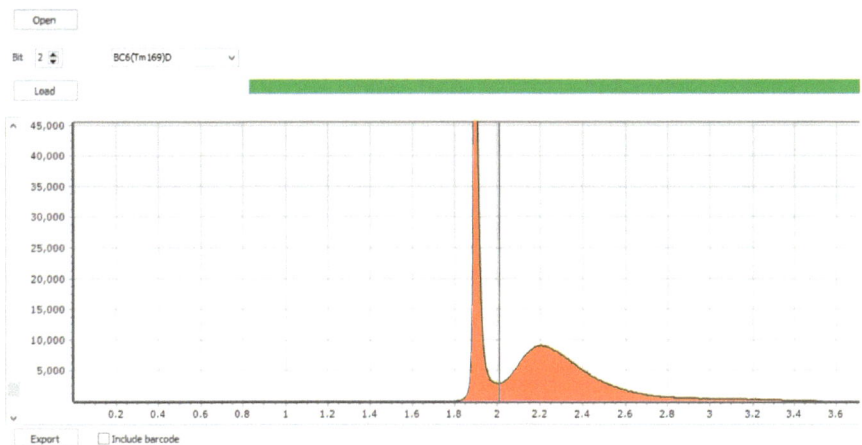

Fig. 6 A dropdown menu lists all the parameters analyzed. BC 1, 2, 3 4, 5, 6, and 7 were the barcodes used in the study by Bodenmiller et al. [3]; each barcode is selected and assigned a bit number. Histograms of each barcode are displayed upon their selection so that a linear gate can be used to select events to the right of the gate and eliminate the noise to the left of the gate

(a) **Staurosporine Concentration**

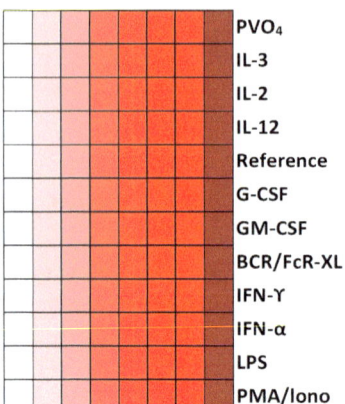

(b)

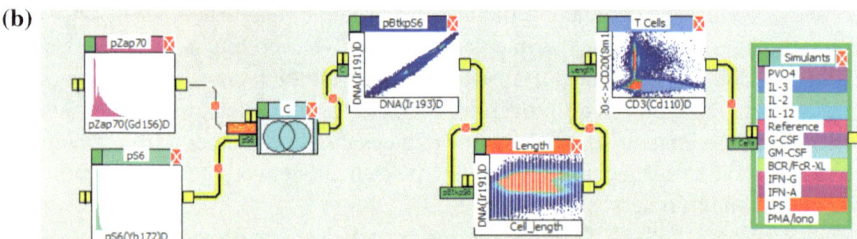

(c)

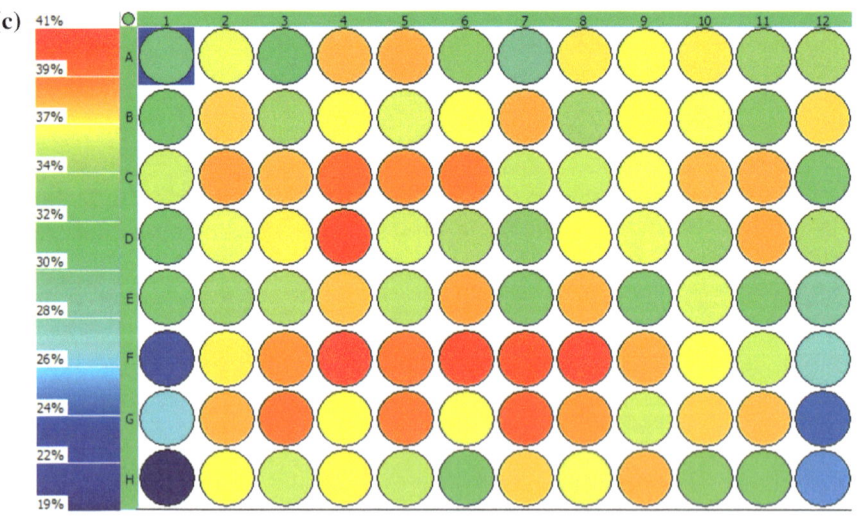

Fig. 7 (continued)

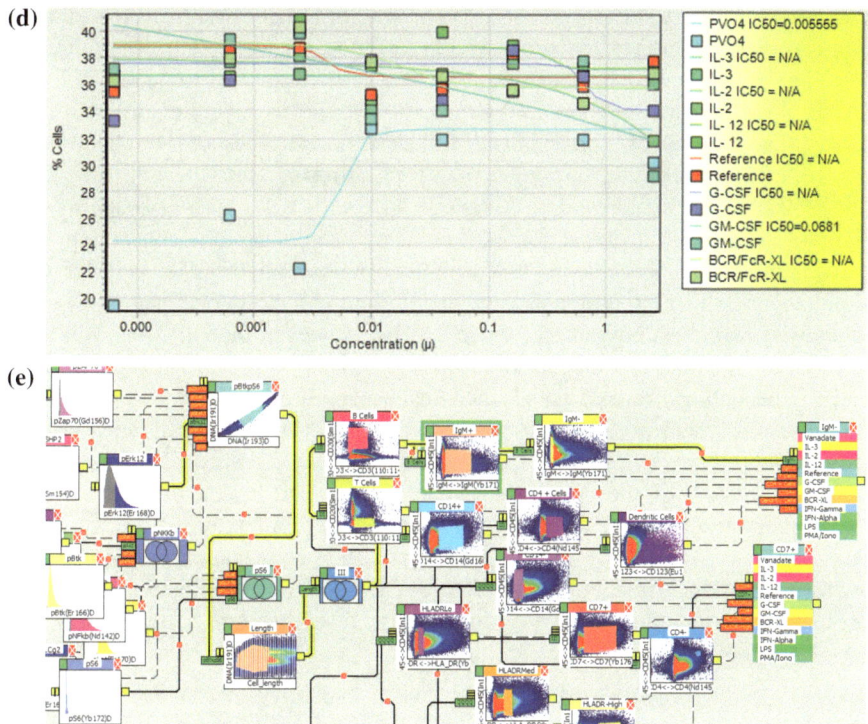

Fig. 7 a Display of the assay layout, indicating the stimulants (*rows*) and the 8-point staurosporine dilution (reproduced from reference [3]) series (*columns*). **b** A simple logic map is constructed to determine phosphorylation of pZap70 and pS6 among live T cells in the presence of certain stimulants and staurosporine. **c** A visual effect of the number of events that fall within the gates determined in the protocol. **d** IC50 curves and values of a 10-point dilution of staurosporine for the gated populations. **e** Significantly more complex logic maps can be created. The screen shot of the logic map shown displays the effect of staurosporine on each cell in the presence of every stimulant. The highlighted path tells the user that data on the population indicated through the highlighted path are being displayed on the IC50 graph and the heat map (both not shown). Data provided by Dr. Bernd Bodenmiller [3]

the "de-barcoding" independently, giving the user more control of the output. PA has a "de-barcoding" feature integrated in the software (Fig. 6), allowing the user to manipulate the extracted CyTOF data independently of the instrument software.

An additional feature of the PA de-barcoding algorithm is that it allows the user to keep or remove the barcode parameters within the de-barcoded data file. If you remove these parameters, the file size is significantly smaller, but you are unable to cross check the de-barcoding process. Once the file is de-barcoded PA appropriately separates the data into the correct well according to the original microtiter-plate assay layout and stores the data from each well as an individual FCS file. Once the data are sorted, the user encounters a similar obstacle as with drug screens and cytokine assays: high-content and multi-parametric data are difficult to analyze on traditional software because such software was originally designed to evaluate a limited number of parameters or events per test tube. An example of the complexity of a data set produced by CyTOF is demonstrated by a study conducted by Dr. Bernd Bodenmiller [3]. The study aimed to examine the effect of different doses of staurosporine on 14 phosphorylation sites in 10 types of peripheral blood mononuclear cells in the presence of 12 stimuli (Fig. 7a). In sum, this would be 2352 8-point dose-response titrations (14 cell types × 14 phosphorylation sites × 12 stimuli). Comparison of each well would be time-consuming using conventional software, making PA's ability to view all the data at once while simultaneously making protocols very useful. Figure 7 shows one method by which to analyze two phosphorylation sites in T cells. However, the logic map can be far more extensive than that displayed in this figure, such as the one in Fig. 7e. At the end of the gating process, the desired results are IC50 curves to evaluate dose response. As discussed in the drug-screen example above, PA can generate IC50 curves without needing to export data into a different software (Fig. 7d). Another important feature of the approach we have developed is that one can visualize changes in any parameter displayed on the logic map in real time. For example, if one wished to evaluate the impact of a drug only on cells that expressed very high levels of an antigen (e.g., only the brightest CD3 cells) one could move the gate of the CD3 cell population and in real time observe dose-response curves for any other set of parameters.

5   Renaming FCS Files

For any analysis software, an important aspect to consider, one which is often overlooked, is the proper naming of parameters. Before data are collected on a flow cytometer, parameter names are created and assigned to one of the cytometer PMTs. These parameter names are then reflected on the software displaying data as they are collected on the instrument, and are integrated into the resulting FCS file. However, if parameters are not named *before* data collection they often remain titled FL1 or FL2, etc. Names such as FL1 or FL2 are occasionally included on papers and presentations without appropriate explanation as to the actual parameter identity—something that is unacceptable in either publications or laboratory records. Another significant disadvantage and source of confusion when parameters remain unnamed is the lack of accurate metadata when data sets are sent to collaborators for analysis. On some occasions familiar to many, digital data files are shelved for some time and notebooks containing details on the assay and parameter names are lost. In such cases, the parameters associated with FL1, FL2, etc. may be forgotten or inaccurately remembered, making analysis of otherwise high-quality data impossible. Currently, there is no software capable of changing parameters names in an FCS file because of the complex relationship between the binary data and the file description information at the beginning of the FCS file format. In addition, this task would be computationally expensive if the data were to be analyzed by the majority of analytical software packages because every FCS file would need to have its parameters re-named during analysis. This is especially problematic when dealing with microtiter-plate assays, which can have hundreds of FCS files per plate. In the FCS file specification, the starting block of data in the binary file is determined when the file is created. Changing any parameter information in the American Standard Code for Information Interchange (ASCII) data file will change that binary starting block and render the file unreadable by any flow cytometry software. Therefore, changing parameter names is not a simple task. The program FCS Rename (Fig. 8) was constructed to rename an FCS file so that when data are visualized on analysis software the correct parameter is listed even if it was not collected when the sample was run on the flow cytometer. With standard flow software, every FCS file of a microtiter plate would need to be renamed prior to analysis because the software gathers parameter information from each FCS file as it is being read. If parameter information were collected from only one FCS file, and applied to the entire plate of data, then one would not need to rename every FCS file, but only the first. This is in fact how PA reads data files, making the application of FCS Rename very practical for data sets consisting of hundreds of FCS files. In the next section, we will explore a program that can allow access to all the details of the FCS file, making certain previously inaccessible information accessible.

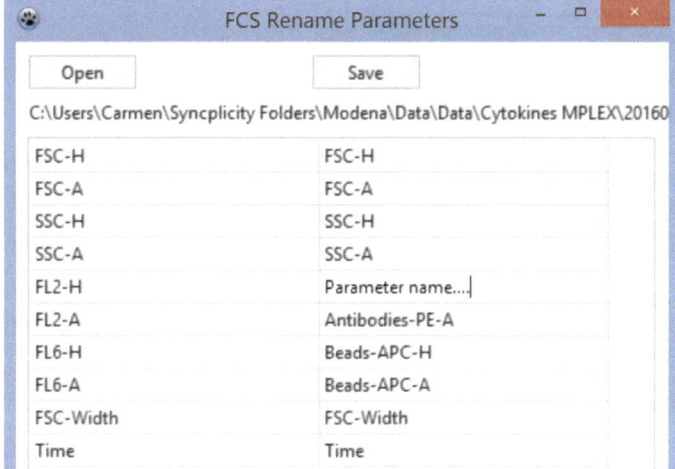

Fig. 8 Parameters from the multiplexed cytokine example described previously are being re-named from the titles indicated on the left-hand side of the column to the titles listed on the right-hand side. The re-named parameters will then be displayed on the axes of graphs and charts made in PA

6   LData: Extracting Raw Data from an FCS File

As seen in the previous section, FCS Rename allows the user to access the parameter names of an FCS file in a very specific manner. On occasion, a deeper level of access to the FCS file is needed to extract raw data.

The program LData (Fig. 9) was developed to transform an FCS file into an ASCII or a.txt file. In this format, raw data from the FCS file can be input to an excel file, parameter names can be manually changed, and the flow cytometer data-collection settings can be viewed. In a clinical laboratory, some information contained in the FCS file must remain confidential as required by the Institutional Review Board (IRB) [19] because it regards a patient's personally identifiable information (PII). For this reason, LData was designed with the ability to erase PII prior to data export, a function called "Incognify."

Using a software such as LData to access information in an FCS file in a form that can easily be changed, exported, and modified is a very powerful tool for flow cytometrists. It can be used in developing software that manipulates data in new and interesting ways, such as PA, FCS Rename, and CytoSpec—software designed for a new generation of flow cytometers discussed in the next section.

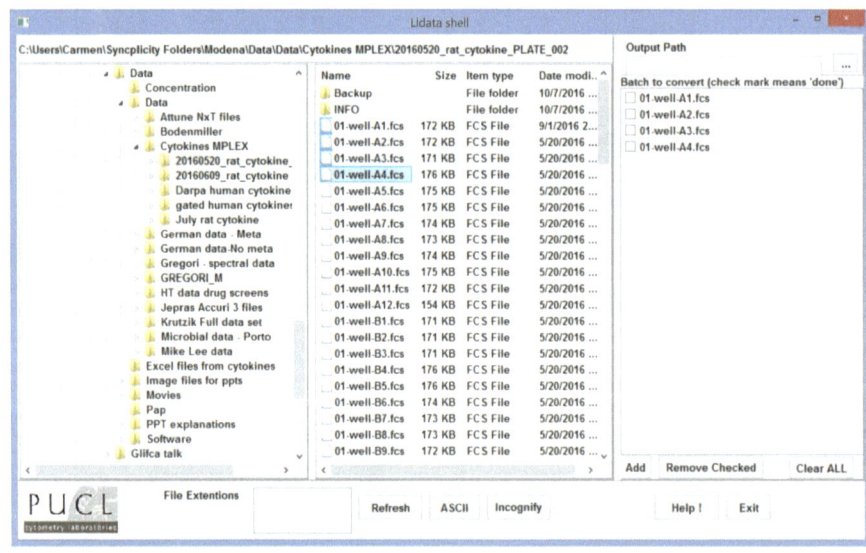

Fig. 9 The LData program interface allows the user to select the file/files to convert to ASCII or comma-separated (.txt) format. Once the data are selected and added into the table on the right, the ASCII option is selected to convert files. The "Incognify" option is specific for FCS files containing personally identifiable information (PII), such as FCS files on patient samples. "Incognify" obscures PII to facilitate data sharing

7 Cytospec: Analyzing Hyperspectral Data

Flow cytometers typically use PMTs to amplify and detect signals with a specific wavelength, each PMT having a single channel. Whereas the early days of flow cytometry used two to three PMTs for two- to three-color detection, recent commercial instruments can detect 10–12 colors using more single-channel PMTs [20, 21]. Though increasing the number of PMTs also increases the degree of multiplexing, there is a limit to the number of colors one can detect simultaneously owing to spectral overlap between fluorophores, the number of lasers available, the fluorophores, and the number of channels. Spectral overlap can to a large degree be resolved through a process of compensation [22]. Increasing the number of PMTs has additional limitations: sensitivity is compromised owing to loss of signal as it passes through more filters, and the design of small instruments becomes more challenging [23].

To address the need of multi-color detection in flow cytometry while overcoming the limitations of traditional systems, a study was conducted in 2004 to construct a prototype hyperspectral flow cytometer in which only one 32-channel PMT was used to detect fluorophores simultaneously per event [4], the equivalent of having a flow cytometer with 32 PMTs. Hyperspectral cytometry places less emphasis on signal intensity at specific wavelengths, and more emphasis on detecting a broad range of wavelengths. This allows for emission profiles of fluorophores to be recorded, not just their intensity at a specific wavelength. In fact, spectral cytometry uses the entire spectrum as a parameter. To maximize the number of fluorophores that could be detected in a single event, an additional study was conducted in 2012 that utilized an approach mathematically similar to compensation: spectral un-mixing [23]. Traditional compensation methods that are still applied in today's flow cytometry software involve mathematically eliminating the fluorescence signal that spills into the wrong channel [24, 25]. However, this process can eliminate potentially useful information. To use the information from spectral spillover in the 2012 study [23], a new and increasingly accepted method based on techniques in image analysis mathematically considers the spilled-over signal as part of the signal of interest. Doing so places more emphasis on measuring a fluorophore spectrum instead of a single emission wavelength and more accurately describes the number of fluorophores present in the analyte (Fig. 10). Approaching spectral un-mixing from the angle of image analysis offers other advantages. The formulas used for spectral un-mixing in image analysis do not assume that the number of detectors equals the number of fluorophores [22]. However, the concept applied to conventional flow cytometry is that the number of PMTs present or "on" is equal to the number of fluorophores being detected. The traditional spectral-compensation approach poses complications for analyzing data produced by new cytometers (where the PMTs are all "on" when the instrument is operating) and hyperspectral cytometry (a single 32-channel PMT). Since hyperspectral cytometry data cannot be analyzed by traditional software because of the constraints needed for compensation, the software *CytoSpec* was created. *CytoSpec*

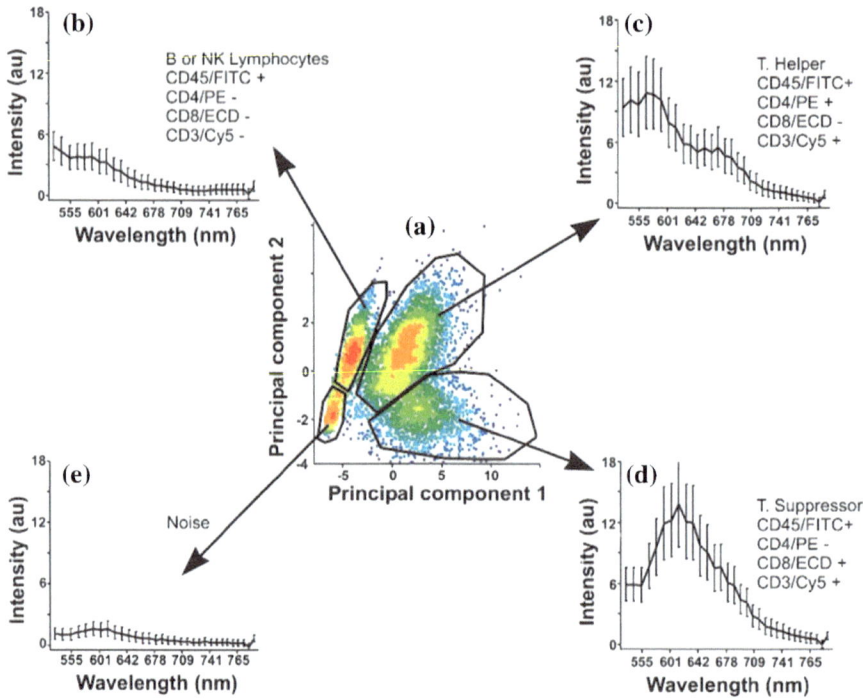

Fig. 10 Image reproduced from Grégori et al. [23]. A spectral analysis of the principal components can be conducted to identify subsets of cells based on the spectra and spatial arrangement characterizing each population

applies spectral un-mixing techniques from image analysis to accommodate the mismatch between the individual PMTs and fluorophores, and utilizes spectral spillover to detect and identify the emission profile of all fluorophores in the analyte.

7.1 Principal Component Analysis

Like most cytometry software, *Cytospec* was designed for not only unmixing spectral data, but analyzing it as well. In multi-parametric studies such as those performed in hyperspectral cytometry, the relationship between parameters is often of interest. To tease apart these relationships, standard flow cytometry data-analysis techniques call for scatter plots comparing each variable or combinations of variables. Given a 10-color data set, 45 scatter plots would be needed to examine different combinations of 2 parameters. This number significantly increases as combinations of parameters are compared, and data analysis becomes challenging for those conducting the study. *Cytospec* provides the user with the option of performing principal component analysis (PCA), a function that analyzes all data points to identify and group observations according to patterns in the data. The output is a list of principal components, each representing an axis of a coordinate system designed to visually identify patterns in the data (Fig. 11). Until recently, software has not incorporated PCA in the toolset for data analysis because flow cytometry software was originally designed for analyzing much simpler 2- or 3-color data sets [26]. However, to accommodate highly multiplexed studies conducted in hyperspectral cytometry and multi-color flow cytometry, PCA was incorporated into *CytoSpec*. Through *CytoSpec*, PCA can be performed on any number of parameters collected by the flow cytometer and generates a list of principal components that can then be compared using scatter plots. This facilitates

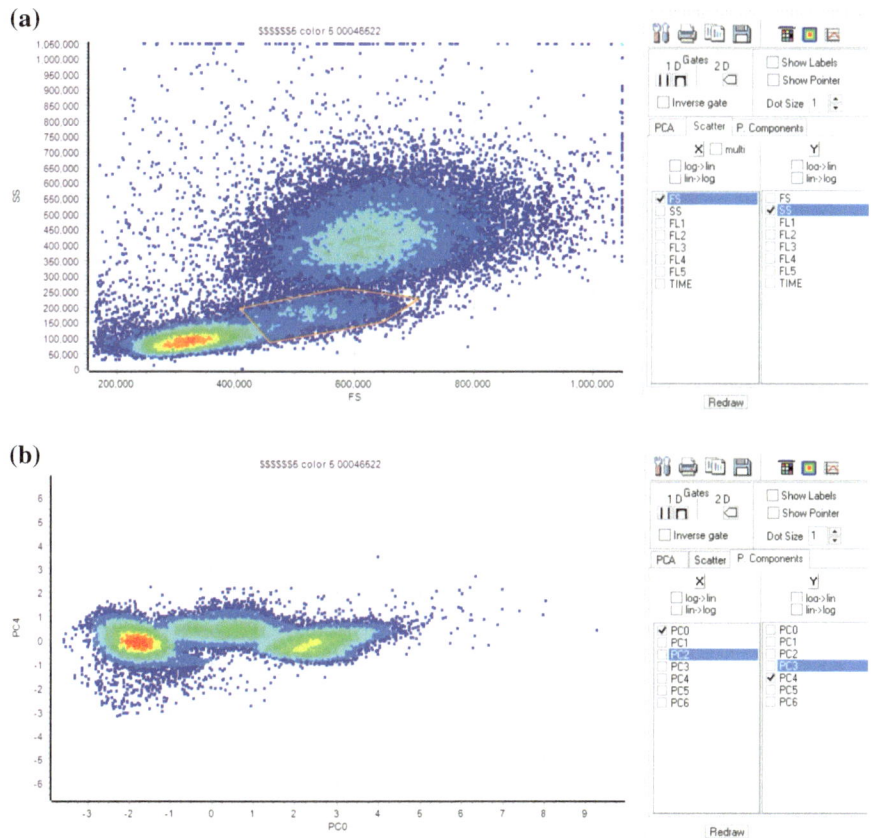

Fig. 11 Human plasma was collected and run on a flow cytometer to observe populations of lymphocytes, monocytes, and granulocytes. **a** CytoSpec was used to analyze the data and create scatterplot displays. **b–d** Cytospec can perform principal component analysis on a selected number of parameters. Different combinations of principal components (PCs) were compared to identify unique groups in the data. Data provided by Dr. J. Paul Robinson's lab group (unpublished)

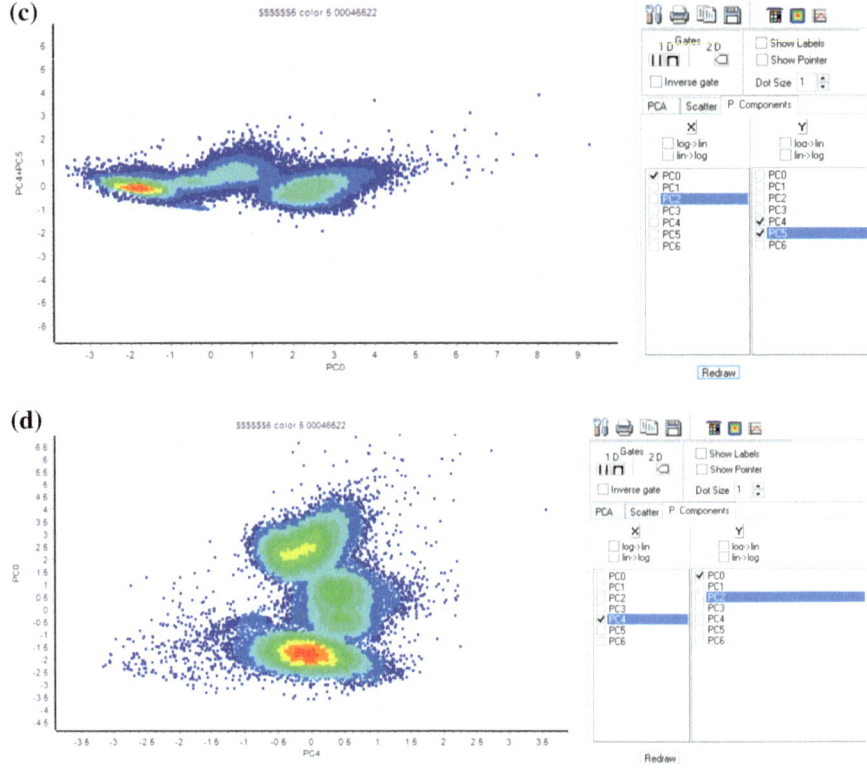

Fig. 11 (continued)

elimination of noise and identification of unique populations that would otherwise have remained invisible.

Elimination of noise is an important part of flow cytometry. Scatter plots of parameters and principal component analyses are both methods by which to discriminate signal from noise. In cases where there is a large degree of noise, populations with low counts can be obscured, and statistics extracted from gated populations can be affected. It is a common feature among analytical software packages to gate out noise, but it may also be desirable to create a new data file on the population of interest that omits most noise or unwanted populations. CytoSpec enables users to generate new FCS files containing only information on a gated set of data points.

In summary, *CytoSpec* is a software tool capable of processing both standard and hyperspectral flow cytometry data using new analytical techniques such as principal component analysis, spectral fingerprint identification, and generation of new FCS files with reduced noise.

8   Conclusion

Instrumentation for fluorescence-based flow cytometry has advanced significantly since its origins in the 1950s. It is noted by Dr. Howard Shapiro that during this time, improvements in flow cytometry automation were out-pacing advancements in data analysis—primarily because computers were largely unavailable to flow cytometrists [27]. As computers became more common, the ability to process the large amount of data produced by flow cytometers caught up to the advancements in instrumentation. Now that the field of flow cytometry has made another great leap forward with automated sample prep, high-throughput sampling, and multi-parametric studies, the supporting analytical tools must again keep pace. This chapter highlights several broad areas of study in which flow cytometry is rapidly advancing: high-throughput drug screens, multi-parametric studies with CyTOF, multiplexed bead assays, and hyperspectral cytometry. Among these areas, six aspects were identified as being of primary importance for the next generation of analytical software:

1. Ability to manipulate an unlimited number of parameters
2. Ability to compare all data parameters in many different combinations
3. Ability to create and customize a data-analysis strategy
4. Ability to see real-time changes in all parameters
5. Ability to integrate advanced statistical tools
6. Rapid reporting toolsets (figures and graphs) with endpoint conclusions.

Based on these six aspects, PlateAnalyzer, *CytoSpec*, FCS Rename, and LData were created to explore a range of software solutions. For high-throughput drug screens, these solutions included simultaneous data visualization, creation and modification of simple-to-extensive logic maps, and generation of IC50 curves, all within the same software. A very similar strategy was taken for analyzing CyTOF data, with the added feature of de-barcoding samples and re-arranging them according to the original microtiter plate array. Multiplexed bead assay analysis was simplified and expedited by using auto-gating systems and ready-to-upload assay databases containing modifiable information on the assay layout. Hyperspectral cytometry posed a new challenge, given that the output, a spectrum of fluorophores, was significantly different from that of conventional cytometers. The software approach tested through *CytoSpec* involved spectral un-mixing and pattern-recognition techniques new to the flow cytometry field. As support for PA and *CytoSpec*, FCS Rename and LData were created to permit the user to change parameter names or extract raw data from FCS files. Information for free download of all four software packages is available on http://www.cyto.purdue.edu/Purdue_software. By creating these data-analysis software options, we hope to form a basis for the next generation of commercial flow cytometry software, which will be geared towards simplifying and expediting high-throughput, multiparametric studies.

References

1. Dean PNBC, Lindmo T, Murphy RF, Salzman GC (1990) Introduction to flow cytometry data file standard. Cytometry 11:321–322. doi:10.1002/cyto.990110302
2. Seamer LC, Bagwell CB, Barden L, Redelman D, Salzman GC, Wood JC, Murphy RF (1997) Proposed new data file standard for flow cytometry, version FCS 3.0. Cytometry 28:118–122
3. Bodenmiller B, Zunder ER, Finck R, Chen TJ, Savig ES, Bruggner RV, Simonds EF, Bendall SC, Sachs K, Krutzik PO, Nolan GP (2012) Multiplexed mass cytometry profiling of cellular states perturbed by small-molecule regulators. Nat Biotechnol 30:67–858. doi:10.1038/nbt.2317
4. Robinson JP (2004) Multispectral cytometry: the next generation. Biophotonics Intl 36–40. doi:10.1017/S1431927605510328
5. Grégori G, Patsekin V, Rajwa B, Jones J, Ragheb K, Holdman C, Robinson JP (2012) Hyperspectral cytometry at the single-cell level using a 32-channel photodetector. Cytometry Part A 81:35–44. doi:10.1002/cyto.a.21120
6. Robinson JP, Durack G, and Kelley S (1991) An innovation in flow cytometry data collection & analysis producing a correlated multiple sample analysis in a single file. Cytometry 12:82–90. doi:10.1002/cyto.990120112
7. Edwards BS, Kuckuck F, Sklar LA (1999) Plug flow cytometry: An automated coupling device for rapid sequential flow cytometric sample analysis. Cytometry 37:156–159. doi:10.1002/cyto.990120112
8. Bocsi J, Tárnok A (2008) Toward automation of flow data analysis. Cytometry Part A 73A:679–680. doi:10.1002/cyto.a.20617
9. Chen X, Hasan M, Libri V, Urrutia A, Beitz B, Rouilly V, Duffy D, Patin É, Chalmond B, Rogge L, Quintana-Murci L, Albert ML, Schwikowski B (2015) Automated flow cytometric analysis across large numbers of samples and cell types. J Clin Immunol 157:249–260. doi:10.1016/j.clim.2014.12.009
10. Finak G, Perez J-M, Weng A, Gottardo R (2010) Optimizing transformations for automated, high throughput analysis of flow cytometry data. BMC Bioinform 11:546. doi:10.1186/1471-2105-11-546
11. Black CB, Duensing TD, Trinkle LS, Dunlay RT (2011) Cell-based screening using high-throughput flow cytometry. Assay and Drug Dev Technol 9:13–20. doi:10.1089/adt.2010.0308
12. Robinson JP, Patsekin V, Holdman C, Ragheb K, Sturgis J, Fatig R, Avramova LV, Rajwa B, Davisson VJ, Lewis N, Narayanan P, Li N, Qualls CW Jr (2013) High-throughput secondary screening at the single-cell level. J Lab Autom 18:85–98. doi:10.1177/2211068212456978
13. Wood J, *Personal Communication on PRISM Processor*, J.P. Robinson, Editor. 2011
14. Elshal MF, McCoy JP (2006) Multiplex bead array assays: performance evaluation and comparison of sensitivity to ELISA. Methods 38:317–323. doi:10.1016/j.ymeth.2005.11.010
15. Horan P, Wheeless L (1977) Quantitative single cell analysis and sorting. Science 198:149–157. doi:10.1126/science.905822
16. BioLegend (2014) Principle of the Assay. Available from: http://www.biolegend.com/legendplex
17. Krutzik PO, Nolan GP (2006) Fluorescent cell barcoding in flow cytometry allows high-throughput drug screening and signaling profiling. NatMethods 3:361–368. doi:10.1038/nmeth872
18. Bendall SC, Simonds EF, Qiu P, Amir el AD, Krutzik PO, Finck R, Bruggner RV, Melamed R, Trejo A, Ornatsky OI, Balderas RS, Plevritis SK, Sachs K, Pe'er D, Tanner SD, Nolan GP (2011) Single-cell mass cytometry of differential immune and drug responses across a human hematopoietic continuum. Science 332:96–687. doi:10.1126/science.1198704

19. US Department of Health and Human Services. Guidance regarding methods for de-identification of protected health information in accordance with the health insurance portability and accountability act (HIPAA) privacy rule. Available from: http://www.hhs.gov/hipaa/for-professionals/privacy/special-topics/de-identification/index.html

20. Bendall SC, Nolan GP, Roederer M, Chattopadhyay PK (2012) A deep profiler's guide to cytometry. Trends Immunol 33:323–332. doi:10.1016/j.it.2012.02.010

21. BioLegend (2014) History of Flow Cytometry. Available from: http://www.biolegend.com/historyofflow

22. Novo D, Grégori G, Rajwa B (2013) Generalized unmixing model for multispectral flow cytometry utilizing nonsquare compensation matrices. Cytometry Part A: J Intl Soc Anal Cytol 83:508–520. doi:10.1002/cyto.a.22272

23. Grégori G, Patsekin V, Rajwa B, Jones J, Ragheb K, Holdman C, Robinson JP (2012) Hyperspectral cytometry at the single-cell level using a 32-channel photodetector. Cytometry Part A 81A:35–44. doi:10.1002/cyto.a.21120

24. Bagwell CB, Adams EG (1993) Fluorescence spectral overlap compensation for any number of flow cytometry parameters. Ann NY Acad Sci 677:84–167. doi:10.1111/j.1749-6632.1993.tb38775.x

25. Roederer M (2002) Compensation in flow cytometry. Curr Protoc Cytom 1–14. doi:10.1002/0471142956.cy0114s22

26. Loken MR, Lanier LL (1984) Three-color immunofluorescence analysis of Leu antigens on human peripheral blood using two lasers on a fluorescence-activated cell sorter. Cytometry 5:151–158. doi:10.1002/cyto.990050209

27. Shapiro HM (2005) History. In: Practical flow cytometry. Wiley, Hoboken, pp 73–100

Photon Detection: Current Status

Masanobu Yamamoto

*I therefore take the liberty of proposing for this hypothetical
new atom, which is not light but plays an essential part of every
process of radiation, the name **Photon**—Lewis [1]*

Abstract Fluorescence analysis at low-level light intensity is important and
inevitable for flow cytometry and cell biology. The photomultiplier (PMT) has been
used as a photon-detection device for many years because of its high sensitivity; it
can amplify a single photoelectron to millions of electrons by a cascade of dynodes
in a vacuum. In addition, the photocathode in the PMT has the advantage of a wide
detection area and wide dynamic range through analog photocurrent detection.
Recently, microelectromechanical system (MEMS)-based PMTs and many
solid-state sensors such as Si photodiodes (PDs), avalanche photodiodes (APDs),
and Si photomultipliers (SiPMs) have been developed and improved in UV to
near-IR wavelengths. Advancements in photosensors especially for photon detection and potential applications are described in this chapter.

1 Single-Photon Detection and Photocurrent Detection

Since Max Planck suggested that "radiation is quantized as quanta" in 1900 and
Albert Einstein proposed the quantum theory of light in *Annus Mirabilis* (Miracle
Year) 1905, the properties of the light particle known as the **photon** have become
well known and commonly understood in this 21st century [2]. From its beginnings
in the 1950s, flow cytometry has developed into an increasingly powerful tool for
cell-population analysis through detection of fluorescence photons from conjugated
cell surfaces. On the other hand, for the past half-century the photon signal has been
detected by an analog photocurrent signal from vacuum PMTs. Based on new

M. Yamamoto (✉)
Basic Medical Science, Purdue University, West Lafayette, IN 47907, USA
e-mail: my@cyto.purdue.edu

M. Yamamoto
CTO, Miftek Corporation, 1231 Cumberland Avenue, West Lafayette, IN 47906, USA

© Springer Nature Singapore Pte Ltd. 2017
J.P. Robinson and A. Cossarizza (eds.), *Single Cell Analysis*,
Series in BioEngineering, DOI 10.1007/978-981-10-4499-1_10

understanding of detection systems, it is likely that the feasibility of single-photon detection by the latest photon sensors and electronics will contribute significantly to next-generation cellular analysis. That is the motivation and objective for this chapter.

Flow cytometry uses "molecules of equivalent soluble fluorochrome" (MESF), a concept formalized by 2004 [3], for sensitivity evaluation. MESF refers to a standard fluorochrome solution and is useful to compare various data obtained under similar detection configurations. But in order to study physical characteristics of photons, it is necessary to go back to the système international d'unités (SI units) to investigate as physical energy-packet characteristics.

Planck and Einstein defined the photon energy of an electric magnetic wave as proportional to frequency, with Planck's constant as the proportionality constant:

$$E = h\nu \tag{1}$$

where h is Planck's constant (6.626×10^{-34} Js) and ν is the radiation frequency (1/s). The wave frequency is calculated as follows:

$$\nu = c/\lambda \tag{2}$$

where c is the speed of light (3.0×10^8 m/s) and λ is the wavelength in meters. Combining (1) and (2),

$$E = h\nu = hc/\lambda \tag{3}$$

where E is the photon energy in Joules (J). Expressing wavelength λ in nm and energy in eV, Eq. (3) is simplified as

$$E = 1240/\lambda \quad eV/nm \tag{4}$$

where 1 eV = 1.60×10^{-19} J.

From Eq. (4), photon energy versus wavelength is calculated in Fig. 1. Values for typical laser wavelengths in flow cytometry are:

405 nm:	740 THz	3.06 eV	488 nm:	614 THz	2.54 eV
532 nm:	563 THz	2.33 eV	594 nm:	504 THz	2.08 eV
633 nm:	473 THz	1.95 eV	780 nm:	384 THz	1.58 eV

Owing to the very small energy per photon, the number of photons per picoW (pW = 10^{-12}W) is a surprisingly large number, expressed as megacounts per second (Mcps):

405 nm: 2.04 Mcps/pW	488 nm: 2.46 Mcps/pW
532 nm: 2.68 Mcps/pW	594 nm: 3.00 Mcps/pW
633 nm: 3.20 Mcps/pW	780 nm: 3.95 Mcps/pW

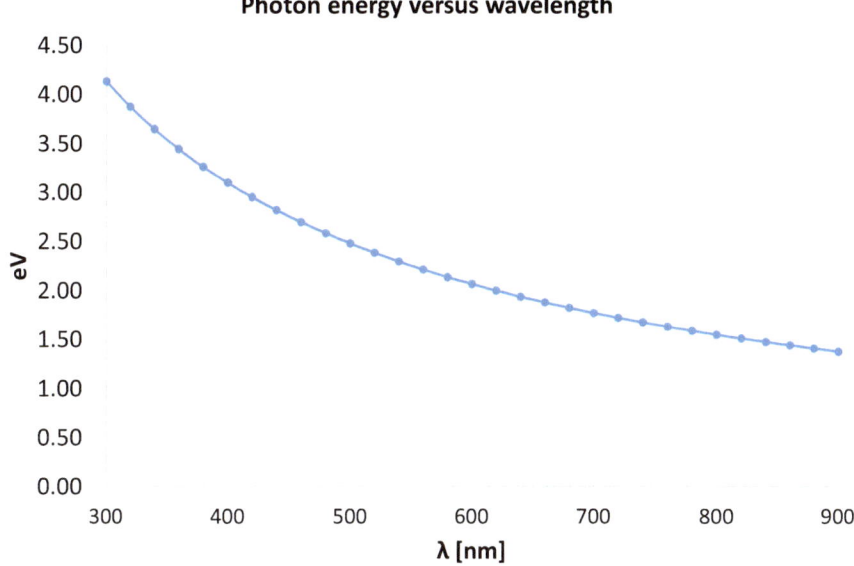

Fig. 1 Single-photon energy measured in eV (1.60×10^{-19} J) as a function of wavelength λ (nm)

In general, one pW is the lowest detection limit of a PMT photocurrent signal.

The photon pulse detected by a sensor is not equal to the incident photon number. The conversion coefficient from photon to detected photoelectron (PE) is called the quantum efficiency (QE). Depending on sensor efficiency and wavelength, the maximum QE lies in the range of 0.2 to ~0.9. Conversely, detection of one photon per second (if possible) is the ultimate in photo detection efficiency. One-photon energy is calculated as follows;

405 nm: 0.48 aW/photon	488 nm: 0.40 aW/photon
532 nm: 0.37 aW/photon	594 nm: 0.33 aW/photon
633 nm: 0.31 aW/photon	780 nm: 0.25 aW/photon

where aW is an attoWatt, equal to 10^{-18} W.

The calculated photon number as a function of optical energy is shown in Fig. 2.

In general, a photon sensor has thermal noise, known as dark count, in the range of 1 cps to ~1 Mcps. Dark count is sensitive to temperature and determines the detection limit. In addition, dark-count standard deviation per second (σ) and coefficient of variation CV(%) are considered as the resolution limit of light intensity. Temperature control of the sensor can improve dark count and its standard deviation.

Another aspect of photon detection is the internal gain of the sensor. A simple model to obtain a 1-mV pulse height across 50 Ω in 1 ns requires a gain $>10^4$. This means that a photodiode with a gain of 1 and an avalanche photodiode with a gain of ~100 have difficulty in obtaining sufficient signal amplitude as a photon detector. Both photodiodes have excellent linearity and dynamic range as photocurrent sensors.

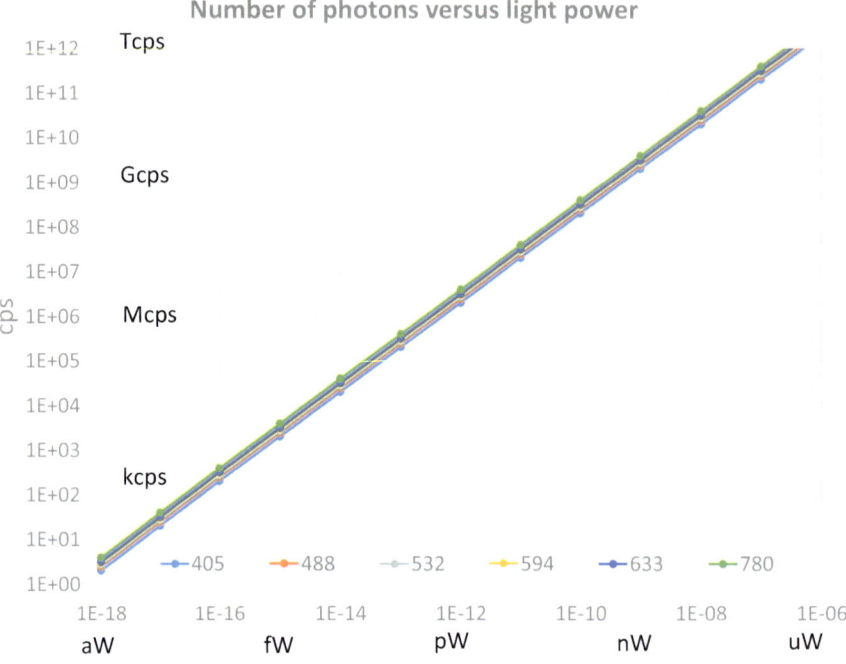

Fig. 2 Number of photons (cps = counts per second) as a function of optical power (W = J/s) at typical wavelengths in flow cytometry

2 Photon Counting, Spectrum Analysis, and Time Domain Analysis

Photon counting is the digital measurement of light intensity with extremely high sensitivity and linearity. If a detected photon pulse is an ideal impulse with pulse width "zero" and dark count "zero," the photo detection system is ideal. Unfortunately, real photon pulses have finite pulse width and waveform. The upper count rate is determined by pulse width and the lower limit by dark-count rate [3]. Owing to pulse overlapping, the true count value N and measured value M are described as follows;

$$N - M = N^* M^* t \qquad (5)$$

where t is the pulse pair resolution. The calculated M/N ratio in the case of t = 1 ns is described in Fig. 3. This graph shows a small error up to 10 Mcps and a gradually larger error for higher count rates. If necessary, it is possible to correct the measured value up to ~1 Gcps by this model. 1 Gcps is equal to 0.5/QE nW at 405 nm and 0.25/QE nW at 780 nm. The figure suggests that it might be possible to achieve over six decades of magnitude and linearity if low dark count were less than 1 kcps and pulse pair resolution 1 ns. But this is quite challenging because conventional

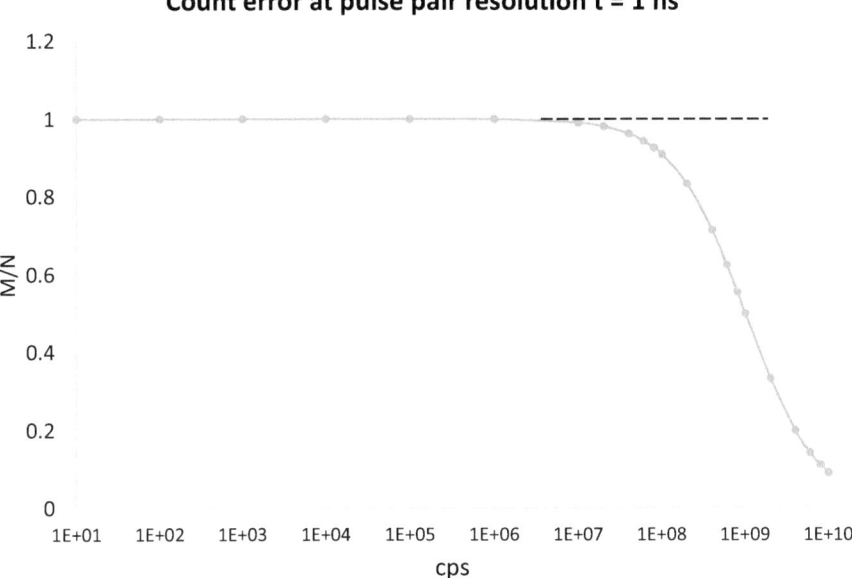

Fig. 3 Calculated count error ratio by photon pulse overlapping at pulse pair resolution 1.0 ns

photon pulse pair resolution is longer than 10 ns. In addition, the above model is based on a continuous pulse train. The actual measured signals have fluctuations that produce counting threshold deviation. This is an additional source of error. Nevertheless, the latest sensor improvement and ultrahigh-speed electronics have potential to achieve the target performance.

Dark count rate (DCR) is another key factor to determine sensitivity limit. In general, dark count is proportional to sensor area. A smaller photocathode or sensor active area usually reduces dark counts. A tenth of sensor area may reduce a tenth of dark count. This is attractive, but there are trade-off with optics design. Conventional flow cytometer optics for fluorescence detection have relatively large aberrations as well as a large spot image due to broad wavelength and high-NA collection lens without compensation. Reflective optics or in combination with optical fiber coupling may resolve this trade-off.

Sensor structure and material can also contribute significantly to dark counts. Dark-count origins include thermal noise in the sensor or photocathode. Materials with higher sensitivity in the IR region have higher dark-count characteristics. For example, comparing bialkali and multi-alkali photocathode materials for detection at extended longer wavelengths, multi-alkali shows a higher dark count.

Dark count and signal deviations can be reduced by temperature control. Peltier cooling is an effective solution to reduce dark count. For cooling purpose, a smaller sensor is easier to implement. If our target is to achieve a six-orders-of-magnitude dynamic range with a maximum 1 Gcps, the dark count must be less than 1 kcps.

Light intensity with theoretical linearity is measured by photon counting in the digital domain. The next question is how should we analyze the photon spectrum or photon energy. Hyperspectral analysis is a very important opportunity for cellular analysis. Photon spectroscopy can be implemented in two ways. One is in combination with a motorized monochromator and photon detector using a long capture time (>1 s). A recent motorized monochromator has a wavelength scanning speed of 500 nm/s. Because photon measurement for flow particles is in the ms to μs time domain, it is necessary to build a parallel photon-detection system. Technically, this is feasible, but cost and implementation as a system may be a hurdle.

Time-domain analysis by a sub-ns photon pulse is an interesting topic and could become an important frontier. Several applications are reported, such as TOF-PET for cancer diagnostics, single-molecule detection, various types of fluorescence resonance-energy transfer (FRET), DNA analysis, fluorescence correlation spectroscopy (FCS), photon imaging, and others.

These analyses are highly dependent upon photon-detection performance, optics, electronics, and software. "What kind of contribution single-photon detection might make to cytometry" is the key question for next-generation cellular analysis.

3 Advancement of Photon Detectors for Cytometry

Biological fluorescence analysis requires high sensitivity and wide dynamic range with linearity in visible wavelengths. Photon detection has ultimate sensitivity if dark count is low. The upper dynamic range of photon detection is mainly determined by maximum count capability per second. The challenges is to obtain the shortest photon pulse width and reduce dead time. A conventional photon sensor is limited by a maximum count rate in the range of 1 to 10 Mcps. Thus a photon pulse width with minimum dead time is the key sensor characteristics for cytometry application.

3.1 The Photo Multiplier Tube (PMT)

The discovery of the photon by Planck and Einstein in 1905 was based on the photoelectric effect. Elster and Geiger invented the photoelectric tube in 1913. Twenty years later, RCA laboratories commercialized the first PMT in 1936 [4]. Inevitably, the PMT became the choice of flow cytometry for fluorescence detection from the very beginning and has continued in use to this day. A new small PMT combined with Si-MEMS structure, the micro PMT, has been developed by Hamamatsu. Unlike a conventional PMT with metal dynode structure, the micro PMT dynode is made by a Si MEMS process, which accurately produces a small and thin structure. When a photon pulse from a micro PMT is amplified by a high-speed preamp, it is possible to obtain a shorter photon pulse of 4–5 ns (Fig. 4). The pulse waveform shows lower noise and distortion from smaller input capacitance owing to short distance and small

Micro PMT structure and detected photon pulse
(Hamamatsu micro PMT Brochure 2016)

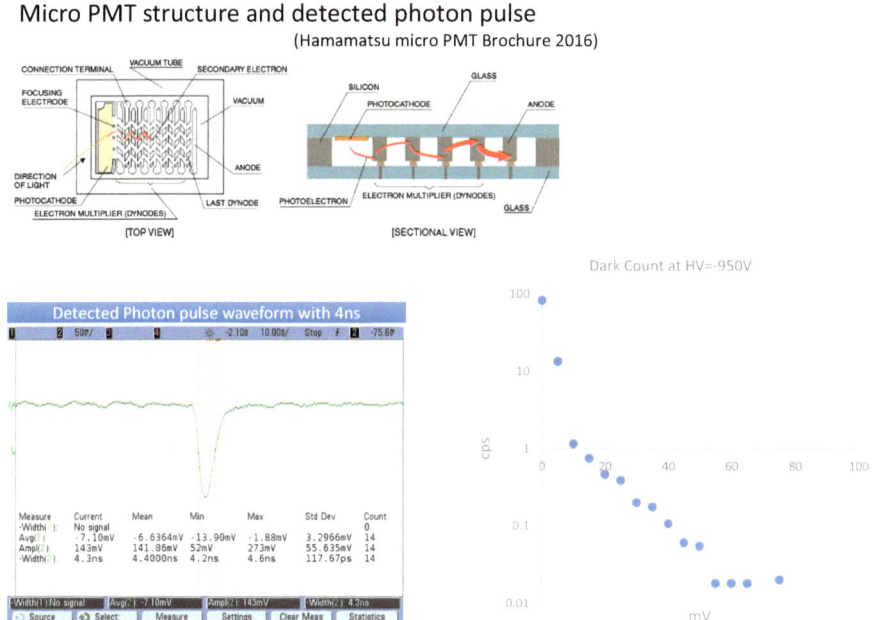

Fig. 4 *Upper* Schematic internal structure of micro PMT by Hamamatsu. *Lower left* Photoelectron pulse waveform amplified by high-speed preamp. *Lower right* Dark count (cps) as a function of comparator level (−mV) at HV = −950 V

area. In addition, a small photocathode area (1 × 3 mm) achieved very low dark-count rates at room temperature. According to the velocity distribution of cascade electrons, the photon pulse for the PMT shows a continuous PE level. The combination of low dark count and narrower pulse width can provide a photon detection system with a wider dynamic range. The drawbacks are the small photocathode area and the cost. In order to collect fluorescence light and couple to the sensor, the utilization of fiber- or aberration-free optics is advisable. PMT output is flexible to control gain and dynamic range for both photocurrent and photon modes. It is necessary to calibrate the detected value to the absolute power level for each measurement condition.

3.2 The Si Photomultiplier (SiPM)

Early development of solid-state single-photon detectors took place in the 1960s and early 70s at RCA and in Russia and Japan. The first single-photon avalanche diode (SPAD) was produced by Perkin-Elmer, and a solid-state Geiger-mode photomultiplier was developed by Rockwell in 1987 [4]. Since then, many developments and improvement have been made for a variety of applications, mainly in nuclear physics, astronomy, elementary particle physics, medical imaging, and the like.

APDs have been proposed for flow cytometry for some time [4, 5]. Recently an avalanche photo diode (APD) has been installed in a commercial compact flow cytometer instead of a PMT [6].

A solid-state photon detector has the advantage of small size, high quantum efficiency, lower bias voltage, light durability, insensitivity to magnetic fields, lower cost, and more, compared to a conventional PMT. In a PMT, an incident photon produces an electron-hole pair in the photocathode, which is an electrical insulator; vacuum and high voltage are mandatory for capture of the induced electron. On the other hand, the electron-hole pair of a solid-state sensor is produced in the p-n junction, which is semi-conductive material. The produced electron moves rapidly and the acceleration depends on the reverse bias electric field. There are three phases of operation: P-I-N mode with gain = 1, linear avalanche mode with gain ~ 100, and Geiger mode over break-down voltage with gain $\sim 10^6$ (see Fig. 5).

Geiger mode is highly sensitive for incident photons owing to a high QE ~ 0.8 and gain $>10^6$. Once a photon has hit, a quenching register is required to recharge electrons. Quenching time, typically 50 ns, is called dead time because the detector will not fire even if struck by a photon. In order to expand dynamic range, a Geiger-mode sensor has a structure arrayed as pixels (Fig. 6). SiPM photon detection efficiency PDE is defined as QE $\times$ εgeo $\times$ εtrig, where εgeo is the geometrical fill-factor and εtrig is the avalanche triggering probability. When a pixel is "fired," a SiPM has secondary fire phenomena called "cross-talk," at adjacent pixels, and "afterpulse," a delayed signal in the fired pixel. Another issue is dark-count rate caused by thermal excitation and field-assisted excitation [5]. These drawbacks of SiPMs have been amended and SiPMs serve in several applications with very high time resolution on the order of picoseconds for TOF-PET, nuclear physics, and the like.

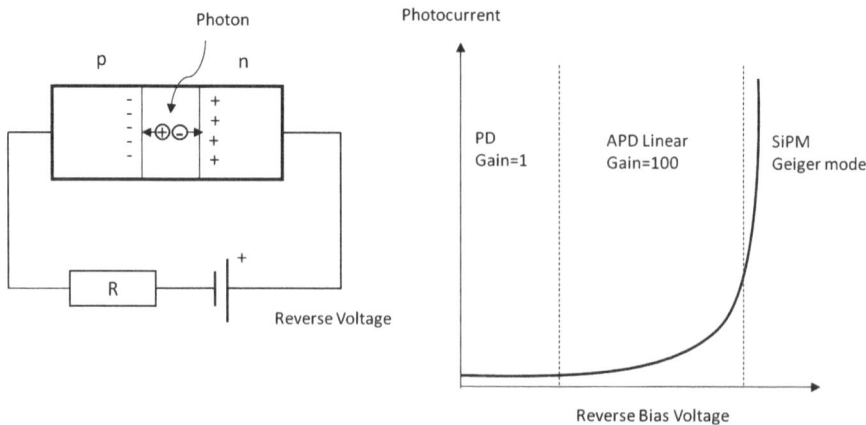

Fig. 5 *Left* Basic configuration of the p-n junction for photon detection. *Right* Photocurrent and gain as a function of reverse bias voltage

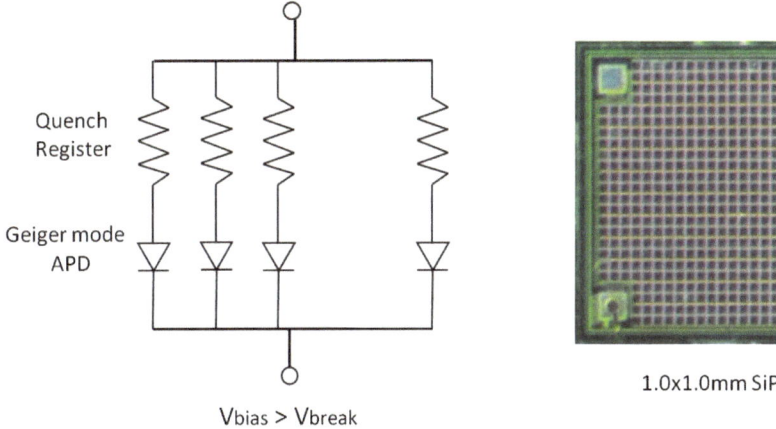

Fig. 6 *Left* Equivalent circuit diagram of SiPM. *Right* Typical 1.0 × 1.0 mm SiPM with 35-μm pixel

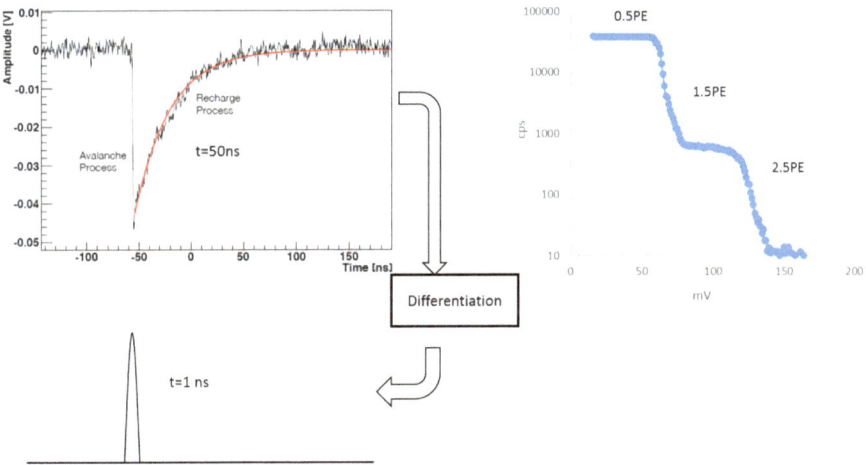

Fig. 7 *Left* Typical photoelectron pulse waveform with 50-ns quenching time and expected pulse waveform after highpass filter and differential circuit. *Right* Photoelectron characteristics at dark condition as a function of comparator level

A typical photon pulse waveform and PE characteristics are shown in Fig. 7. Owing to the micron-length distance in the p-n junction, avalanche electrons are accelerated in a short path and the SiPM pulse-height distribution shows steps designated 0.5 PE and 1.5 PE. The dynamic range of a SiPM, typically three orders of magnitude, is limited by the number of pixels and the quench time. In order to be useful in biological applications, a wider dynamic range is clearly required.

Detection of multiple photon pulse even during quenching time

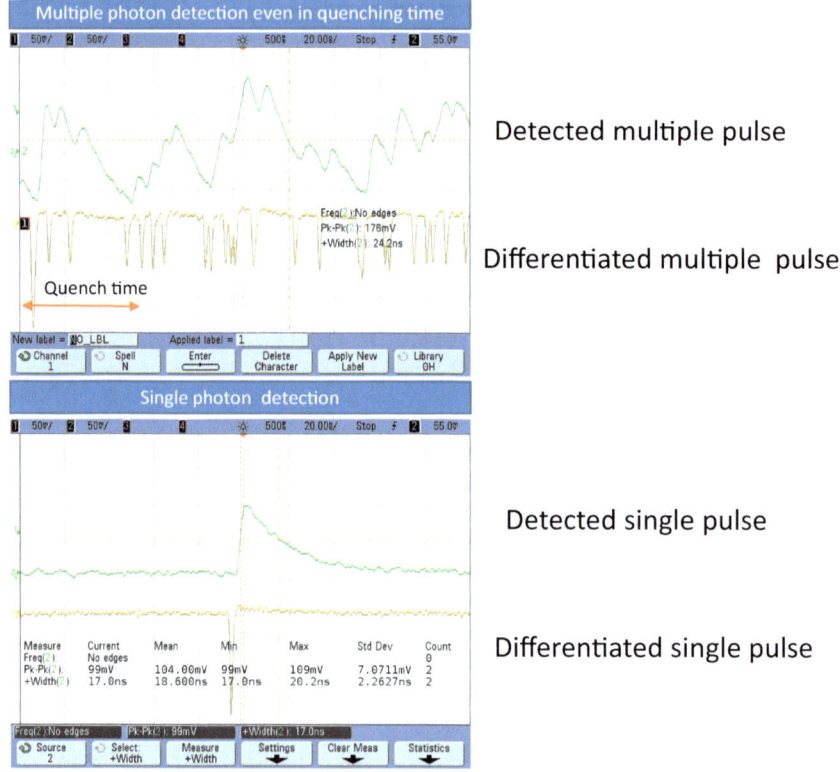

Fig. 8 *Lower* Observed photoelectron pulse waveform before and after differentiation for a single photon incident. *Upper* Observed photoelectron pulses before and after differentiation for multiple photon incidents during the quenching period

The SiPM output waveform has a very fast avalanche process measured in picoseconds and a slow recharge process measured in nanoseconds. It is suggested that a highpass filter and signal differentiation may detect only the avalanche process. Experiments using an ultrahigh-speed differentiating circuit with GHz-bandwidth amp or equivalent differentiation in pixels show the feasibility of multiple photon detection even during quenching dead time. This is a significant breakthrough for wide-dynamic-range photon detection. This method combined with signal processing we have termed "*differential Geiger mode*." Multiple photon detection during quenching is shown in Fig. 8.

4   Performance of Differential Geiger Mode and Preliminary Evaluation

Differential Geiger mode with signal processing is a new approach for photon detection by SiPMs. It is possible to expand the dynamic range of detection by multiple photon detection during quenching time. The current status of detection performance and a preliminary evaluation of basic material for flow cytometry are described.

4.1   Linearity Between Incident Light Intensity and Counting Rate

Sensing linearity is evaluated by a 10-mW 405-nm laser and optics as shown in Fig. 9. Laser power is adjusted using a rotational polarization beam splitter (PBS) and neutral density filters (ND2 & ND3), and coupled to a 0.6 mm—core optical fiber through a lens. The Q8230 (Advantest) optical power meter is calibrated to 405 nm with 0.01 nW resolution. The SiPM and power meter are connected to the optical fiber by an SMA connector to eliminate ambient light. The incident 405-nm light is adjusted by the power meter and the photon count measured. Measurement results show excellent linearity to 400 pW, after which the line curves as the system gradually becomes saturated. Photon counting sensitivity is described as count per pW. The 405-nm and 1-pW light includes 2.04 Mcps photons. The PDE of the tested sensor is 0.24 and the estimated photoelectron number is 489 kcps/pW. The measured rate, 358 kcps/pW, is reasonable considering the coupling efficiency ($\sim 75\%$) from the fiber to the 1×1-mm sensor area. In addition, the measured rate is confirmed to be constant within the operational range of the bias voltage. As shown in the figure, the maximum count rate is over 250 Mcps and the measured time pair resolution is 1.20 ns. In order to achieve a target of 1 Gcps and 1 ns, it is necessary but feasible to improve ultrahigh-speed signal processing by applying the latest and highest-bandwidth ICs.

4.2   Dark Count Rate by Cooling and Dynamic Range

Dark-count rate (DCR) is the lower limit of sensitivity and determines the dynamic range of photon counting. The main cause of high dark counts is thermal noise in p-n junctions. Thermal electrons are amplified in a similar manner to incident photons. Unlike photocurrent noise (thermal noise, shot noise, amp noise, and so on), the dark count can easily be subtracted from the evaluated count rate. Peltier cooling for SiPMs is very effective to reduce dark count. Figure 10 shows DCR temperature dependence of a 1-mm^2 sensor. At temperatures higher than 25 °C, the DCR is over 100 kcps, but is reduced to 2 kcps at −10 °C. The approximate

Light intensity–count linearity

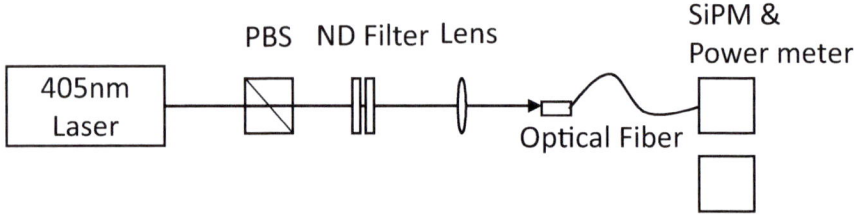

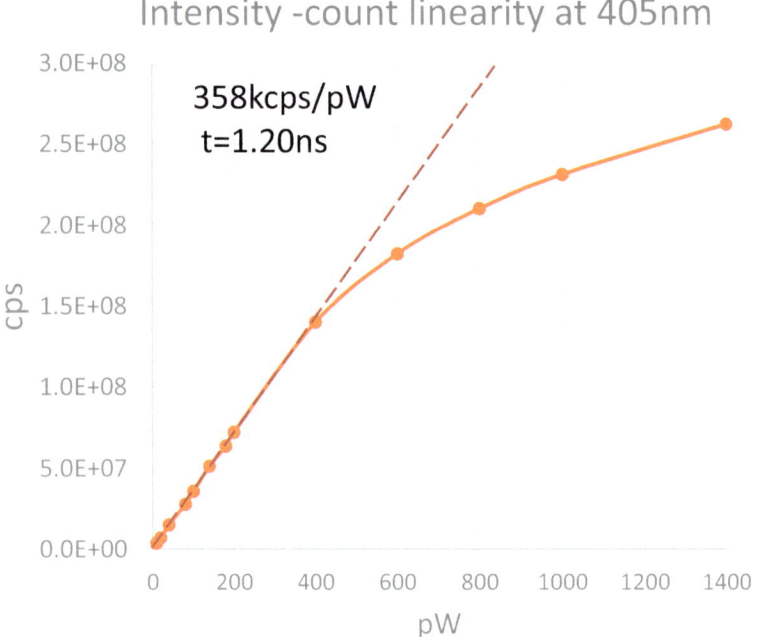

Fig. 9 *Upper* Schematic layout of the linearity measurement between photoelectron counts and optical power meter. *Lower* Plotted photoelectron counts per second as a function of incident power (pW)

curve suggests that a target of 1 kcps would be possible at temperatures lower than −20 °C. Our experimental evidence showed that a DCR of 100 cps is confirmed by dry-ice cooling at −50 °C.

A counting range from 1 kcps to 1 Gcps means a six-orders-of-magnitude dynamic range with theoretical linearity in the digital environment. Further development is required, but current performance is very close to achieving the targeted function. DCR standard deviation and CV % are considered as the resolution of light intensity difference by sensor and statics. Dark count is sensitive to

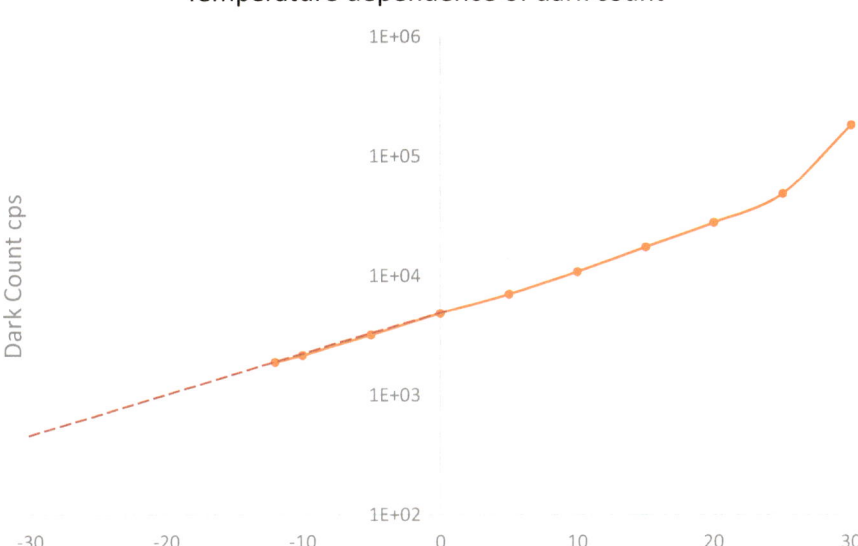

Fig. 10 Example of temperature dependence of dark count for a 1-mm^2 SiPM

temperature change, so temperature control is certainly required for biological measurements for data accuracy, stability, and reproducibility. The measured value of DCR controlled at 4 °C is 5 kcps, the standard deviation per second, σ, is 10 to 50 cps, and the coefficient of variation is 0.2–1.0%, respectively.

4.3 Excited Fluorescence Evaluation by Photon Counting

Photon counting is the ultimate high-sensitivity approach for photon measurement. In the case of 10 kcps photons at 405 nm, the intensity is approximately 5 femtoW (fW). This is roughly 1000 times higher sensitivity than is obtained with a conventional photocurrent approach. It is very difficult to analyze the causes in photocurrent noise; photon counting can distinguish each cause and opens a new frontier in cellular analysis. In order to evaluate basic material in flow cytometry and biology, we have configured autofluorescence (AFL) evaluation optics as shown in Fig. 11. An excitation wavelength of 405 nm is the shortest wavelength in the visible region with excitation energy that provides a full visible spectrum longer than the laser wavelength. A laser spot 100 μm in diameter illuminates the sample behind a bandpass filter to remove stimulated spontaneous emission in the laser beam. Excited fluorescence photons are collected by an NA-0.125 lens and coupled to an NA-0.22 optical fiber through a 405-nm notch filter and a longpass filter to

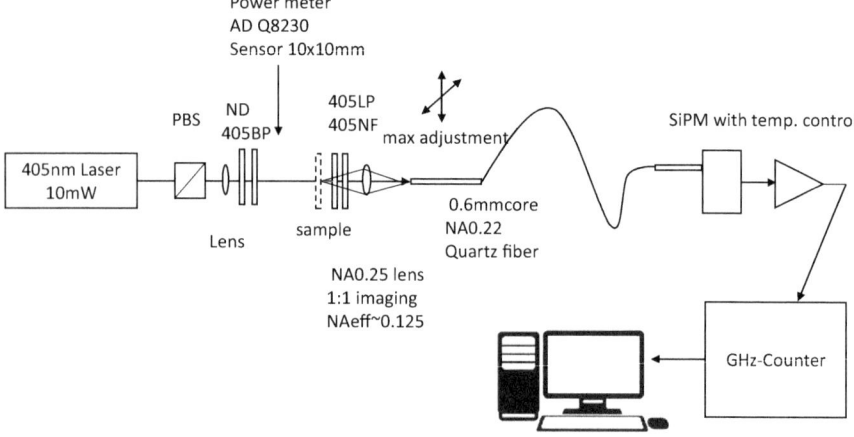

Fig. 11 Schematic layout of fluorescence evaluation: 405-nm laser, fiber-coupled Si photon detector, and GHz counter

remove excitation photons. Detection occurs at 420–900 nm and count is given as total number of photoelectrons (PEs) without detection-efficiency correction. Assuming that excited fluorescence is emitting uniformly to any solid angle, the total number of emitted photons is estimated as 1000 times the measured PE number because of lens collection efficiency (1/250) and sensor PDE (1/4).

Excited autofluorescence intensity is basically proportional to illuminating power and sample thickness under fixed optics. In order to compare autofluorescence, an excitation coefficient k is defined as the detected PE number per μW illumination for a 1-mm sample thickness (PEcps/μWmm). Interestingly, quartz, glass, and many materials show AFL and photobleaching (Fig. 12). It is necessary to check the excitation coefficient before and after photobleaching. Photobleaching is difficult to observe with the noncoherent light source in a conventional fluorescence spectrometer. Lasers can provide very high illumination intensity, over 10^6 J/m^2, which is not proportional to total illumination energy (J). Measurement is first dark count, system AFL, and dry vial for liquid, and finally the sample to calculate a count only for that sample.

As an example, an excitation coefficient of 1000 cps/μWmm means that the total number of emitted photons is estimated as 1 Mcps (1 k × 1 kcps) under measurement conditions. 1 Mcps photons is about 1 pW. Illumination at 405 nm/1 μW contains 2.04 Gigaphotons. This equals 1 Mcps/2 Gcps, ∼ 1/2000; it takes 2000 illuminating photons to produce a single emitted photon. Using a measured excitation coefficient, it is possible to estimate the AFL from the illumination level. A material with k = 1000 cps/μW mm emits 1 k × 1 kcps = 1 Mcps, ∼ 1 pW AFL under 1 mW illumination.

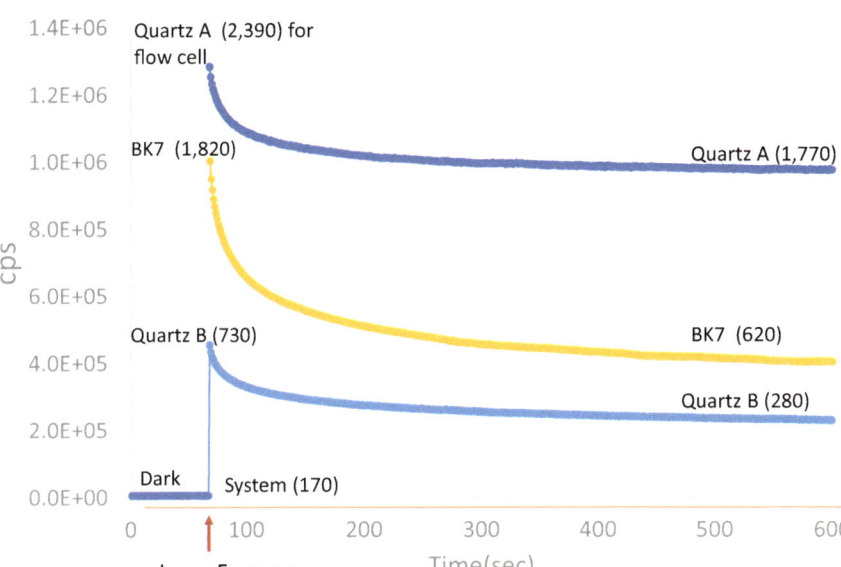

Fig. 12 Autofluorescence from high-quality quartz and optical glass with photobleaching as a function of exposure time at 500 μW of 100-μm φ spot at 405 nm

Several material-evaluation results are shown in Table 1. We have to recognize that every material for flow cytometry exhibits autofluorescence, even distilled water, sheath, or clean beads. Yellow-Green dye diluted in water for flow check in a cytometer has a count of 1.2 Mcps/μWmm, meaning roughly one emitted photon for every two excitation photons in the illuminated volume. In order to reduce the influence of AFL from the tube, a trapping method involving illumination along the coaxial direction has been developed for liquid-sample evaluation.

Table 1 Measured excitation coefficient per 1 μW/405 nm exposure and 1-mm sample thickness for basic materials in flow cytometry

Excitation coefficients at 405 nm/1 μW/100-μm spot/NA 0.125		
Material	Excitation coefficient k [cps/uW mm]	Remark
Flow cell quartz	3500–1500	Photobleach before/after
Highest grade quartz	700–300	Photobleach before/after
Distilled water	250	Tube trapping
Sheath	1120	Tube trapping
Clean polymer bead 1 μm in water	2210	Tube trapping
Clean silica bead 1 μm in water	10,700	Tube trapping
Y-G highlighter	1,180,000	Tube trapping
Slide glass borosilicate	4700	1 mmt
Cover glass borosilicate	3200 (530)	(0.17 mmt)

Many researchers may recognize the influence of autofluorescence at the time of measurement, but it is difficult to understand quantitatively. Photon counting can provide a method for order estimation. The influence of excited AFL depends on optics configuration. In-line detection to laser incident direction may be highly affected, but perpendicular fluorescence detection as in a conventional flow cytometer may be minimal. In the case of confocal fluorescence imaging or flow cytometry with longer gate period for small particles, optics AFL must be reduced for higher contrast and accurate detection. AFL reduction for optical glass and coatings is quite challenging, but it may be possible to improve after establishing an evaluation approach.

5 Conclusion and Discussion

We have developed a wide-dynamic-range Si photon-detection system, differential Geiger mode, and confirmed the potential performance for flow cytometry and cellular analysis. Photon counting with sensitivity over three orders higher than that of conventional photocurrent detection has made clear that optical and biological materials are the limit of background noise such as autofluorescence. The next step is single-photon spectrum analysis for identifying the origins of autofluorescence and for the highest sensitivity analysis on biological assays. Photon counting is simply intensity analysis with time deviation. In addition, single-photon detection with picosecond time resolution could open new frontiers for live single-cell and population analysis in flow cytometry and cellar analysis.

Acknowledgements I would like to acknowledge the substantial contributions of Keegan Hernandez on electronics and software development, Kathy Ragheb, Jennifer Sturgis, and Eva Biela on biological samples, Gretchen Lawler on chapter editing, and J. Paul Robinson on the opportunity and support.

References

1. Lewis GN (1926) Letters to the editor; conservation of photons. Nature 118:874
2. Okun LB (2006) Formulae E=MC^2 in the year of physics. Acta Physica Polonica B 37:1127–1132
3. Schwartz A, Gaigalas AK, Wang L, Marti GE, Vogt RF, Fernandez-Repollet E (2004) Formalization of the MESF unit of fluorescence intensity. Cytometry B Clin Cytom 57:1–6
4. Shutao Z, Xiaodong W, Yongqin C, Ce W, Tang Y (2011) High gain avalanche photodiode (APD) arrays in flow cytometer opitical system. In: 2011 international conference on multimedia technology
5. Lawrence WG, Varadi G, Entine G, Podniesinski E, Wallace PK (2008) Enhanced red and near infrared detection in flow cytometry using avalanche photodiodes. Cytom Part A: J Intl Soc Anal Cytol 73:767–776
6. Tung J, Condello D, Donnenberg AD, Duggan E, Lemus JA, Nolan J, Ragheb K, Sturgis J, Robinson JP, Jing Z, Fenoglio D, Huang Y Scibelli P(2015) CytoFLEX system performance evaluation. In: CYTO2015. Wiley, Liss, Glasgow

Identification of Small-Molecule Inducers of FOXP3 in Human T Cells Using High-Throughput Flow Cytometry

Rob Jepras, Poonam Shah, Metul Patel, Steve Ludbrook,
Gregory Wands, Gary Bonhert, Andrew Lake, Scott Davis
and Jonathan Hill

Abstract The increasing use of disease-relevant human primary cells in cellular target assays in drug discovery and phenotypic screening is driving the requirement for more sensitive high-throughput technologies that derive maximum information from fewer and fewer cells. It is widely recognized that heterogeneous primary cell populations are more suited to high-content single-cell analysis techniques such as flow cytometry. A robust flow cytometry assay was developed to identify compounds that up-regulate FOXP3 in primary human T cells. Over 70,000 small-molecule compounds were screened in single shot and several thousand hits were followed up at full dose response. Several compounds were identified that had a profound effect on FOXP3 expression in T cells; one in particular had an EC50 of <100 nM. These novel tool compounds can be

R. Jepras (✉) · P. Shah · M. Patel
GlaxoSmithKline Medicines Research Centre, Gunnelswood Road, Stevenage SG12NY, UK
e-mail: robert.i.jepras@gsk.com

P. Shah
e-mail: poonam.c.shah@gsk.com

M. Patel
e-mail: metul.2.patel@gsk.com

S. Ludbrook
GlaxoSmithKline Medicines Research, Gunnelswood Road, Stevenage SG12NY, UK
e-mail: steve.b.ludbrook@GSK.com

G. Wands · G. Bonhert · S. Davis
Tempero Pharmaceuticals, A GSK Company, Cambridge, MA 02139, USA
e-mail: gregory.d.wands@gsk.com

G. Bonhert
e-mail: gary.j.bohnert@gsk.com

S. Davis
e-mail: scott.p.davis@gsk.com

A. Lake · J. Hill
Surface Oncology, 50 Hampshire Street, 8th Floor, Cambridge, MA 02139, USA
e-mail: alake@surfaceoncology.com

J. Hill
e-mail: jhill@surfaceoncology.com

© Springer Nature Singapore Pte Ltd. 2017
J.P. Robinson and A. Cossarizza (eds.), *Single Cell Analysis*,
Series in BioEngineering, DOI 10.1007/978-981-10-4499-1_11

used for dissecting signaling pathways upstream of FOXP3 and provide a molecular target that may lead to therapeutic intervention for autoimmune and inflammatory disease. However, a good lead starting point was not found. In order to increase chances of identifying lead molecules, greater compound structural diversity is required, meaning significantly larger compound libraries containing millions of compounds. The high cost and limited availability of human tissues pose challenges for attempts to screen large compound libraries.

Keywords Regulatory T cells · Tregs · FOXP3 · High-throughput flow cytometry · Phenotypic screening

With its ability to make single-cell high-content measurements, flow cytometry is an important platform for drug discovery. It is used at every stage of the drug-discovery cycle both in pharma and in biotech companies [1–3]. The increasing use of phenotypic human primary cell–based assays in drug discovery [4] is driving more and more flow cytometry drug-screening capability and applications. Since the advent of faster automated sampling technologies [5, 6] flow cytometry has been accepted as a technique for drug-screening applications, and automated high-throughput platforms are present in many drug-screening laboratories [7–9]. 1536-well [6] microtiter-sampling flow cytometers are commercially available for rapid sampling of microliters of cell suspension. The use of integrated robotics, peripheral automation, and liquid-handling systems is critical to drug-screening applications, and automation-friendly instruments are a pre-requisite [8].

The majority of flow-cytometry applications employed in drug discovery are associated with cellular target–based assays. These are defined as assays in which the measured effect is directly dependent on a known molecular target or pathway such as receptor-ligand interactions, either by monitoring direct binding of ligand to receptor, e.g., competition between drug and ligand for receptor [10, 11], or by measuring a cell functional response as a result of interfering with a receptor/ligand interaction. Many of these assays serve as routine drug-discovery critical-path assays important for making decisions on drug progression. For example, automated microtiter plate–sampling flow cytometry can be used to identify small-molecule inhibitors that target kinase signaling pathways in human primary cells. One assay used routinely is pAkt activity in expanded T cells following treatment with PI3 kinase-targeted compounds. The phosphatidylinositide 3-kinase (PI3K) pathway is a signal-transduction pathway that promotes survival and growth in response to extracellular signals [12]. Monitoring the activity of proteins involved in PI3K signalling, such as Akt, is important in understanding the mechanism of action of drugs. Figure 1 shows an example where automated high-throughput flow cytometry is used to identify compounds that inhibit the PI3 kinase pathway. These data are robust enough to rank compound activities, and the additional information obtained, such as cell toxicity, helps with compound hit progression decisions. When these data are used in conjunction with cell-free biochemical enzyme activity, information on cell penetration and off-target effects by the compound is also obtained (Fig. 2).

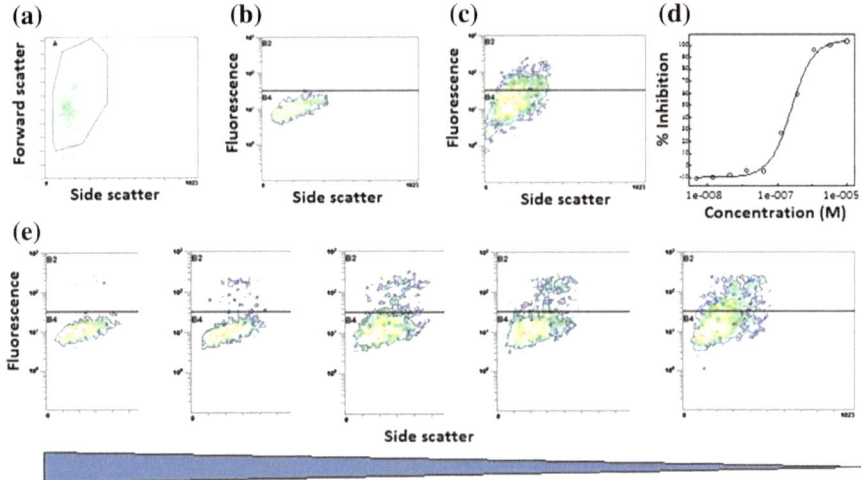

Fig. 1 Flow cytometry plots depicting the effects of the PI3 kinase inhibitor LY294002 on single human T-cell Akt phosphorylation. **a–c** show the effects of compound on single-cell Akt phosphorylation, and **d** is a 10-point concentration-response curve produced by the inhibitor. Plots **e** depict the concentration effects of LY292004 on pAkt in cell populations. Human expanded T cells were treated for 30 min with compound and then co-stimulated with anti-CD3 and anti-CD28 for 20 min. Cells were fixed using a 4% solution of formaldehyde and then permeabilized using ice-cold 90% methanol before intracellular pAkt labeling using fluorescently conjugated phospho-specific monoclonal antibodies. Flow cytometry sampling was automated using a plate-based sampling flow cytometer integrated with a refrigerated plate stacker

Fig. 2 Correlation between single-cell intracellular pAkt activity determined using flow cytometry and free enzyme inhibition. Such data provide information on **a** off-target compound effects and **b** potential compound cell-penetration issues (see Fig. 1 legend for further information)

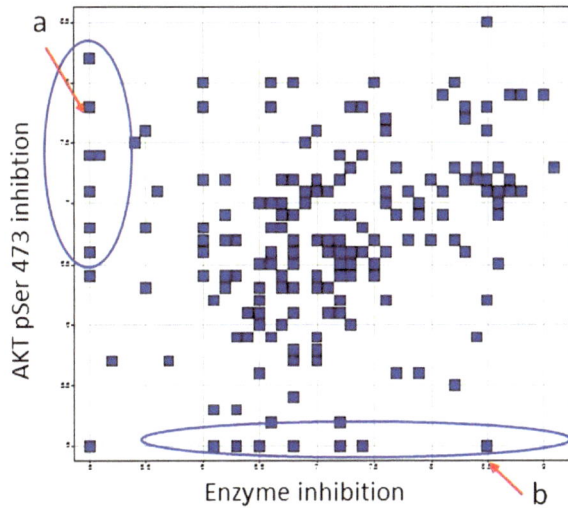

Like high-content imaging [13], flow cytometry profoundly benefits phenotypic drug-discovery efforts. A phenotypic assay is defined as an assay that measures a desired endogenous effect in a biological system, that effect being agnostic of a defined target. The ability of flow cytometry to rapidly analyze and identify large numbers of individual cells in complex heterogeneous populations makes it a powerful technique for phenotypic drug discovery, particularly when using physiologically relevant primary human cells [4, 6, 7].

A high-throughput flow cytometry screen was established to identify compounds that influence the transcriptional factor FOXP3 (forkhead boxP3) expression in $CD4^+$ T cells. Several signaling pathways common to all T cells are involved in FOXP3 expression; it is uncertain what pathways control its turnover or half-life, this primarily owing to their plasticity [14, 15]. This makes prioritizing a single 'signaling-pathway' target difficult. Therefore, a phenotypic screening approach was taken. FOXP3 functions as a master regulator of the regulatory pathways in the development and function of regulatory T cells [16]. Regulatory T cells (Tregs) have an important role in maintaining immune homeostasis; they suppress autoreactive lymphocytes, and control innate and adaptive immune responses. FOXP3 is necessary for the development of Tregs and plays a critical role in coordinating the expression of several genes involved in immune regulation [17]. Mutations in the *FoxP3* gene lead to the inherited multi-organ autoimmune disease IPEX (immune dysregulation, polyendocrinopathy, enteropathy, X-linked) syndrome [18], and a similar disease in the Scurfy mouse [19]. There is considerable interest in strategies to increase $FOXP3^+$ Treg cells in patients with chronic GVHD (graft-versus-host disease) and other autoimmune disorders. Most strategies are currently focused on the adoptive transfer of Treg cells that have been purified and expanded ex vivo. These cellular therapies have been difficult to standardize in humans, but measurable clinical benefit has been seen in studies of GVHD [20]. Recent evidence also indicates that pharmacological control of Treg cells may be possible in patients by the administration of low-dose IL-2, a growth factor critical for Treg cells [21, 22]. The goal of the screen was to identify small-molecule pharmacological agents that can modulate the immune response by increasing the frequency of FOXP3-positive T cells, thus leading to an effective therapy for inflammatory diseases such as GVHD.

One of the main issues with running large screens using human cells is ensuring that enough cells are available to complete the screening campaign and follow-up studies. Screening large compound collections using human primary cells taken from single individual living donors is, at present, difficult, unless cell populations have been expanded following isolation. Although it is advantageous to gain insights into donor-to-donor variation during a drug-discovery process, this actually poses problems in hit identification during screening campaigns, where day-to-day or week-to-week stability is required. To ensure that enough cells were available and to avoid donor-to-donor variation, PBMCs from at least ten donors were isolated, pooled, and frozen and stored at $-140\ ^\circ$C using controlled-rate cryopreservation [23]. Using pooled frozen PBMCs, a robust flow cytometry protocol was developed for identifying compounds that upregulate FOXP3 in human T cells.

Several focused compound library screening sets [24], totaling approximately 70,000 compounds, were tested in the FOXP3 assay at a single concentration with hits confirmed in 10-point concentration-response experiments. These sets were enriched with small molecules that have been identified as hits in other target- or cell-based screens. Other compound sets screened contained published kinase inhibitors, marketed drugs, natural products, and lead-like molecules.

The screen consisted of a complex 96-well, 6-day assay (Fig. 3) in which human PBMCs were cultured in the presence of anti-CD3 and low concentrations of IL2 (1 ng/ml) and TGFβ (2 ng/ml). In the first instance, a concentration of IL-2 (50 ng/ml) was used as a standard positive control for assay quality-control purposes. This was replaced during the course of the screen once a small-molecule compound was identified. This compound had an EC50 of <100 nM.

The high-content nature of the data can be used to provide useful information on the compounds being tested. A simple parameter such as viable cell number, based

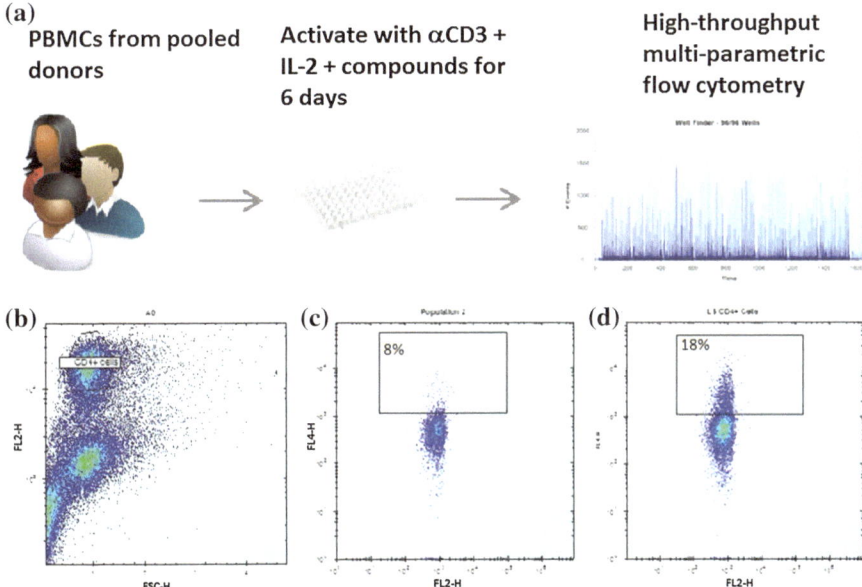

Fig. 3 High-throughput FOXP3 flow cytometry screening assay. **a** Frozen PBMCs from 10 donors were thawed and allocated into wells of a 96-well plate. Bar-coded plates were pre-stamped with the compound libraries. Final assay concentration of compound was 1 μM. Cells were activated with anti-CD3 in the presence of 1 ng/ml IL-2 and 2 ng/ml TGFβ. Columns 11 and 12 of the plate were used as positive (50 ng/ml IL-2) and negative controls respectively. Plates were sampled using a HyperCyt (Intellicyt) integrated to a Beckman Coulter NX (robotic plate/liquid handler) and an Accuri C6 (BD Biosciences) or CyAn (Beckman Coulter) flow cytometer. **b–d** Typical flow cytometry plots (IntelliCyt Hyperview software) representing the total number of cells sampled from the plate 'all-well data.' FOXP3$^+$ cells were identified as a % of CD4$^+$ T cells (**b**). **c** Unstimulated. **d** Stimulated (positive control) with anti-CD3, 50 ng/ml IL-2, and 2 ng/ml TGFβ

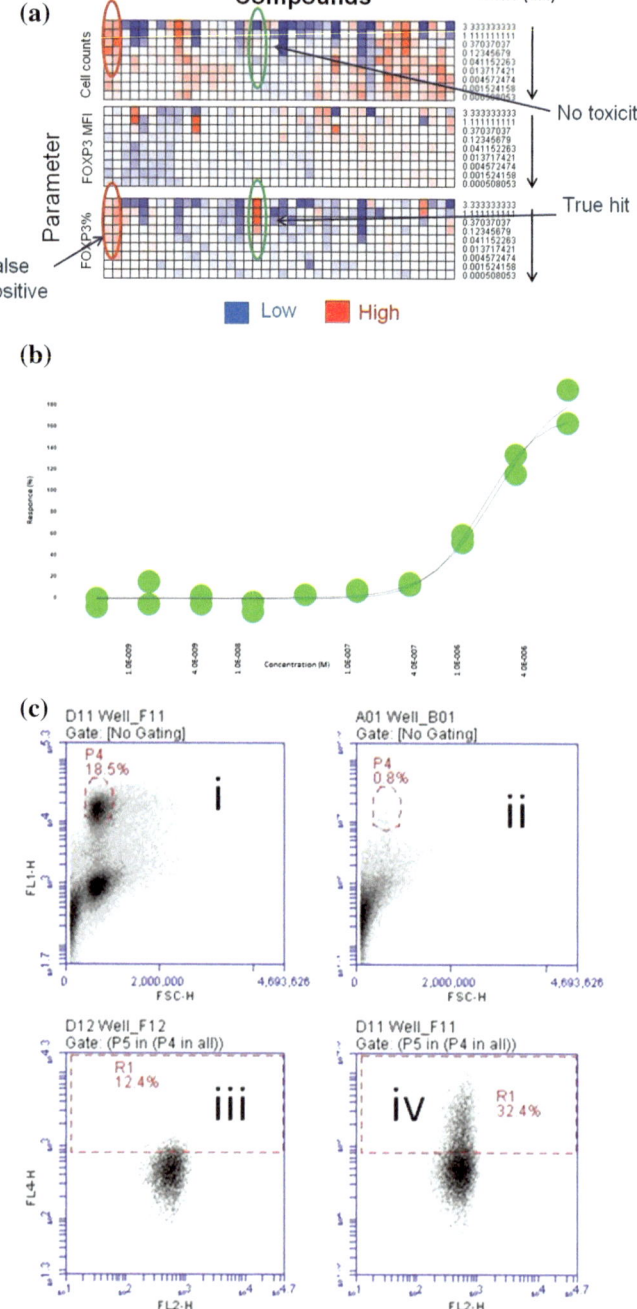

Fig. 4 The high-content nature of flow cytometry allows the ability to separate true hits from false positives. **a** Heat plot containing selected data depicting the effects of compound on cell counts, FOXP3%, and mean fluorescence intensity. **b** Concentration response curve of a representative screening compound. Here the compound appears to be highly effective at inducing FOXP3 up-regulation but is in fact a false positive hit as demonstrated by its effects on CD4$^+$ cells, where CD4$^+$ cells have been reduced from 18.5 to 0.8% respectively (**c** i and **c** ii). **c** iii and **c** iv depict representative data of FOXP3% in un-stimulated and a positive control compound respectively

on cell counts and light scattering, provides important information on compound toxicity, thus helping to make decisions on compound progression (Fig. 4).

Several compounds had profound effects on both the percentage of FOXP3$^+$ cells in the population and the mean fluorescence intensity (MFI) of FOXP3$^+$ cells. Some compounds had a larger effect on the numbers of FOXP3$^+$ cells in the population than on MFI and vice versa, i.e., the amount of single-cell expression of FOXP3 (Fig. 5). Compound activity against both these parameters is important, as there is evidence indicating that FOXP3 cell frequency and individual single-cell level of FOXP3 expression can affect the antiinflammatory effects of regulatory T cells [25].

Similar compound series showed a distinct structure/activity relationship (Fig. 6). Identifying both active and inactive analogs of compounds is important when running phenotypic screens if target identification is required, particularly if chemo proteomics is used to aid drug target identification [26].

Traditional flow cytometry has an important role throughout the drug-development cycle and is represented at each phase in some capacity. High-through put flow cytometry is becoming an important technology for drug discovery, particularly for phenotypic screening. The introduction of faster sampling

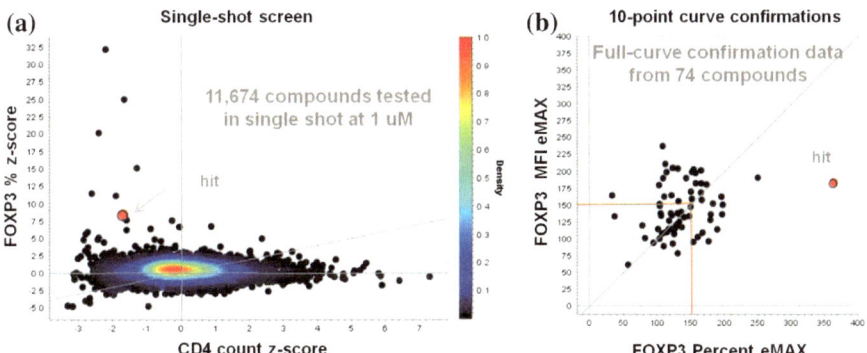

Fig. 5 Example of screening data taken from a subset of compounds and consequent hit selection. **a** Effects of 1 μM concentration of compounds on FOXP3% and CD4$^+$ cell counts. **b** Hits from **a** confirmed in full curve concentration response. FOXP3 MFI maximum response is plotted against FOXP3% maximum response

Fig. 6 Structure/activity
relationship between
compound analogs on
FOXP3% in CD4$^+$ T cells

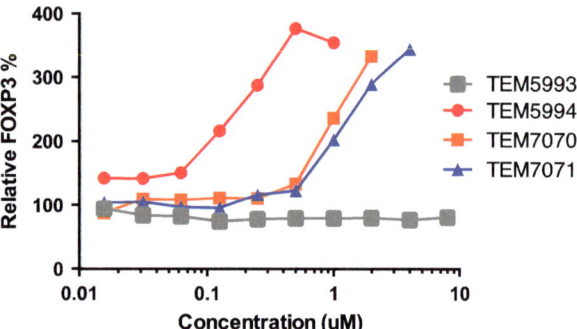

technologies and the use of peripheral automation and liquid handling have
transformed flow cytometry as a drug-screening tool, and more and more screening
laboratories are using it. The use of high-throughput flow cytometry for routine
drug-discovery critical-path cellular-target assays is helping to make critical decisions on drug progression. High-throughput flow cytometry, like high-content
imaging, is an extremely powerful technique and is well placed for running large
complex phenotypic screens, such as identifying small-molecule inducers of CD4$^+$
T-cell FOXP3 in PBMC populations.

Even though strides are being made in flow cytometry technology, other
important factors require thought, such as assay costs and cell supply, and in
particular finding ways to reduce numbers of rare and expensive cells. Where a cell
supply is readily available, often it is the biology that limits the throughput, particularly if prolonged cell culture is required. Cells are often cultured for days or
weeks before they differentiate into the required phenotype. This biological complexity makes it difficult to miniaturize assays, and replication of 96-well assays in a
384-well format is difficult without altering the biology of the assay. A greater
understanding of cell function and behavior is also required in artificial micro- or
nanoscale environments. Efforts are being made to advance miniaturization of
cell-based assays using nanowells and microfluidics [27–29], particularly for
single-cell transcriptomics, which is helping us understand the complexity and
heterogeneity of cell systems [30].

Although some useful tool molecules were identified in this screen that can be
used to aid new target identification, none of the compounds were progressable as
drug leads. In order to increase chances of identifying lead molecules, larger
compound structural diversity is required, meaning that significantly larger compound libraries containing millions of compounds are required. Even using
designed libraries based on quantitative models relating chemical structural similarity to biological activity, screening collections of at least 800,000 compounds are
required [31]. The high cost and limited availability of human tissues poses challenges when attempting to screen large compound libraries.

Note: "The human biological samples were sourced ethically and their research
use was in accord with the terms of the informed consents."

References

1. Ramanathan M (1997) Flow cytometry applications in pharmacodynamics and drug delivery. Pharm Res 14(9):1106–1114
2. Litwin V, Marder P (ed) (2010) Flow cytometry in drug discovery and development. Wiley, London
3. Luu YK, Rana P, Duensing TD, Black C, Will Y (2012) Profiling of toxicity and identification of distinct apoptosis profiles using a 384-well high-throughput flow cytometry screening platform. J Biomol Screen 17(6):806–812
4. Feng Y, Mitchison TJ, Bender A, Young DW, Tallaricon JA (2009) Multi-parameter phenotypic profiling: using cellular effects to characterize small-molecule compounds. Nat Rev Drug Discov 8:566–578
5. Edwards BS, Young SM, Saunders MJ, Bologa C, Oprea TI, Ye RD, Prossnitz ER, Graves SW, Sklar LA (2007) High-throughput flow cytometry for drug discovery. Exp Opin Drug Discov 2(5):1–12
6. Edwards BS, Sklar LA (2015) Flow cytometry: impact on early drug discovery. J Biomol Screen 20(6):689–707
7. Black CB, Duensing TD, Trinkle LDLS, Dunlay RT (2011) Cell-based screening using high-throughput flow cytometry. Assay Drug Dev Technol 9(1):13–20
8. Jepras R, Ludbrook S (2013) Evolution of flow cytometry as a drug screening platform. Drug Discov World (Spring 13):43–50
9. Arenas A, Hernandez-Campo P, Primo D, Barrio S, Ayala RM, Ballesteros J, Martínez-López J (2013) High throughput screening, with a flow cytometry automated platform (Ex vivo Biotech), to identify potential combination partners for the JAK 2 inhibitor ruxolitinib. Blood 122:2534
10. Edwards BS, Young SM, Oprea TI, Bologa CG, Prossnitz ER, Sklar LA (2006) Biomolecular screening of formylpeptide receptor ligands with a sensitive, quantitative, high-throughput flow cytometry platform. Nat Protoc 1(1):59–66
11. Kim K, Wang L, Hwang I (2009) A novel flow cytometric high throughput assay for a systematic study on molecular mechanisms underlying T cell receptor-mediated integrin activation. PLoS ONE 4(6):1–8
12. Vanhaesebroeck B, Stephens L, Phillip Hawkins P (2012) PI3K signalling: the path to discovery and understanding. Nat Rev Immunol 13:195–203
13. Kang J, Hsu C-H, Wu Q, Liu S, Coster AD, Posner BA, Altschuler SJ, Wu L (2016) Improving drug discovery with high-content phenotypic screens by systematic selection of reporter cell lines. Nat Biotechnol 34: 70–77
14. Feuerer M, Hill JA, Mathis D, Benoist C (2009) Foxp3$^+$ regulatory T cells: differentiation, specification, subphenotypes. Nat Immunol 10:689–695
15. DuPage M, Bluestone JA (2016) Harnessing the plasticity of CD4$^+$ T cells to treat immune-mediated disease. Nat Rev Immunol 149–163
16. Rudensky AY (2011) Regulatory T cells and Foxp3. Immunol Rev 241(1):260–226
17. Xie X, Stubbington MJT, Nissen JK, Anderson KG, Hebenstreit D, Teichmann SA, Betz AG (2015) PLOSOne genetics. Published: June 24, pp 1–32
18. Powell BR, Buist NR, Stenzel P (1982) An X-linked syndrome of diarrhea, polyen-docrinopathy, and fatal infection in infancy. J Pediatr 100(5):731–737
19. Sharma R, Sung SS, Fu SM, Ju ST (2009) Regulation of multi-organ inflammation in the regulatory T cell-deficient scurfy mice. J Biomed Sci 16(1):20–28
20. Theil AS, Oelschlägel U, Maiwald A, Döhler D, Oßmann D, Zenkel A, Wilhelm C, Middeke JM, Shayegi N, Trautmann-Grill K, von Bonin M, Platzbecker U, Ehninger G, Bonifacio E, Bornhäuser M (2015) Adoptive transfer of allogeneic regulatory T cells into patients with chronic graft-versus-host disease. Cytotherapy 17(4):473–86

21. Koreth J, Matsuoka K-I, Haesook TK, McDonough SM, Bindra B, Alyea III., Armand P, Cutler C, Ho VT, Treister NS, Bienfang DC, Prasad S, Tzachanis D, Joyce RM, Avigan DE, Antin JH, Ritz J, Soiffer RJ (2011) Interleukin-2 and regulatory T cells in graft-versus-host disease. N Engl J Med 365:2055–2066
22. Klatzmann D, Abbas AK (2015) The promise of low-dose interleukin-2 therapy for autoimmune and inflammatory diseases. Nat Rev Immunol 15:283–294
23. Ramos TV, Mathew AJ, Thompson ML, Ehrhardt RO (2014) Standardized cryopreservation of human primary cells. Curr Protoc in Cell Biol 2, 64:A.3I.1–8
24. Valler MJ, Green D (2000). Screening versus focussed screening in drug discovery. DDT 5(7) 286–293
25. Chauhan SK, Saban DR, Lee HK, Dana R (2009) Levels of Foxp3 in regulatory T cells reflect their functional status in transplantation. J Immunol 182(1):148–153
26. Moellering RE, Cravatt BF (2012) How chemoproteomics can enable drug discovery and development. Chemical Biology 19(1):11–22
27. Mazutis L, Gilbert J, Ung LW, Weitz DA, Griffiths AD, Heyman JA (2013) Single-cell analysis and sorting using droplet-based microfluidics. Nat Protoc 8:870–891
28. Love KR, Bagh S, Choi J, Love JC (2013) Microtools for single-cell analysis in biopharmaceutical development and manufacturing. Trends Biotechnol 31:280–286
29. Dura U, Dougan SK, Barisa M, Hoehl MM, Lo CT, Ploegh HL, Voldman J (2015) Profiling lymphocyte interactions at the single-cell level by microfluidic cell pairing. Nat Commun 6 (5940):1–13
30. Stubbington MJ, Lönnberg T, Proserpio V, Clare S, Speak AO, Dougan G, Teichmann SAT (2016) Cell fate and clonality inference from single-cell transcriptomes. Nat Methods 13: 329–332
31. Harper G, Pickett SD, Green DV (2004) Design of a compound screening collection for use in high throughput screening. Comb Chem High Throughput Screen 7(1):63–70

Cancer Stem Cells and Multi-drug Resistance by Flow Cytometry

Jordi Petriz

Abstract This chapter will focus on the study of stem cells, reviewing and updating a series of methods and protocols mainly applied in the author's laboratory to learn more from the hematopoietic system, one of the best-studied compartments so far. The lack of specific markers for stem cells makes the physical identification and isolation of this compartment difficult; the author will introduce different methodologies aimed to identify as well as to isolate very rare stem cells and their subpopulations using functional flow cytometry, which has advantages over many other methods for identifying very primitive precursor cells. With all this information, researchers interested in the identification of cancer or leukemic stem cells should take into account that flow cytometric–based experiments must be carefully evaluated and planned. If not, these elusive cells can be mismeasured, which results in frustrating and unsuccessful time-consuming efforts.

Keywords Stem cells · Multidrug resistance · Flow cytometry · CD34 antigen · Side population · ABCB1 · ABCG2

Hematopoietic stem cells express the CD34 antigen and are able to self-renew and to differentiate into progenitors of all hematopoietic lineages [1, 2]. These processes are regulated by a large number of cytokines [3] to give rise to red cells, white cells, and platelets [4, 5]. Flow cytometry is a very powerful methodology, allowing the identification and isolation of rare cells. In general, CD34+ cells can be found in steady-state peripheral blood, representing less than 0.1% of leukocytes and less than 1.5% of total white cells in the marrow. In 1998, Bhatia et al. demonstrated the existence of CD34neg stem cells [6]; two years previously, Goodell and collaborators demonstrated that bone-marrow cells contain CD34neg progenitors [7] with a clonogenic capacity 100-fold greater compared to that of CD34+ cells, based on clonogenic culture assays. Although flow cytometry CD34+ counting methods are

J. Petriz (✉)
Josep Carreras Leukaemia Research Institute, Badalona, Barcelona, Spain
e-mail: jpetriz@carrerasresearch.org

© Springer Nature Singapore Pte Ltd. 2017
J.P. Robinson and A. Cossarizza (eds.), *Single Cell Analysis*,
Series in BioEngineering, DOI 10.1007/978-981-10-4499-1_12

in general well accepted, calibrated, and standardized [8], new protocols moving from the bench to the clinics will provide more sensitive measurements of hematopoietic stem cells, having an impact on research in a broad range of scientific and clinical disciplines, and could also finally clarify the number of stem cells needed for engraftment. In the meanwhile, new assays will need to pass different stages prior to validation and application in daily practice.

In general, many cell tissues in the adult can express a series of mechanisms of protection against cell damage, specifically to protect very sensitive tissues through the hematoencephalic or the hematotesticular barriers. One of the most studied mechanisms of multidrug resistance was discovered using cell lines highly resistant to mitotic blocking agents. In 1979, P-glycoprotein (P-gp) was identified as the mechanism to efflux drugs out from cells [9] and was classified as a member of the ATP binding cassette (ABC) multidrug transporters [10, 11]. Classically, this mechanism of pleiotropic resistance, commonly to cationic and lipophilic drugs, was also known as the product of the *MDR*1 gene [12] (mdr, multidrug resistance). Since then, more than 50 ABC multidrug transporters have been identified [13] and are involved in many physiological functions, including protection against anti-cancer drugs [14].

Multidrug resistance–associated proteins (MRPs) [15] and the breast cancer–resistance (BCRP) transporter [16, 17] are also well known members of the ABC family. Like P-glycoprotein, these other transporters were also identified by studying cells highly refractory against anticancer agents and all together are considered as major drivers of therapy refractoriness. However, other mechanisms of drug resistance have been described, for example those related to direct mechanisms such as the imbalance of topoisomerases and anthracycline resistance [18, 19], nucleoside transporters [20], vault proteins [21, 22], or promotion of cell cycle quiescence [23]. Indirect mechanisms of drug resistance are associated with mechanisms that make difficult the access of drugs to interact with target cells, such as those mediated by interactions of tumor cells and vascular tissue through regulatory angiogenic factors, among others [24].

With the passing of the years and with the increased knowledge of ABC transporters, many of them have been renamed and nowadays are called in accordance with the ABC transporter classification database. Thus, the group of ABC transporters better studied so far in human disease includes ABCB1 (P-gp), ABCC1 (MRP1), and ABCG2 (BCRP1). ABC transporters use energy to actively efflux drugs, and based on this ability many strategies to reverse drug resistance exploited the use of inhibitors or molecule blockers. Reversal efficacy against ABC transporters has been demonstrated in the laboratory, and different multicenter trials were designed to develop clinical strategies based on blocking their activity. However, the use of these blocking therapies was quickly discontinued based on the significant side effects of the reversal agents [14, 25].

Over the years, a growing concept has evolved around the stem cell compartment with new questions about these intriguing cells [26]. Are stem cells a phenotype or a function? And what is the role of ABC transporters expressed in stem cells? Although many markers and functional mechanisms have been associated with stem cells [27, 28], true stem cells still remain elusive and unidentified. One of the well-known markers to identify hematopoietic progenitor cells is the CD34 antigen. This marker is expressed in a rare subset of hematopoietic cells and is used to monitor the quality of the graft for stem-cell transplantation, as well as to detect cells with aberrant CD34 expression in human leukemia. Interestingly, CD34+ cells overexpress functional ABCB1 transporter, and its highest activity is associated with the most primitive cells within this compartment [27]. Moreover, CD34-neg stem-like cells overexpress ABCG2, which appears to be a key regulator of hematopoietic repopulating activity [29]. Malignant hematopoietic cells can also overexpress these transporters, having an important impact, especially in leukemic drug-resistant cells [30, 31].

Flow cytometry has advantages over many other methods to identify very primitive precursor cells. First of all, a large number of cells can be analyzed in order to detect the presence of subsets of primitive cells in very low numbers. Second, flow cytometry is a multiparametric technology that allows the combination of phenotypic and functional markers, resulting in the development of more sensitive measurements of CD34+ and CD34neg stem cells, which should have an impact on research in a broad range of scientific and clinical disciplines, and could also clarify the number of stem cells needed for engraftment after stem-cell transplantation. Thus, hematopoietic stem cells can be purified based on the efflux of fluorescent dyes such as rhodamine 123 [27] and Hoechst 33342 [32, 33]. Although expression of ATP binding cassette transporters has been assumed to contribute to the dye-efflux component of the stem-cell compartment, little is known about which transporters are expressed or what function(s) they might be conferring [29]. Nowadays, we know that CD34+ cells express mainly ABCB1, whereas CD34neg stem cells express the ABCG2 transporter. Many drugs are good fluorescent substrates. These fluorescent substances can intercalate into the DNA and accumulate in the nucleus [34]. The reduced drug-accumulation pattern present in drug-resistant cells can also be reversed by preincubation with some reversal agents, such as verapamil [35] or cyclosporine A [36]. Flow cytometry is increasingly used for quantitation of cellular fluorescent-drug retention, the effect of reversal agents, and the expression of drug resistance–related cellular surface markers.

To achieve optimal growth conditions, cells are seeded in medium in multiwell tissue-culture plates before drug uptake/retention experiments. The drugs are added to the culture medium, in the presence or absence of either verapamil or cyclosporine A. The concentration of the fluorescent substrates has previously been optimized to allow appropriate electronic signal compensation. After 1 h of incubation, cells are washed, fed with drug-free culture medium, cultured for 1 h at 37 °C, again in the presence or absence of the reversing agents to evaluate their effects on drug retention, and monitored on a flow cytometer [37, 38]. Multiple substrates can be used to identify drug-resistant cells in the green channel, such as rhodamine 123 and Syto 13 dyes, or anticancer drugs such as doxorubicin and daunorubicin in the orange channel, or mitoxantrone in the red channel. Optimal staining and concentration of these drugs should be determined and combined with immunophenotyping [39]. Gating of viable cells is then based on light-scatter parameters and simultaneous staining with propidium iodide, 7-AAD, or DAPI. Analysis of a minimum of 10,000 cells per sample is carried out in the fluorescence/count log histogram, collecting the autofluorescence signal in the first decade. Cancer and leukemic cells can simultaneously express different multidrug-resistant transporters, which can be easily distinguished by combining fluorescent probes and multidrug reversal agents. Cyclosporine A is generally specific for P-glycoprotein and probenecide for ABCC1. By using calcein-AM, a molecular probe that can be excluded by these two transporters, different fluorescent profiles can be observed, indicating that cells can evade anticancer treatment consisting of multiple drug combinations [40, 41]. Flow cytometry can also take advantage of other techniques based on the monitoring of drug efflux over time to easily estimate maximum transport velocity, information that can be very helpful for clinicians, because it will have special impact on a reduced drug effect. The author's laboratory uses stable ABCG2 transfectants preloaded with mitoxantrone, a fluorescent drug that can be excited with red lasers, showing that specific transporter mutations consisting of a single amino acid substitution are crucial for changes in the velocity of the transporter, making cells more resistant to anticancer drugs [38]. Multidrug resistance in human leukemia and lymphoma can also be detected by using monoclonal antibodies, such as MRK16 [42], MRK20 [43], and UIC2 [44]. The immature CD34 phenotype is strongly linked with P-gp [27]. An important advantage of flow cytometry for the detection of multidrug-resistant cells seems to be the potential to identify multiple subsets of hematopoietic cells, using lineage-specific or differentiation antigens. Those cells can be separated using positive or negative sorting procedures for gene expression analysis.

Studies on rare cells are especially challenging, and methods for cell sample preparation can be a significant source of perturbation of the assay. Moreover, the observation of rare events will also depend on space and time. Therefore, sampling can be crucial to detect and isolate cells that exist at a very low frequency in blood or in the marrow. Absolute numbers of target cells can be extremely low, representing less than 10 circulating tumor cells/mL whole blood in patients with metastatic disease, whereas 1 mL blood contains several million leukocytes and billions of erythrocytes.

Commonly used hematopoietic stem-cell surface markers can comprise CD34, CD90, CD93, CD105, CD110, CD111, CD117, CD133, CD135, CD150, CD184, CD202b, CD271, CD243, CD309, and CD338, among others. Some other stem cell markers, such as bromodeoxyuridine (BrdU), are not specific markers of stemness. BrdU can be toxic to the cells, thus raising the issue of artifactual stem cell detection [45]. More than 99% of cells that retain BrdU are not stem cells, whereas other markers that are good for embryonic stem-cell detection are expressed at low levels in adult stem cells [46]. To complicate the issue, CD133 can be aberrantly overexpressed in gliomas [47] and astrocytomas [48].

Rare-cell analysis involves adequate sampling, sample preparation, and counting statistics. Data processing, system noise, fluidic stability, antigen density, subset analysis, and/or pre-enrichment of the sample are crucial for success in the identification of target cells. Although all these factors can be accurately controlled and precisely monitored, Heisenberg's uncertainty principle will determine our observations, making it difficult to be sure about the identification of true target cells. Fortunately, polychromatic flow cytometry allows the combination of phenotypic and functional markers that can be very helpful to enrich very primitive stem cells. A previous elegant assay used three-color flow cytometry to analyze ABCB1 function in CD34+ cells by combining CD34 staining, 7-AAD to exclude necrotic cells, and rhodamine 123 (rho123), a probe commonly used for mitochondrial membrane potential assessment and a good fluorescent substrate to monitor the activity of ABCB1. By adding the parameter time, kinetic analysis can easily be performed on a flow cytometer, providing additional information about the functional status of primitive hematopoietic progenitors. Thus, most CD34+/7-AADneg cells fit to a sigmoid rho123 fluorescent profile over time, allowing one to distinguish different CD34 subsets based on rho123 intensities. Using this assay, CD34+/7-AADneg/rho123low cells are enriched in primitive progenitors, indicating high ABCB1 functional activity, and low mitochondrial membrane potential, as occurs in quiescent cells [39].

CD34+/CD38neg cells are also enriched in primitive progenitors; this combination has been used over time in multiple studies to learn more about the biology of this stem-cell subset, even using clonogenic cell culture and stem-cell transplantation and immunodeficient mice models. In parallel, the cell cycle can be also very informative to explore the proliferative potential of stem cells in association with immunophenotyping studies. CD34+/CD38+/HLA-DR+, CD34+/CD38+/HLA-DRneg, and CD34+/CD38neg/HLA-DRneg cells show different cell-cycle distribution, with $14.5 \pm 1.8\%$, $3 \pm 0.7\%$, and $0.5 \pm 0.4\%$ S+G2M cells, respectively, indicating that the progressive lack or decrease of CD38 and HLA-DR is associated with the quiescent status of CD34+ cells. However, in order to distinguish G0 from G1 cells, classical cell-cycle studies based on propidium iodide or DAPI staining are not valid. One more elegant assay used simultaneous RNA and DNA staining to differentiate G0 and G1 cells by combining viable staining of cells in the presence of pyronin Y (RNA) and Hoechst 33342 (DNA) [49]. Flow FISH–based approaches have also been used to study the CD34+ compartment after immunomagnetic separation of these cells, showing that a rare fraction of these cells, comprising less than 0.07% of isolated CD34+ cells, have extremely large telomeres [50]. Altogether, all the above methods can be used for primitive cells, which exist at low frequencies in bone marrow and peripheral blood. However, perturbation of the assay by use of inadequate methods can complicate the detection of very rare subsets.

Red-cell lysing agents are widely used to deplete erythrocytes in leukocyte immunophenotyping. The ISHAGE guidelines for CD34+ cell counting recommend the use of these solutions to deplete red cells. Although these solutions have been designed with the above-mentioned purpose, depletion of leukocytes by lysing reagents cannot be ruled out. In the author's laboratory, several specific strategies to avoid either red-cell lysing or density gradient separation have been developed [51, 52, 53]. These strategies are based on leukocyte discrimination by collecting violet side scatter [54] or by staining viable nucleated cells and are applied for rare-cell subset discrimination (Fig. 1). In the absence of a gold standard, these methods can be widely used and routinely applied. Only minor modification of the original optical filter setup on the flow cytometer will be needed for violet side-scatter configuration, or for adapting the staining protocol aimed to discriminate erythrocytes using viable DNA dyes. Moreover, the multilaser capabilities of ultimate flow cytometers can allow multicolor analysis with minimum or no color compensation and reserving phycoerythrin (PE) and other PE-tandem dyes for yellow-green laser excitation. An additional problem for rare-event analysis and detection is also related to subjectivity and gating, which can result in flow cytometric variability, as demonstrated by "in silico" multicenter trials for CD34+ counting (Table 1). With all this information, researchers interested in the identification of cancer or leukemic stem cells should take into account that flow cytometric–based experiments must be

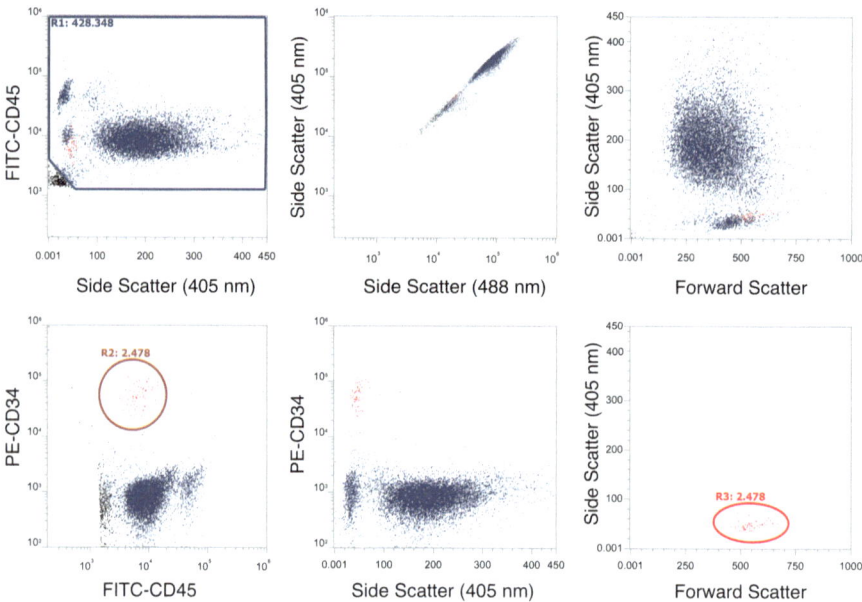

Fig. 1 Single-platform assay for absolute CD34+ cells without beads. Analysis of human CD34+ cells using the Attune NxT cytometer configuration with the 488-nm and 561-nm lasers. Utilizing a fluorescent threshold on CD45, the CD45-stained cells appear on the bivariate plots as a region of dots clustered in three main populations, consisting of lymphocytes, monocytes, and polymorphonuclear cells. RBCs have been eliminated by the CD45 threshold. A SSC versus CD45 dot plot is used to discriminate leukocytes from erythrocytes and debris. A polygonal region R1 is drawn to enclose the CD45-positive cells, and subsequent bivariate plots are generated based on this gate. CD34+CD45 dim cells are shown in R2. Forward-scatter versus violet side-scatter plot is used again to record at least 100 CD34+ cells (R3). Note that red blood cells are excluded as a consequence of an acquisition threshold set on CD45 fluorescence. Importantly, this no-lyse/no-wash staining procedure is used to prevent loss of cells due to red-cell lysing and washing. This strategy exploits the difference in light-scattering properties between red blood cells and leukocytes. This protocol is ideally suited to study the numbers of circulating CD34+ cells

Table 1 Multicenter "in silico" trial on the subjectivity of data analysis from flow cytometric rare-event counting studies

	Aph	CB	BM	PB	Total
CD34%	11.53	47.55	33.38	21.98	28.61
CD45%	6.85	27.73	7.44	8.50	12.63
CD34µL	11.61	53.11	33.14	21.84	29.92
CD45µL	5.35	7.21	5.88	12.30	7.67
					n = 21

A reference laboratory prepared different representative CD34 data files for independent and blind software analysis by 50 participating laboratories. FCS files for leukapheresis (Aphe), cord blood (CB), bone marrow (BM), and mobilized peripheral blood (PB) were successfully analyzed by 21 centers. Coefficients of variation between laboratories showed unexpected high values even for CD45 and CD34+ cells, indicating that the more abundant populations such as CD45+ cells are also outnumbered based on subjective gating

carefully evaluated and planned. If not, these elusive cells can be mismeasured, which results in frustrating and unsuccessful time-consuming efforts.

There is a growing interest in side-population (SP) cells. The SP constitutes an important hallmark for the definition of the stem cell compartment, and can be detected and isolated only by flow cytometry and fluorescence-activated cell sorting. SP cells are primitive and undifferentiated cells that can be found in almost all studied tissues and that show high plasticity and great potential therapeutic value. SP cells are CD34 negative and were discovered using ultraviolet lasers, based on the efflux of Hoechst 33342, a fluorescent DNA stain commonly used to study the cell-cycle phases of viable cells. Under UV excitation, the Hoechst dye shows dual wavelength emission, in both red and blue. SP cells express the ABCG2 transporter, which is responsible for the active decrease in Hoechst concentration inside the SP cells. Based on ABCG2 functional activity, SP cells display less fluorescence over time, which results in a dim tail of cells when the red and blue Hoechst 33342 emissions are displayed. Hoechst 33342 varies from red to blue as a consequence of the ABCG2 activity, which results in this specific tail pattern in association with a chromatic shift in the fluorescence of the cells. SP cells represent a small fraction of bone-marrow cells (0.02-0.03%), and the fluorescent profile of the SP can be reverted in the presence of ABCG2 inhibitors, such as fumitremorgin C. Sample manipulation heavily affects ABCG2 function, making the identification of SP cells difficult, especially when the SP represents less than 0.01% of cells, as can occur in cancer or leukemia, and more especially in chronic myelogenous leukemia [55].

Despite the therapeutic potential of SP cells in tissue regeneration, it has been shown that cells with SP phenotype can drive many types of cancer and leukemia. It is controversial if stem or stem-like cells constitute this rare population. However, the high prevalence of SP cells in cancer and leukemic cell lines in the laboratory has been demonstrated, especially under low oxygen concentrations. SP cells have also been identified with high prevalence in human cancer and leukemia, which show high SP cell counts, with a median of 0.04% (range, 0–13.88) [56].

However, little is known about the physiological role of SP cells in humans, with the exception of their high prevalence in human cancer and leukemia. SP cells are extremely difficult to detect or undetectable in peripheral blood compared to bone marrow, making difficult the development of studies to enlighten their role in human hematopoietic repopulation. One of the limitations of the classic use of Hoechst 33342 for SP identification and isolation is ultraviolet excitation. With the passing of the years, flow cytometers have become more potent tools for multicolor analysis, with the incorporation of solid-state lasers. Many cytometers all over the world are now commonly equipped with violet lasers, even in the clinics, allowing the simultaneous analysis of eight to ten colors simultaneously for leukemia immunophenotyping. Vybrant DyeCycle Violet stain (DCV) is an alternative dye that can be used to identify SP cells and is excited with violet lasers [55, 57]. Violet lasers (405 nm) are less expensive, consume less power, and last longer than ultraviolet lasers. This availability of violet lasers will bring in the near future excellent opportunities for clinical studies using SP cells in combination with DCV and immunophenotyping for rare subset identification. DCV is a DNA-selective and cell membrane–permeant stain that can also be used to discriminate nucleated from non-nucleated cells, allowing whole-blood studies with minimum manipulation, and to distinguish different subsets of leukocytes in steady-state peripheral blood, bone marrow, leukapheresis products, cord blood, and in leukemia. Moreover, DCV effectively identifies subsets and orders CD34+ cells for the definition of different functional properties that can be related to the characterization, resolution, and purification of primitive hematopoietic stem cells in combination with specific useful markers for multicolor flow cytometric measurements [53].

In vitro experiments with stem cells need oxygen concentrations comparable to that of their normal niche. Tissue-cell incubators operating at 21% oxygen will result in hyperoxic conditions leading to aberrant expression of multiple genes. In vitro physioxia will mimic physiological conditions, as none of our cells live in our body under harmful hyperoxic atmospheric conditions. ABCG2 is regulated by oxygen concentration; low oxygen promotes ABCG2 expression and activity, mediating increased drug resistance in tissues and tumors with high hypoxia [58]. High hypoxia is also one of the most important driving stimuli for tumor angiogenesis [59], which can also lead to poor tumor vascularization and increased drug resistance by limiting the delivery and penetration of anticancer drugs into the stem cell tumor microenvironment [24].

Recent studies have highlighted the relevance of the expression of functional ABC multidrug-resistance transporters and their association with stemness. This field of investigation is providing new clues about how ABC pumps may protect specific signaling pathways highly conserved in stem cells, such as Sonic-Hedgehog [60], Notch [61], or PI3-Akt [62]. In fact, these studies have suggested that multidrug transporters in stem cells may act as firewalls to avoid having most of these important pathways for stem-cell maintenance inhibited or altered by other molecules with potential roles in pathway activation or inhibition. This exciting mechanism of regulation may help to explain why most primitive stem-like side population cells express high levels of ABCG2 functional transporter, and how this mechanism of protection against differentiation also contributes to pleiotropic drug resistance in cancer and leukemia, which can also trigger the maintenance and expansion of highly refractory malignant stem cells.

Some enzyme activity can also be applied to identify primitive cells. The aldehyde dehydrogenase (ALDH) assay is widely used for healthy and cancer stem-cell isolation using flow cytometry [63]. Alkaline phosphatase is also a promising marker for stem cell detection and can be very easily adapted in flow cytometry experiments in combination with antibodies. The alkaline phosphatase (AP) is a hydrolase enzyme responsible for removal or transfer of phosphoryl groups. It is most effective in an alkaline environment (pH = 10), and purified sources undergo three types of activity: (i) hydrolytic, (ii) phosphotransferase, and (iii) pyrophosphatase. Measurement of serum alkaline phosphatase has diagnostic use; AP can be altered in malignant and non-malignant disease. AP activity has been used to identify pluripotent stem cells, and is expressed in embryonic stem cells, embryonic germ cells, and embryonic carcinoma cells [64, 65]. AP expression can be studied with flow cytometry and antibodies against the enzyme, and more recently with nontoxic, cell-permeant and fluorogenic substrates with fluorescence emission at 530 nm [66]. Live AP staining is specific and stained cells can be propagated further. Moreover, this live AP stain can be used as a quick method to test its activity in primitive progenitors in combination with commonly available antibodies against stem cells. AP staining is stable over time and this fluorogenic substrate is not effluxed by multidrug transporters, making this stain feasible for the identification of differential expression within stem-like SP cells, especially under high hypoxic conditions [67].

One more fascinating application of AP live stain is associated with early detection of highly refractory leukemic cells in association with classic panels used for leukemia immunophenotyping at diagnosis, during treatment, follow-up, and relapse. Interestingly, blast populations identified with well-standardized panels used in leukemia may appear as apparently clonal populations at diagnosis and during relapse. However, by adding AP live stain, different subsets of malignant cells can be identified with unprecedented levels of sensitivity compared to classic flow cytometry immunophenotyping panels, suggesting that high levels of alkaline phosphatase activity in leukemic blast cells can also be associated with highly refractory cells that sustain leukemogenesis over time [67].

The emergence of new, powerful flow cytometers equipped with multiple laser sources will help in the near future to integrate panels designed for the simultaneous analysis of immunophenotyping and functional mechanisms that occur in living cells. These studies will be fundamental to understand the complex biology of stem cells, but also to better understand other pathologies, such as autoimmune disease. More colors will involve more possible combinations of simultaneously analyzed markers, giving rise to the following question: Is more better? Five-color immunophenotyping of leukemia in one single tube will generate 15 possible combinations of the markers used. Many hematology laboratories involved in leukemia diagnosis perform one of the most potent flow cytometry applications to evaluate minimal residual disease (MRD) in adults and in pediatric patients, with the aim to both identify and quantify persistent or new malignant phenotypes. The MRD method should be the most sensitive one possible. Theoretically, in order to achieve a sensitivity of 0.01%, one million total events will be required. Unfortunately, among other factors, it is not uncommon that sample degeneration, hemodilution, or low cell count will negatively affect accurate MRD assessment. In these cases, software analysis can be very helpful to better identify abnormal phenotypes. However, abnormal cell populations will be identified only when rare events are observed in both space and time. Therefore, sampling and time point measurements, together with expertise in normal and abnormal cell maturation, will be crucial to gather the valid information needed for the accurate identification of rare-cell subsets in cancer and leukemia.

References

1. Berenson RJ, Andrews RG, Bensinger WI, Kalamasz D, Knitter G, Buckner CD, Bernstein ID (1988) Antigen CD34+ marrow cells engraft lethally irradiated baboons. J Clin Invest 81:951–955. doi:10.1172/JCI113409
2. Srour EF, Brandt JE, Briddell RA, Leemhuis T, van Besien K, Hoffman R (1991) Human CD34+ HLA-DR-bone marrow cells contain progenitor cells capable of self-renewal, multilineage differentiation, and long-term in vitro hematopoiesis. Blood Cells 17:287–95
3. Bernstein ID, Andrews RG, Zsebo KM (1991) Recombinant human stem cell factor enhances the formation of colonies by CD34+ and CD34+ lin-cells, and the generation of colony-forming cell progeny from CD34+ lin-cells cultured with interleukin-3, granulocyte colony-stimulating factor, or granulocyte-macrophage colony-stimulating factor. Blood 77:2316–2321
4. Mitjavila MT, Natazawa M, Brignaschi P, Debili N, Breton-Gorius J, Vainchenker W (1989) Effects of five recombinant hematopoietic growth factors on enriched human erythroid progenitors in serum-replaced cultures. J Cell Physiol 138:617–623. doi:10.1002/jcp.1041380324
5. Debili N, Hegyi E, Navarro S, Katz A, Mouthon MA, Breton-Gorius J, Vainchenker W (1991) In vitro effects of hematopoietic growth factors on the proliferation, endoreplication, and maturation of human megakaryocytes. Blood 77:2326–2338
6. Bhatia M, Bonnet D, Murdoch B, Gan OI, Dick JE (1998) A newly discovered class of human hematopoietic cells with SCID-repopulating activity. Nat Med 4:1038–1045. doi:10.1038/2023
7. Goodell MA, Brose K, Paradis G, Conner AS, Mulligan RC (1996) Isolation and functional properties of murine hematopoietic stem cells that are replicating in vivo. J Exp Med 183:1797–1806
8. Sutherland DR, Anderson L, Keeney M, Nayar R, Chin-Yee I (1996) The ISHAGE guidelines for CD34+ cell determination by flow cytometry. International Society of Hematotherapy and Graft Engineering. J Hematother 5:213–226. doi:10.1089/scd.1.1996.5.213
9. Juliano RL, Ling V (1976) A surface glycoprotein modulating drug permeability in Chinese hamster ovary cell mutants. Biochim Biophys Acta 455:152–162. doi:10.1016/0005-2736(76)90160-7 [pii]
10. Gerlach JH, Endicott JA, Juranka PF, Henderson G, Sarangi F, Deuchars KL, Ling V (1986) Homology between P-glycoprotein and a bacterial haemolysin transport protein suggests a model for multidrug resistance. Nature 324:485–489. doi:10.1038/324485a0

11. Cornwell MM, Tsuruo T, Gottesman MM, Pastan I (1987) ATP-binding properties of P glycoprotein from multidrug-resistant KB cells. FASEB J 1:51–54

12. Shen DW, Fojo A, Chin JE, Roninson IB, Richert N, Pastan I, Gottesman MM (1986) Human multidrug-resistant cell lines: increased mdr1 expression can precede gene amplification. Science 232:643–645

13. Alexander SP, Kelly E, Marrion N, Peters JA, Benson HE, Faccenda E, Pawson AJ, Sharman JL, Southan C, Davies JA (2015) The concise guide to pharmacology 2015/16: transporters. Br J Pharmacol 172:6110–6202. doi:10.1111/bph.13355

14. Gottesman MM, Fojo T, Bates SE (2002) Multidrug resistance in cancer: role of ATP-dependent transporters. Nat Rev Cancer 2:48–58. doi:10.1038/nrc706

15. Szczypka MS, Wemmie JA, Moye-Rowley WS, Thiele DJ (1994) A yeast metal resistance protein similar to human cystic fibrosis transmembrane conductance regulator (CFTR) and multidrug resistance-associated protein. J Biol Chem 269:22853–22857

16. Doyle LA, Yang W, Abruzzo LV, Krogmann T, Gao Y, Rishi AK, Ross DD (1998) A multidrug resistance transporter from human MCF-7 breast cancer cells. Proc Natl Acad Sci U S A 95:15665–15670

17. Ross DD, Yang W, Abruzzo LV, Dalton WS, Schneider E, Lage H, Dietel M, Greenberger L, Cole SP, Doyle LA (1999) A typical multidrug resistance: breast cancer resistance protein messenger RNA expression in mitoxantrone-selected cell lines. J Natl Cancer Inst 91:429–433

18. Glisson B, Gupta R, Hodges P, Ross W (1986) Cross-resistance to intercalating agents in an epipodophyllotoxin-resistant Chinese hamster ovary cell line: evidence for a common intracellular target. Cancer Res 46:1939–1942

19. Glisson B, Gupta R, Smallwood-Kentro S, Ross W (1986) Characterization of acquired epipodophyllotoxin resistance in a Chinese hamster ovary cell line: loss of drug-stimulated DNA cleavage activity. Cancer Res 46:1934–1938

20. Aronow B, Allen K, Patrick J, Ullman B (1985) Altered nucleoside transporters in mammalian cells selected for resistance to the physiological effects of inhibitors of nucleoside transport. J Biol Chem 260:6226–6233

21. Slovak ML, Ho JP, Cole SP, Deeley RG, Greenberger L, de Vries EG, Broxterman HJ, Scheffer GL, Scheper RJ (1995) The LRP gene encoding a major vault protein associated with drug resistance maps proximal to MRP on chromosome 16: evidence that chromosome breakage plays a key role in MRP or LRP gene amplification. Cancer Res 55:4214–4219

22. Scheffer GL, Wijngaard PL, Flens MJ, Izquierdo MA, Slovak ML, Pinedo HM, Meijer CJ, Clevers HC, Scheper RJ (1995) The drug resistance-related protein LRP is the human major vault protein. Nat Med 1:578–582

23. Sullivan DM, Latham MD, Ross WE (1987) Proliferation-dependent topoisomerase II content as a determinant of antineoplastic drug action in human, mouse, and Chinese hamster ovary cells. Cancer Res 47:3973–3979

24. Hedley DW (1993) Flow cytometric assays of anticancer drug resistance. Ann N Y Acad Sci 677:341–353

25. Holohan C, Van Schaeybroeck S, Longley DB, Johnston PG (2013) Cancer drug resistance: an evolving paradigm. Nat Rev Cancer 13:714–726. doi:10.1038/nrc3599 [pii]

26. Blau HM, Brazelton TR, Weimann JM (2001) The evolving concept of a stem cell: entity or function? Cell 105:829–841. doi:10.1016/S0092-8674(01)00409-3 [pii]

27. Chaudhary PM, Roninson IB (1991) Expression and activity of P-glycoprotein, a multidrug efflux pump, in human hematopoietic stem cells. Cell 66:85–94. doi:0092-8674(91)90141-K [pii]

28. Bunting KD (2002) ABC transporters as phenotypic markers and functional regulators of stem cells. Stem Cells 20:11–20. doi:10.1634/stemcells.20-3-274

29. Zhou S, Schuetz JD, Bunting KD, Colapietro AM, Sampath J, Morris JJ, Lagutina I, Grosveld GC, Osawa M, Nakauchi H, Sorrentino BP (2001) The ABC transporter Bcrp1/ABCG2 is expressed in a wide variety of stem cells and is a molecular determinant of the side-population phenotype. Nat Med 7:1028–34. doi:10.1038/nm0901-1028nm0901-1028 [pii]

30. Steinbach D, Legrand O (2007) ABC transporters and drug resistance in leukemia: was P-gp nothing but the first head of the Hydra? Leukemia 21:1172–1176. doi:10.1038/sj.leu.2404692 [pii]

31. Eechoute K, Sparreboom A, Burger H, Franke RM, Schiavon G, Verweij J, Loos WJ, Wiemer EA, Mathijssen RH (2011) Drug transporters and imatinib treatment: implications for clinical practice. Clin Cancer Res 17:406–15. doi:10.1158/1078-0432.CCR-10-2250 [pii]

32. Van Zant G, Fry CG (1983) Hoechst 33342 staining of mouse bone marrow: effects on colony-forming cells. Cytometry 4:40–46. doi:10.1002/cyto.990040106

33. Goodell MA, Rosenzweig M, Kim H, Marks DF, DeMaria M, Paradis G, Grupp SA, Sieff CA, Mulligan RC, Johnson RP (1997) Dye efflux studies suggest that hematopoietic stem cells expressing low or undetectable levels of CD34 antigen exist in multiple species. Nat Med 3:1337–1345

34. Glazer AN, Rye HS (1992) Stable dye-DNA intercalation complexes as reagents for high-sensitivity fluorescence detection. Nature 359:859–861. doi:10.1038/359859a0

35. Beck WT (1991) Modulators of P-glycoprotein-associated multidrug resistance. Cancer Treat Res 57:151–170

36. Gupta S, Tsuruo T, Gollapudi S (1992) Multidrug resistant gene 1 product in human T cell subsets: role of protein kinase C isoforms and regulation by cyclosporin A. Adv Exp Med Biol 323:39–47

37. Petriz J, Sanchez J, Bertran J, Garcia-Lopez J (1997) Comparative effect of verapamil, cyclosporin A and SDZ PSC 833 on rhodamine 123 transport and cell cycle in vinblastine-resistant Chinese hamster ovary cells overexpressing P-glycoprotein. Anticancer Drugs 8:869–875

38. Garcia-Escarp M, Martinez-Munoz V, Sales-Pardo I, Barquinero J, Domingo JC, Marin P, Petriz J (2004) Flow cytometry-based approach to ABCG2 function suggests that the transporter differentially handles the influx and efflux of drugs. Cytometry A 62:129–138. doi:10.1002/cyto.a.20072

39. Petriz J, Garcia-Lopez J (1997) Flow cytometric analysis of P-glycoprotein function using rhodamine 123. Leukemia 11:1124–1130

40. Feller N, Kuiper CM, Lankelma J, Ruhdal JK, Scheper RJ, Pinedo HM, Broxterman HJ (1995) Functional detection of MDR1/P170 and MRP/P190-mediated multidrug resistance in tumour cells by flow cytometry. Br J Cancer 72:543–549

41. Feller N, Broxterman HJ, Wahrer DC, Pinedo HM (1995) ATP-dependent efflux of calcein by the multidrug resistance protein (MRP): no inhibition by intracellular glutathione depletion. FEBS Lett 368:385–388. doi:10.1016/0014-5793(95)00677-2 [pii]

42. Hamada H, Tsuruo T (1986) Functional role for the 170- to 180-kDa glycoprotein specific to drug-resistant tumor cells as revealed by monoclonal antibodies. Proc Natl Acad Sci USA 83:7785–7789

43. Sugawara I, Ohkochi E, Hamada H, Tsuruo T, Mori S (1988) Cellular and tissue distribution of MRK20 murine monoclonal antibody-defined 85-kDa protein in adriamycin-resistant cancer cell lines. Jpn J Cancer Res 79:1101–1110

44. Mechetner EB, Roninson IB (1992) Efficient inhibition of P-glycoprotein-mediated multidrug resistance with a monoclonal antibody. Proc Natl Acad Sci USA 89:5824–5828

45. Kiel MJ, He S, Ashkenazi R, Gentry SN, Teta M, Kushner JA, Jackson TL, Morrison SJ (2007) Haematopoietic stem cells do not asymmetrically segregate chromosomes or retain BrdU. Nature 449:238–242. doi:10.1038/nature06115 [pii]

46. Zangrossi S, Marabese M, Broggini M, Giordano R, D'Erasmo M, Montelatici E, Intini D, Neri A, Pesce M, Rebulla P, Lazzari L (2007) Oct-4 expression in adult human differentiated cells challenges its role as a pure stem cell marker. Stem Cells 25:1675–80. doi:10.1634/stemcells.2006-0611 [pii]

47. Zeppernick F, Ahmadi R, Campos B, Dictus C, Helmke BM, Becker N, Lichter P, Unterberg A, Radlwimmer B, Herold-Mende CC (2008) Stem cell marker CD133 affects clinical outcome in glioma patients. Clin Cancer Res 14:123–129. doi:10.1158/1078-0432. CCR-07-0932 [pii]

48. Christensen K, Schroder HD, Kristensen BW (2008) CD133 identifies perivascular niches in grade II-IV astrocytomas. J Neurooncol 90:157–170. doi:10.1007/s11060-008-9648-8

49. Darzynkiewicz Z, Juan G, Srour EF (2004) Differential staining of DNA and RNA. Curr Protoc Cytom Chapter 7:Unit 7 3. doi:10.1002/0471142956.cy0703s30

50. Garcia-Escarp M, Martinez-Munoz V, Barquinero J, Sales-Pardo I, Domingo JC, Marin P, Petriz J (2005) A rare fraction of human hematopoietic stem cells with large telomeres. Cell Tissue Res 319:405–412. doi:10.1007/s00441-004-1022-3

51. Fornas O, Garcia J, Petriz J (2000) Flow cytometry counting of CD34 + cells in whole blood. Nat Med 6:833–836. doi:10.1038/77571

52. Alvarez-Larran A, Jover L, Marin P, Petriz J (2002) A multicolor, no-lyse no-wash assay for the absolute counting of CD34+ cells by flow cytometry. Cytometry 50:249–253. doi:10. 1002/cyto.10129

53. Nunez-Espinosa C, Garcia-Godoy MD, Ferreira I, Rios-Kristjansson JG, Rizo-Roca D, Rico LG, Rubi-Sans G, Palacio C, Torrella JR, Pages T, Ward MD, Viscor G, Petriz J (2016) Vybrant DyeCycle violet stain discriminates two different subsets of CD34+ cells. Curr Stem Cell Res Ther 11:66–71. doi:CSCRT-EPUB-67694 [pii]

54. Ost V, Neukammer J, Rinneberg H (1998) Flow cytometric differentiation of erythrocytes and leukocytes in dilute whole blood by light scattering. Cytometry 32:191–197. doi:10.1002/(SICI)1097-0320(19980701)32:3<191::AID-CYTO5>3.0.CO;2-N [pii]

55. Petriz J (2013) Flow cytometry of the side population (SP). Curr Protoc Cytom Chapter 9 (Unit9):23. doi:10.1002/0471142956.cy0923s64

56. Avendano A, Sales-Pardo I, Garcia-Godoy MD, Rico LG, Marin P, Petriz J (2014) Accuracy and reproducibility of stem cell side population measurements on clinically relevant products. Curr Stem Cell Res Ther 9:526–534. doi:CSCRT-EPUB-60274 [pii]

57. Telford WG (2010) Stem cell side population analysis and sorting using DyeCycle violet. Curr Protoc Cytom Chapter 9:Unit9 30. doi:10.1002/0471142956.cy0930s51

58. Krishnamurthy P, Ross DD, Nakanishi T, Bailey-Dell K, Zhou S, Mercer KE, Sarkadi B, Sorrentino BP, Schuetz JD (2004) The stem cell marker Bcrp/ABCG2 enhances hypoxic cell survival through interactions with heme. J Biol Chem 279:24218–24225. doi:10.1074/jbc. M313599200M313599200 [pii]

59. Pugh CW, Ratcliffe PJ (2003) Regulation of angiogenesis by hypoxia: role of the HIF system. Nat Med 9:677–684. doi:10.1038/nm0603-677nm0603-677 [pii]

60. Balbuena J, Pachon G, Lopez-Torrents G, Aran JM, Castresana JS, Petriz J (2011) ABCG2 is required to control the sonic hedgehog pathway in side population cells with stem-like properties. Cytometry A 79:672–683. doi:10.1002/cyto.a.21103

61. Takebe N, Miele L, Harris PJ, Jeong W, Bando H, Kahn M, Yang SX, Ivy SP (2015) Targeting Notch, Hedgehog, and Wnt pathways in cancer stem cells: clinical update. Nat Rev Clin Oncol 12:445–464. doi:10.1038/nrclinonc.2015.61 [pii]

62. Bleau AM, Hambardzumyan D, Ozawa T, Fomchenko EI, Huse JT, Brennan CW, Holland EC (2009) PTEN/PI3 K/Akt pathway regulates the side population phenotype and ABCG2 activity in glioma tumor stem-like cells. Cell Stem Cell 4:226–235. doi:10.1016/j. stem.2009.01.007 [pii]

63. Greve B, Kelsch R, Spaniol K, Eich HT, Gotte M (2012) Flow cytometry in cancer stem cell analysis and separation. Cytometry A 81:284–293. doi:10.1002/cyto.a.22022

64. Takahashi K, Okita K, Nakagawa M, Yamanaka S (2007) Induction of pluripotent stem cells from fibroblast cultures. Nat Protoc 2:3081–3089. doi:10.1038/nprot.2007.418 [pii]

65. Takahashi K, Tanabe K, Ohnuki M, Narita M, Ichisaka T, Tomoda K, Yamanaka S (2007) Induction of pluripotent stem cells from adult human fibroblasts by defined factors. Cell 131:861–872. doi:10.1016/j.cell.2007.11.019 [pii]
66. Singh U, Quintanilla RH, Grecian S, Gee KR, Rao MS, Lakshmipathy U (2012) Novel live alkaline phosphatase substrate for identification of pluripotent stem cells. Stem Cell Rev 8:1021–1029. doi:10.1007/s12015-012-9359-6
67. Rico LG, Junca J, Ward MD, Bradford J, Petriz J (2016) Is alkaline phosphatase the smoking gun for highly refractory primitive leukemic cells? Oncotarget 7:72057–72066. doi:10.18632/oncotarget.12497 [pii]